Thin Films and Nanostructures

Advances in Plasma-Grown Hydrogenated Films

Volume 30

Recent volumes in this serial appear at the end of this volume

Thin Films and Nanostructures

Advances in Plasma-Grown Hydrogenated Films

Edited by

Maurice H. Francombe

Department of Physics and Astronomy
Georgia State University
Atlanta, Georgia

VOLUME 30

ACADEMIC PRESS

An Elsevier Science Imprint

San Diego San Francisco New York Boston
London Sydney Tokyo

The articles in this book are selected from the Academic Press multi-volume work titled *Handbook of Thin Film Materials*, edited by Hari S. Nalwa, and are uniquely arranged to focus on current advances in surface science.

Academic Press
An Imprint of Elsevier Science
525 B Street, Suite 1900, San Diego, California 92101-4495, USA
http://www.academicpress.com

Academic Press
Harcourt Place, 32 Jamestown Road, London NW1 7BY, UK
http://www.academicpress.com

International Standard Book Number: 0-12-533030-8

International Standard Serial Number: 1079-4050

PRINTED IN THE UNITED STATES OF AMERICA
01 02 03 04 05 CO 9 8 7 6 5 4 3 2 1

Contents

Methods of Deposition of Hydrogenated Amorphous Silicon for Device Applications

Wilfried G. J. H. M. van Sark

Growth, Structure, and Properties of Plasma-Deposited Amorphous Hydrogenated Carbon–Nitrogen Films

D. F. Franceschini

Contributors

Methods of Deposition of Hydrogenated Amorphous Silicon for Device Applications: Wilfried G. J. H. M. van Sark, Debye Institute, Utrecht University, NL-3508 TA Utrecht, The Netherlands

Growth, Structure, and Properties of Plasma-Deposited Amorphous Hydrogenated Carbon–Nitrogen Films: D. F. Franceschini, Instituto de Física, Universidade Federal Fluminense, Avenida Litorânea s/n, Niterói, RJ, 24210-340, Brazil

Preface

Starting with Volume 30, the title of the series, *Thin Films*, is being changed to *Thin Films and Nanostructures*. We feel that this new title reflects more accurately the rapidly growing inclusion of research and development efforts on nanostructures, especially in relation to novel solid-state device formats.

The material presented in Volume 30 of this series complements and extends the subject matter published in Volume 29, which reviewed *Non-Crystalline Films for Device Structures* and dealt primarily with gate-dielectric and passivating layers. Volume 30, titled *Advances in Plasma-Grown Hydrogenated Films*, contains two chapters covering in detail the plasma preparation and properties of hydrogenated amorphous semiconductor layers and their emerging applications in optoelectronic devices and arrays. As was the case with Volume 29, the review articles included in this volume were selected from the new multi-volume work, *Handbook of Thin Film Materials*, edited by Dr. Hari Singh Nalwa for publication by Academic Press.

Investigation of amorphous semiconductor films for use in a wide variety of applications has received considerable attention for approximately the past four decades. Since the mid-1960s interest has focused on hydrogenated amorphous silicon, initially for potential use in low-cost large-area solar-cell installations. Chapter 1 offers a complete and well-documented survey, authored by W. G. J. H. M. van Sark, discussing the deposition of these films for a number of important device applications. The main deposition approach considered involves several versions of plasma enhances chemical vapor deposition (PECVD), using multi-chambered, load-locked vacuum systems, affording the ability to grow multi-layered (junction) structures under UHV background conditions. Material aspects are treated, including the role of hydrogen in the plasma for passivation and reduction of defects in the *a*-Si:H layers. Further sections include detailed descriptions of equipment and techniques, the physics and chemistry of PECVD, plasma modeling and analysis, modifications of PECVD, and discussions of applications in active matrix displays, light and chemical sensors, loudspeakers, etc. The critical need to improve material properties and intrinsic stability of *a*-Si:H, e.g., by working with discharge conditions that are "at the edge of crystallinity," is stressed.

Chapter 2, authored by D. F. Franceschini, addresses the growth, structure, and properties of plasma-deposited amorphous hydrogenated carbon–nitrogen films. The approach seeks to exploit the diamond-like properties of the hypothetical compound β-C_3N_4 (with the β-Si_3N_4 structure) by incorporating nitrogen in plasma-deposited hard amorphous hydrogenated carbon (*a*-C:H) films. Films in the chemical system a-C(N):H have previously been shown to be useful in a number of applications, for example, adherent antireflective coatings for Ge-based infrared detectors and overcoat layers on electron-field emission tips in flat-panel display devices. This review deals first with the structure and plasma deposition of a-C:H films, followed by the determination of composition and modeling of surface processes and film growth. Structure characterization, using optical, electron, X-ray photoelectron, and nuclear magnetic resonance spectroscopy to determine both structure and bonding aspects, is discussed. Control of optical and electrical characteristics of the layers is still somewhat limited by the low solubility of nitrogen and its increasing tendency to bond preferentially to hydrogen at higher nominal concentrations. However, the evolution of new plasma-growth conditions may help to extend the potential applications for these novel film materials.

MAURICE H. FRANCOMBE

Methods of Deposition of Hydrogenated Amorphous Silicon for Device Applications

WILFRIED G. J. H. M. VAN SARK

Debye Institute, Utrecht University, NL-3508 TA Utrecht, The Netherlands

1.1. Introduction

This chapter describes the deposition of hydrogenated amorphous silicon (*a*-Si:H) and related materials by employing a low-temperature, low-density plasma. The method basically is a special form of chemical vapor deposition (CVD), which is known as plasma-enhanced chemical vapor deposition (PECVD) or plasma CVD. Essentially, silane gas (SiH_4) is excited by a radiofrequency (RF, 13.56 MHz) plasma, which causes silane molecules to dissociate. Subsequently, dissociation products are deposited on heated substrates and form a layer. Most research and industrial reactor systems consist of two parallel electrodes in a stainless steel chamber. Because of the relative ease of depositing *a*-Si:H uniformly over large areas, the original parallel plate geometry with RF excitation frequency is commonly used in industry, and has not changed much over the past two or three decades [1–13]. Material and device optimization is mostly done empirically, and so-called *device quality* *a*-Si:H layers having excellent uniformity are made by PECVD.

Nevertheless, several modifications have evolved since the first demonstration of deposition of *a*-Si:H, such as the use of higher excitation frequencies [from

THIN FILMS AND NANOSTRUCTURES
Vol. 30
ISBN 0-12-533030-8
ISSN 1079-4050/02 $35.00

VHF (50–100 MHz) up to the gigahertz range], the use of remote excitation of the plasma, plasma beams, and modulation of the plasma in time or frequency. Even methods without the assistance of a plasma have evolved, such as the hot-wire CVD (HWCVD) method.

The material properties of layers deposited in a PECVD reactor strongly depend on the interaction between the growth flux and the film surface. Therefore, a central theme in this chapter is the relation between material properties and deposition parameters. Considering the plasma as a reservoir of species, we can distinguish neutrals, radicals, and ions, which can be either positive or negative. In a typical RF discharge, which is weakly ionized, the neutral species are the most abundant, having a concentration of about 10^{16} cm^{-3}, while the concentrations of radicals and ions are only about 10^{14} and 10^{10} cm^{-3}, respectively. The energies of these species may differ considerably. The neutrals, radicals and ions *within* the plasma are not energetic at all (below 0.1 eV). The ions that reach substrates and reactor walls are much more energetic (1–100 eV). This can have enormous consequences for the effect of species on a growing film. For example, the amount of energy that is present in the plasma amounts to 10^{13} eV/cm^3 for radicals and 10^{12} eV/cm^3 for ions. This shows that although ions are much less present in the plasma, their effect may be comparable to that of radicals.

In this chapter we will treat the common RF PECVD method for *a*-Si:H deposition, with emphasis on intrinsic material. First, a short introduction on the material properties of hydrogenated amorphous silicon is given. Subsequently, details are given on experimental and industrial deposition systems, with special emphasis on the UHV multichamber deposition system ASTER (Amorphous Semiconductor Thin Film Experimental Reactor) at Utrecht University [14, 15]. This is done not only because many experimental results presented in this chapter were obtained in that system, but because in our opinion it can also be seen as a generic multichamber deposition system. Then, a thorough description of the physics and the chemistry of the discharge is presented, followed by plasma modeling and plasma analysis results. Subsequently, relations will be formulated between discharge parameters and material properties, and models for the deposition of *a*-Si:H are presented.

In further sections extensions or adaptations of the PECVD method will be presented, such as VHF PECVD [16], the chemical annealing or layer-by-layer technique [17], and modulation of the RF excitation frequency [18]. The HWCVD method [19] (the plasmaless method) will be described and compared with the PECVD methods. The last deposition method that is treated is expanding thermal plasma CVD (ETP CVD) [20, 21]. Other methods of deposition, such as remote-plasma CVD, and in particular electron cyclotron resonance CVD (ECR CVD), are not treated here, as to date these methods are difficult to scale up for industrial purposes. Details of these methods can be found in, e.g., Luft and Tsuo [6].

As all these methods are used in research for improving material properties with specific applications in mind, a summary of important applications for which *a*-Si:H is indispensable is given in the last section.

It will be clear throughout this chapter that it is biased toward research performed at or in collaboration with the research group at the Debye Institute at Utrecht University. However, numerous references to other work are presented in order to put this research in a much broader perspective.

For further reading one may find excellent material in the (edited) books by (in chronological order) Pankove [1], Tanaka [2], Street [3], Kanicki [4, 5], Luft and Tsuo [6], Bunshah [7], Bruno et al. [8], Machlin [9, 10], Schropp and Zeman [11], Searle [12], and Street [13].

1.1.1. HISTORICAL OVERVIEW

The research on amorphous semiconductors in the 1950s and 1960s was focused on the chalcogenides, i.e. materials containing group VI elements (sulfur, selenium, and tellurium), such as As_2Se_3. These glasses are formed by cooling from the melt, their structure being similar to oxide glasses. Of particular interest was the relation between disorder of the structure and its electronic properties. This still is a question not fully answered; see for example Overhof in his Mott Memorial Lecture [22].

Amorphous silicon (and germanium) was prepared in those days by thermal evaporation or sputtering (see e.g. [23]). This unhydrogenated material was highly defective, which inhibited its use as a semiconductor. Research on incorporating hydrogen as a passivating element was pursued by introducing hydrogen in the sputtering system, which indeed improved the electronical properties [24–26].

In 1965 it was discovered that deposition of amorphous silicon employing the glow discharge technique yielded a material with much more useful electronic properties [27, 28]. The deposition occurs on a moderately heated substrate (200–300°C) through reactions of gas radicals with the substrate. At that time the infrared absorption bands of silicon–hydrogen bonds present in the deposited material were observed, but they were not recognized as such. Some years later, Fritsche and co-workers in Chicago confirmed that *a*-Si produced from a glow discharge of SiH_4 contains hydrogen [29, 30]. A recent personal account of the early years in amorphous silicon research shows the struggle to reach this conclusion [31]. Spear and co-workers in Dundee succeeded in improving the electrical properties, and finally, in 1975, a boost in research activities occurred when it was shown that *a*-Si:H could be doped *n*- or *p*-type by introducing phosphine (PH_3) or diborane (B_2H_6) in the plasma [32]: a range of resistivity of more than 10 orders of magnitude could be reached by adding small amounts of these dopants.

The achievement of Spear and LeComber [32] immediately paved the way for practical amorphous silicon devices. In fact, it is argued that the research field became "polluted by the applications" [33]. Carlson and Wronski at RCA Laboratories started in 1976 with the development of photovoltaic devices [34]. The first *p–i–n* junction solar cell was reported by the group of Hamakawa [35, 36]. In 1980, Sanyo was the first to market devices: solar cells for hand-held calculators [37]. Also, considerable research effort was directed towards amorphous silicon photoconductors for application in photocopying machines and laser printers [38, 39]. The *a*-Si:H photoconductor is, amongst other materials, used as the light-sensitive component in the electrophotographic process.

The first field effect transistors were also reported [40–42] at about this time. These thin film devices take advantage of the capability to deposit and process *a*-Si:H over large areas. It took only a few years before these thin-film transistors (TFTs) were utilized in active matrix liquid crystal displays (AMLCDs) by various companies. Active matrix addressing can also be used in printer heads. Combining the photoconductive properties and the switching capabilities of *a*-Si:H has yielded many applications in the field of linear sensor arrays, e.g., 2D image sensors and position-sensitive detectors of charged particles, X-rays, gamma rays, and neutrons [43–47].

1.1.2. MATERIAL ASPECTS OF HYDROGENATED AMORPHOUS SILICON

1.1.2.1. Atomic Structure

Hydrogenated amorphous silicon is a disordered semiconductor whose optoelectronic properties are governed by the large number of defects present in its atomic structure. The covalent bonds between the silicon atoms in *a*-Si:H are similar to the bonds in crystalline silicon. The silicon atoms have the same number of neighbors and *on average* the same bond lengths and bond angles. One can represent the disorder by the atom pair distribution function, which is the probability of finding an atom at a distance r from another atom. A perfect crystal is completely ordered to large pair distances, while an amorphous material only shows short-range order. Because of the short-range order, material properties of amorphous semiconductors are similar to their crystalline counterparts.

Amorphous silicon is often viewed as a continuous random network (CRN) [48, 49]. In the ideal CRN model for amorphous silicon, each atom is fourfold coordinated, with bond lengths similar (within 1% [50]) to that in the crystal. In this respect, the short-range order (<2 nm) of the amorphous phase is similar to that of the crystalline phase. Amorphous silicon lacks long-range order because the bond angles deviate from the tetrahedral value (109.5°). The average bond-angle variation $\Delta\Theta$ reflects the degree of structural disorder in the random network. Raman spectroscopy is used to determine the vibrational density of states. The transverse optical (TO) peak region has been used to determine $\Delta\Theta$. Beeman et al. [51] have related the half width at half maximum, $\Gamma/2$, of the TO peak to $\Delta\Theta$. Values for $\Gamma/2$ range from 33 to 50 cm^{-1}, which translates into values of 8–13° for the average bond-angle variation.

The CRN may contain defects, but the crystalline concepts of interstitials or vacancies are not valid here. Instead, in the CRN one identifies a *coordination defect* when an atom has too few or too many bonds. In *a*-Si:H a silicon atom can have too few bonds to satisfy its outer sp^3 orbital. It is the common view that the dominant defect in amorphous silicon is a threefold-coordinated silicon atom. This structural defect has an unpaired electron in a nonbonding orbital, called a *dangling bond*. Pure amorphous silicon has a high defect density, 10^{20} cm^{-3} (one dangling bond for every ~500 Si atoms), which prevents photoconductance and doping. The special role of hydrogen with regard to amorphous silicon is its ability to passivate defects. Hydrogenation to a level of ~10 at.% reduces the defect density by four to five orders of magnitude.

The incorporation of phosphorus yields fourfold-coordinated P atoms, which are positively charged, as phosphorus normally is threefold coordinated. This substitutional doping mechanism was described by Street [52], thereby resolving the apparent discrepancy with the so-called $8 - N$ rule, with N the number of valence electrons, as originally proposed by Mott [53]. In addition, the incorporation mechanism, because charge neutrality must be preserved, leads to the formation of deep defects (dangling bonds). This increase in defect density as a result of doping explains the fact that *a*-Si:H photovoltaic devices are not simple *p–n* diodes (as with crystalline materials): an intrinsic layer, with low defect density, must be introduced between the *p*- and *n*-doped layers.

1.1.2.2. Microstructure and Hydrogen Content

The structure on a scale from the atomic level to about 10 nm is called *microstructure*. Many structural aspects are represented there, such as hydrogen bonding configurations (SiH, SiH_2, SiH_3, and polysilane $[(SiH_2)_n]$ groups), multivacancies, internal surfaces associated with microvoids, density fluctuations, the unbonded hydrogen distribution (isolated and bulk molecular hydrogen), and the bonded hydrogen distribution (clustered and dispersed). Columnar structure, void size, and volume fraction are also included in this concept of microstructure, although these are of much larger scale.

The hydrogen content C_H greatly influences structure and consequently electronic and optoelectronic properties. An accurate measurement of C_H can be made with several ion-beam-based methods; see e.g. Arnold Bik et al. [54]. A much easier accessible method is Fourier-transform infrared transmittance (FTIR) spectroscopy. The absorption of IR radiation is different for different silicon–hydrogen bonding configurations. The observed absorption peaks have been indentified [55–57] (for an overview, see Luft and Tsuo [6]). The hydrogen content can be determined from the absorption peak at 630 cm^{-1}, which includes the rocking modes of all possible silicon–hydrogen bonding configurations. The hydrogen content is now defined as

$$C_{\mathrm{H}} = A_{630} \int_{-\infty}^{+\infty} \frac{\alpha_{630}(\omega)}{\omega}\, d\omega \qquad (1)$$

with $A_{630} = 2.1 \times 10^{19}$ cm^{-2} the proportionality constant determined by Langford et al. [58]. Alternatively, one can determine the hydrogen content by integration of the absorption peaks of the stretching modes (2000–2090 cm^{-1}), albeit by using the proportionality constant $A_{2000\text{–}2090}$ determined by Beyer and Abo Ghazala [59], $A_{2000\text{–}2090} = 1.1 \times 10^{20}$ cm^{-2}. Typical hydrogen contents are 9–11%.

In order to distinguish between isolated silicon–hydrogen bonds in a dense network and other bonding configurations, such as clustered monohydride and dihydride bonds, bonds on internal void surfaces, and isolated dihydride bonds, Mahan et al. [60] have defined the microstructure factor R^* as

$$R^* = \frac{I_{2070\text{–}2100}}{I_{2070\text{–}2100} + I_{2000}} \qquad (2)$$

where I_{2000} is the integated absorption band due to isolated Si—H bonds. For all other bonds the IR absorption shifts to 2070–2100 cm^{-1}. Isolated bonds are preferred, so R^* is close to zero.

The refractive index of amorphous silicon is, within certain limits, a good measure for the density of the material. If we may consider the material to consist of a tightly bonded structure containing voids, the density of the material follows from the void fraction. This fraction f can be computed from the relative dielectric constant ϵ. Assuming that the voids have a spherical shape, f is given by Bruggeman [61]:

$$f = \frac{(\epsilon_d - \epsilon)(2\epsilon_d + 1)}{(\epsilon_d - 1)(2\epsilon_d + \epsilon)} \qquad (3)$$

with the effective refractive index $n = \sqrt{\epsilon}$, and $n_d = \sqrt{\epsilon_d}$ the refractive index of the dense fraction.

The presence of a dense material with a varying void fraction results in compressive stress, with typical values of 500 MPa. Compressive stress can be determined conveniently by comparing the curvature of a crystalline silicon wafer before and after deposition of an *a*-Si:H film.

Hydrogenated amorphous silicon is not a homogeneous material. Its structure is thought to consist of voids embedded in an amorphous matrix [62, 63]. The size and number density of the voids depend on the deposition conditions. Poor-quality material can have a void fraction around 20%, while *device quality a*-Si:H has been shown to contain fewer voids, ~1%, with a diameter of ~10 Å [64–66]. The surfaces of the voids may be partly covered with hydrogen atoms [62, 67], and the voids are also thought to be filled with H_2 [68–72]. The influence of voids on the defect density in *a*-Si:H is the subject of much debate.

1.1.2.3. *Electronic Structure and Transport*

The preservation of the short-range order results in a similar electronic structure of the amorphous material to that of the crystalline one: bands of extended mobile states are formed (defined by the conduction and valence band edges, E_C and E_V), separated by the energy gap E_g, more appropriately termed the *mobility gap*. Figure 1 shows schematically the density of states distribution. The long-range atomic disorder broadens the densities of energy states, resulting in band tails of localized states that may extend deep into the bandgap. Coordination defects (dangling bonds) result in electronic states deep in the bandgap, around midgap. As electronic transport mostly occurs at the band edges, the band tails largely determine the electronic transport properties. The deep defect states determine electronic properties by controlling trapping and recombination.

Normally the defect density is low, and electronic transport is considered to occur predominantly at the mobility edges. For electron transport one can write the following expression for the conductivity σ:

$$\sigma(T) = \sigma_0 \exp\left(-\frac{E_C - E_F}{k_B T}\right) \tag{4}$$

where E_F denotes the Fermi energy, σ_0 the prefactor, k_B Boltzmann's constant, and T the temperature. Of particular interest for devices is the conductivity in the dark, σ_d, and the photoconductivity σ_{ph}. For typical undoped material E_F lies at midgap and σ_d equals about $10^{-10}\ \Omega^{-1}\,\mathrm{cm}^{-1}$ or less.

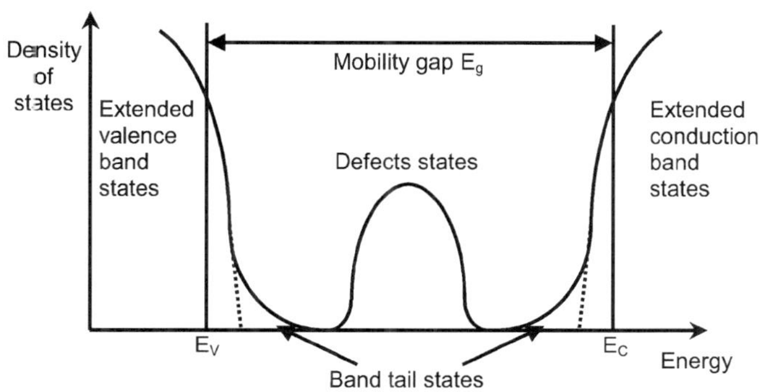

FIG. 1. Schematic density of states distribution. Bands of (mobile) extended states exist due to short-range order. Long-range disorder causes tails of localized states, whereas dangling bonds show up around midgap. The dashed curves represent the equivalent states in a crystal.

The activation energy E_A, defined as $E_C - E_F$ for the conduction band (and analogously for the valence band), can be used to assess the presence of impurities. Due to their presence, either intentional (B or P dopant atoms) or unintentional (O or N), the Fermi level shifts several tenths of an electron volt towards the conduction or the valence band. The activation energy is determined from plots of $\log\sigma(T)$ versus $1/T$, with $50 < T < 160°C$. For undoped material E_A is about 0.8 eV. The Fermi level is at midgap position, as typically E_g is around 1.6 eV.

The photoconductivity usually is determined by illuminating the material with light of known spectral content (AM1.5, 100 mW/cm^2). It has been shown [73] that the photoconductivity can be expressed as

$$\sigma_{\text{ph}} = e\mu\Delta n = \frac{e\eta_g\mu\tau(1 - R_f)F_0}{d}\big[1 - \exp\big(-\alpha(\lambda)d\big)\big] \tag{5}$$

with e the elementary charge, μ the mobility of photogenerated majority carriers (i.e. electrons), Δn their density, η_g the quantum efficiency for generation of carriers, τ their lifetime, R_f and d the reflectance and thickness of the material under study, F_0 the illumination intensity (photons/cm^2 s), and α the wavelength-dependent absorption coefficient. For good intrinsic a-Si:H one usually finds $\sigma_{\text{ph}} = 10^{-5}\ \Omega^{-1}\ \text{cm}^{-1}$ or higher, which is 5–6 orders of magnitude higher than the dark conductivity. One uses the photoresponse (or photosensitivity) $\sigma_{\text{ph}}/\sigma_{\text{d}}$ as a measure of optoelectronic quality.

Besides σ_{ph} and the photoresponse, the quantum-efficiency–mobility–lifetime product, $\eta_g\mu\tau$, is used as a figure of merit. Usually this product is measured at a wavelength of 600 nm, and typical values of $(\eta_g\mu\tau)_{600}$ are 10^{-7} cm^2/V or higher.

1.1.2.4. Optical Properties

The optical properties of a-Si:H are of considerable importance, especially for solar-cell applications. Because of the absence of long-range order, the momentum $\vec{k}$ is not conserved in electronic transitions. Therefore, in contrast to crystalline silicon, a-Si:H behaves as though it had a direct bandgap. Its absorption coefficient for visible light is about an order of magnitude higher than that of c-Si [74]. Consequently, the typical thickness (sub-micrometer) of an a-Si:H solar cell is only a fraction of that of a c-Si cell.

In general one can divide the absorption behavior into three ranges:

1. For absorption coeffient α larger than 10^3 cm^{-1}, absorption takes place between extended states and is described by [75]

$$\big(\alpha(E)n(E)E\big)^{1/(1+p+q)} = B(\hbar\omega - E_g) \tag{6}$$

with $\alpha(E)$ the energy-dependent absorption coefficient, $n(E)$ the refractive index, p and q constants related to the shape of the band edges,

B a proportionality constant, $\hbar$ Planck's constant, and ω the frequency. Tauc has argued that the density of states near the band edges has a square-root dependence on energy, as is the case for crystalline semiconductors [75]. This results in $p = q = 1/2$. Thus, extrapolating $(\alpha(E)n(E)E)^{1/2}$ versus the photon energy to $\alpha(E) = 0$ for $\alpha(E) \geq 10^3\ \text{cm}^{-1}$ yields the Tauc gap. Klazes et al. [76] have proposed the density of states near the band edges to be linear, i.e. $p = q = 1$. Plotting $(\alpha(E)n(E)E)^{1/3}$ versus E yields a so-called cubic gap, which is about 0.1–0.2 eV smaller than the Tauc gap. The cubic plot is linear over a larger energy range than the Tauc gap, and fitting it yields more accurate values. Nevertheless, the Tauc gap is still widely used. For intrinsic a-Si:H the Tauc gap typically is 1.7 eV or smaller. The cubic gap is 1.6 eV or smaller. The refractive index at 600 nm is around 4.3; the absorption coefficient $4 \times 10^4\ \text{cm}^{-1}$.

2. In the range of $1\text{–}10 < \alpha < 10^3\ \text{cm}^{-1}$ absorption takes place between subbandgap states. An exponential dependence of α exists [23]:

$$\alpha = \alpha_0 \exp\left(\frac{E}{E_0}\right) \tag{7}$$

with α_0 a prefactor, and E_0 the so-called Urbach energy. This reciprocal logarithmic slope of the edge, E_0, depends both on the temperature and on the disorder in the material. The slope of the absorption edge is mainly determined by the slope of the valence band tail. Typically, E_0 amounts to 50 meV. The slope of the conduction band tail is about half this value [77].

3. In the low-energy range α depends on the defect density, doping level, and details of the preparation process. Sensitive subbandgap spectroscopy is used to measure α and relate it to the defect density in the material [78, 79].

The optical properties of a-Si:H are influenced both by the hydrogen concentration and bonding in the film, and by the disorder in the a-Si:H network [80]. A linear relation between bandgap and hydrogen concentration is reported [81]: $E_g = 1.56 + 1.27C_{\text{H}}$. However, Meiling et al. [63, 82] have identified three regions in the relation between E_g and C_{H}. The bandgap first increases linearly as a function of hydrogen content, up to a value of about 0.1. A further increase up to $C_{\text{H}} \approx 0.22$ does not lead to a change in the bandgap. For higher values of C_{H} the bandgap further increases linearly, but with a smaller slope than in the first region. This is explained as follows: in the first region ($C_{\text{H}} < 0.1$) hydrogen is present only in monohydride bonds. The microstructure parameter is low; hence there are no voids in the material [62, 83]. In the middle region ($0.1 < C_{\text{H}} < 0.22$) the increasing amount of hydrogen is no longer incorporated as isolated SiH bonds.

The microstructure parameter increases from 0 to 0.6, along with a reduction in silicon density. The material in this region consists of two phases, one which contains only SiH bonds, and one with SiH_2 bonds (chains) and voids. In the third region ($C_H > 0.22$) the material mainly consists of chains of SiH_2 bonds, and the material density is much lower.

Berntsen et al. [84, 85] have separated the effect of hydrogen content and bond-angle variation. The structural disorder causes broadening of the valence and conduction bands and a decrease of the bandgap by 0.46 eV. Hydrogenation to 11 at.% results in an independent increase of the bandgap with 0.22 eV.

For undoped *a*-Si:H the (Tauc) energy gap is around 1.6–1.7 eV, and the density of states at the Fermi level is typically 10^{15} eV^{-1} cm^{-3}, less than one dangling bond defect per 10^7 Si atoms. The Fermi level in *n*-type doped *a*-Si:H moves from midgap to approximately 0.15 eV from the conduction band edge, and in *p*-type material to approximately 0.3 eV from the valence band edge [32, 86].

1.1.2.5. Metastability

An important drawback of *a*-Si:H is its intrinsic metastability: the electronic properties degrade upon light exposure. This was discovered by Staebler and Wronski [87, 88], and is therefore known as the Staebler–Wronski effect (SWE). This effect manifests itself by an increase in the density of neutral dangling bonds, N_{db}, upon illumination, according to $N_{db}(t) \propto G^{0.6}t^{1/3}$, where G is the generation rate and t the illumination time [89]. The excess defects are metastable; they can be removed by annealing the material at temperatures above ~150°C for some hours. The presence of these excess defects in concentrations up to 10^{17} cm^{-3} leads to a reduction of free carrier lifetime, and hence to a lower conversion effiency for solar cells.

The SWE is an intrinsic material property; it is also observed in very pure *a*-Si:H, with an oxygen concentration as low as 2×10^{15} cm^{-3} [90]. At impurity concentrations above 10^{18} cm^{-3} a correlation between SWE and impurity level has been established [91]. The hydrogen concentration, the bonding structure of hydrogen in the silicon network, and the disorder in the silicon network together affect the SWE. It has been long assumed that the origin of the SWE was local. Stutzmann et al. [89] have proposed a model in which photogenerated charge carriers recombine nonradiatively at weak silicon–silicon bonds (e.g., strained bonds). The release of the recombination energy can be used to break the bond, and the two dangling bonds thus formed are prevented from recombining by the passivation of one of the bonds by a back-bonded hydrogen atom. Recent experimental results contradict the local nature of this model. Electron spin resonance (ESR) experiments have revealed that there is no close spatial correlation between the presence of hydrogen and light-induced dangling bonds [92]. Further, it has been found that the rate of dangling-bond creation is independent of temperature in the range of 4 to 300 K [93]. As hydrogen is immobile at 4 K, diffusion can be ruled out.

The SWE has been found to depend on the hydrogen microstructure. The amount of hydrogen bound to silicon in the dilute phase, i.e., monohydride bonds, determines the saturated density of metastable defects [94]. In addition, only in regions of low hydrogen density have light-induced defects been found in ESR experiments [92]; it may well be that in these regions only monohydride bonds up to the solubility limit of 2–4% of hydrogen in silicon [69, 95] are present. The presence of clustered hydrogen (high microstructure parameter R^*) leads to faster defect creation kinetics [96, 97]. These and other observations have led to the realization that the SWE extends over large regions, because configurational defects are transported over long distances. Branz proposed the hydrogen collision model [98]: the defect-creating recombination takes place at a Si—H bond, and the recombination energy is used to lift the hydrogen to a mobile energy level. The mobile hydrogen atom diffuses interstitially through the material. This mobile hydrogen atom is represented by a mobile complex of a Si—H bond and a dangling bond [99]. As the complex diffuses through the silicon network, it breaks one Si—Si bond after another, but each broken bond is re-formed after the complex has passed it. If two mobile hydrogen complexes collide, an immobile metastable two-hydrogen complex is formed. The net result is that two dangling bonds are created that are not spatially correlated to hydrogen atoms.

1.1.2.6. Alloys

The glow discharge technique is especially suitable for controlling various material properties by introducing other precursor gases in the plasma. As was first demonstrated by Anderson and Spear, incorporation of carbon or nitrogen in *a*-Si:H results in material with a large bandgap [100]. A linear relation between bandgap and carbon fraction x_C in *a*-SiC:H has been reported [101, 102]: $E_g = 1.77 + 2.45x_C$. On the other hand, by diluting the plasma with germane (GeH_4), material is obtained with small bandgaps [103]: from 1.7 eV (*a*-Si:H) to 1.0 eV (*a*-Ge:H). Here also a linear relation between bandgap and Ge fraction (x_{Ge}) has been reported [104], in which in addition the hydrogen fraction x_H is included: $E_g = 1.6 + x_H - 0.7x_{Ge}$.

The incorporation of elements as carbon, nitrogen, or germanium, however, leads to material with a low mobility and lifetime of charge carriers. This would limit the application of these alloys, and a large research effort has been undertaken to find ways around this problem.

The addition of an extra feedstock gas (GeH_4 or CH_4) in large quantities (compared to dopant gases) adds an extra degree of freedom to the already complex chemistry of the discharge. It is therefore even more complicated to relate material properties to deposition conditions [105].

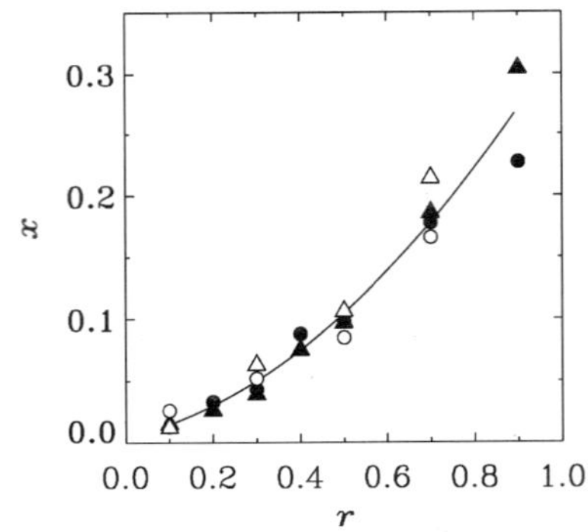

FIG. 2. The dependence of the carbon fraction $x = [C]/([Si] + [C])$ on the gas-flow ratio $r = [CH_4]/([SiH_4] + [CH_4])$ for films deposited in the ASTER system [AST1 (filled circles) and AST2 (filled triangles)] and for films deposited in a similar system (ATLAS) [ATL1 (open circles) and ATL2 (open triangles)]. (From R. A. C. M. M. van Swaaij, Ph.D. Thesis, Universiteit Utrecht, Utrecht, the Netherlands, 1994, with permission.)

As an example, GeH_4 is less stable than SiH_4 in the glow discharge, and Ge is preferentially deposited from SiH_4–GeH_4 mixtures: even a low GeH_4 fraction in the gas results in a high Ge fraction in the solid. Moreover, optimum conditions for *a*-Si:H deposition (low power density) cannot be translated simply to *a*-Ge:H deposition, in which case high power density is required to obtain good-quality material [106]. The deposition of *a*-SiGe:H requires elaborate fine tuning of deposition parameters. Other possibilities have been pursued to improve the properties of *a*-SiGe alloys, such as the use of strong dilution of SiH_4–GeH_4 mixtures with hydrogen [107], and the use of fluorinated reactants such as SiF_4 [108]. Modulation of the discharge (see also Section 1.8.3) has been demonstrated to increase the amount of Ge in the alloy by a factor of 2–10, depending on discharge conditions, compared to a continuous discharge [109].

As SiH_4 is less stable than CH_4, a low C fraction in the solid is obtained for a high CH_4 fraction in the gas. This is illustrated in Figure 2, which shows the dependence of the carbon fraction $x = [C]/([Si] + [C])$ on the gas flow ratio $r = [CH_4]/([SiH_4] + [CH_4])$ for films deposited in the ASTER system and a system with a similar reactor (ATLAS) [101, 102, 110]. These films were deposited in the so-called *low-power* deposition regime [111, 112], which is defined as the regime in which the applied power density is lower than the threshold power density required for the decomposition of methane [112]. In this regime the deposition of films is dominated by the decomposition of silane and is not dependent on the methane concentration [111, 112]. Incorporation of carbon results from chemical reactions between methane molecules and silane species that are created by the plasma. The deposition rate in this regime is almost indepen-

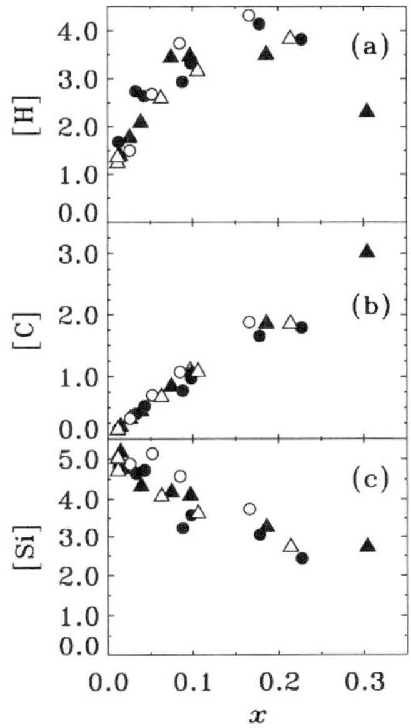

FIG. 3. Absolute atomic concentrations in units of 10^{22} at./cm^3, determined by ERD, RBS, and optical reflection and transmission spectroscopy, of (a) hydrogen, (b) carbon, and (c) silicon as a function of the carbon fraction x. Results are presented for the series AST1 (filled circles), AST2 (filled triangles), ATL1 (open circles), and ATL2 (open triangles). (From R. A. C. M. M. van Swaaij, Ph.D. Thesis, Universiteit Utrecht, Utrecht, the Netherlands, 1994, with permission.)

dent of the gas flow ratio r, provided the gas in the reactor is not depleted in the discharge [101, 102, 113].

The dependence of the absolute atomic concentration of hydrogen, silicon, and carbon on the carbon fraction is shown in Figure 3 [101, 102]. The atomic concentrations were determined by using elastic recoil detection (ERD) and Rutherford backscattering spectrometry (RBS) [114]. The most striking observation is the rapid increase of the hydrogen concentration upon carbon alloying (Fig. 3a), which can be ascribed to the incorporation of CH_n groups in the material during deposition [112, 115]. It can be inferred from the data that up to a carbon fraction of about 0.1, three hydrogen atoms are incorporated per carbon atom. Above $x = 0.15$ the rate of increase of the hydrogen concentration becomes smaller. For higher x-values the hydrogen concentration even tends to decrease.

In Figure 3b and c the absolute atomic concentrations of carbon and silicon, respectively, are shown as a function of the carbon fraction. As expected, the carbon concentration increases upon alloying. In contrast, the silicon content decreases rapidly, which implies that the material becomes less dense. As it was reported that the Si$-$Si bond length does not change upon carbon alloying [116], it thus can be inferred that the a-SiC:H material contains microscopic voids, which is in agreement with small-angle X-ray scattering (SAXS) results [62].

Much more can be said about Ge and C alloying, but that is not within the scope of this chapter. We refer to the books by Kanicki [5] and Luft and Tsuo [6].

1.1.2.7. *Device Quality Characteristics*

Over the past decades the term *device quality* has come to refer to intrinsic PECVD hydrogenated amorphous silicon that has optimum properties for application in a certain device. Of course, depending on the type of device, different optimum values are required; nevertheless the properties as listed in Table I are generally accepted, e.g. [6, 11]. Many of these properties are interrelated, which has to borne in mind when attempting to optimize only one of them.

Optimum properties for p- and n-type doped a-Si:H have been identified [6, 11]. For p-type doping boron is used as the dopant element. Due to alloying with silicon, the bandgap is reduced. This can be compensated by adding carbon. Typically silane (SiH_4), methane (CH_4), and diborane (B_2H_6) are used, with silane and carbon in about equal amounts and diborane a factor of thousand lower. This yields p-type material with $E_g = 2.0$ eV, $E_A = 0.5$ eV, $\sigma_d = 10^{-5}\ \Omega^{-1}\,\text{cm}^{-1}$, and $\alpha_{600} = 10^4\ \text{cm}^{-1}$. Adding phosphine ($PH_3$) to silane in the ratio 0.025 : 1 yields n-type material of good quality: $E_g = 1.8$ eV, $E_A = 0.3$ eV, $\sigma_d = 10^{-3}\ \Omega^{-1}\,\text{cm}^{-1}$, and $\alpha_{600} = 4 \times 10^4\ \text{cm}^{-1}$.

TABLE I
SELECTED PROPERTIES OF *Device Quality* HYDROGENATED AMORPHOUS SILICON FILMS

Property	Symbol	Value	Unit
Optical bandgap (Tauc)	E_g	1.8	eV
Optical bandgap (cubic)	E_g	1.6	eV
Refractive index	n	4.3	
Absorption coefficient (600 nm)	α_{600}	4×10^4	cm^{-1}
Urbach energy	E_0	50	meV
Dark conductivity	σ_d	10^{-10}	$\Omega^{-1}\,\text{cm}^{-1}$
Photoconductivity (AM1.5)	σ_{ph}	10^{-5}	$\Omega^{-1}\,\text{cm}^{-1}$
Activation energy	E_A	0.8	eV
Mobility–lifetime product	$\mu\tau_{600}$	10^{-7}	$\text{cm}^2\,\text{V}^{-1}$
Hydrogen content	C_H	8–12	%
Microstructure factor	R^*	0–0.1	
HWHM of TO Raman peak	$\Gamma/2$	33	cm^{-1}
Intrinsic stress	σ_i	400–500	MPa
Defect density	N_s	10^{15}	$\text{cm}^{-3}\,\text{eV}^{-1}$

1.2. Research and Industrial Equipment

1.2.1. GENERAL ASPECTS

A plasma deposition system usually consists of several subsystems, each providing different functions [117]. The gas handling system includes process gas storage in high-pressure cylinders, mass flow controllers to measure and control the different gases to the reactor, and tubing. The vacuum system comprises pumps and pressure controllers. The plasma reactor is operated between 10^{-4} and 10 Torr.[1] A much lower background pressure, in the ultrahigh vacuum (UHV) range (10^{-9} mbar), is often required to ensure cleanliness of the process. High-vacuum rotary pumps are used in combination with turbomolecular pumps [118].

The deposition setup as shown in Figure 4a is the central part of the most commonly used planar diode deposition system. The power to the reactor system is delivered by means of a power supply connected to the reactor via appropriate dc or RF circuitry (matchboxes). Power supplies can consist of generator and amplifier combined in one apparatus, with a fixed RF frequency. More flexible is to have an RF generator coupled to a broadband amplifier [119, 120].

Using the planar diode geometry, *p–i–n* devices have been made in single-chamber reaction systems, either with [121] or without load lock [122]. In such single-chamber systems cross-contamination occurs. After deposition, small amounts of (dopant) gases will remain in the reactor due to adsorption on the walls. If no load lock is used, water vapor and oxygen are also adsorbed upon sample loading. These residual gases will desorb during subsequent depositions and may be built in as impurities in the amorphous layer [123, 124]. Additionally, the desorption of gases after a deposition may cause contaminated interfaces, e.g., a graded boron concentration profile may be present in the first monolayers of the i-layer of a p^+–i interface [125, 126]. Contaminated interfaces combined with (too) high impurity levels in the layers lead to worse electrical properties of the layer [127, 128], which results in low efficiency of solar cells.

By using a multichamber system [129], exchange of residual gases between successive depositions will be strongly decreased, and very sharp interfaces can be made. Furthermore, the use of a load-lock system ensures high quality of the background vacuum, and thus low levels of contaminants in the bulk layers. Multichamber reactor systems have been used for the fabrication of solar cells, and considerable improvements in energy conversion efficiency have been achieved [130, 131].

Most of the gases used are hazardous: they can be corrosive, flammable, explosive, and/or highly toxic. Silane is pyrophoric; the dopant gases diborane, phos-

---

[1] Throughout this chapter the torr, millibar, and pascal are used as units of pressure, according to the original data rather then converting the first two to the SI unit. Note than 1 Pa = 0.01 mbar = 0.0076 Torr = 7.6 mTorr.

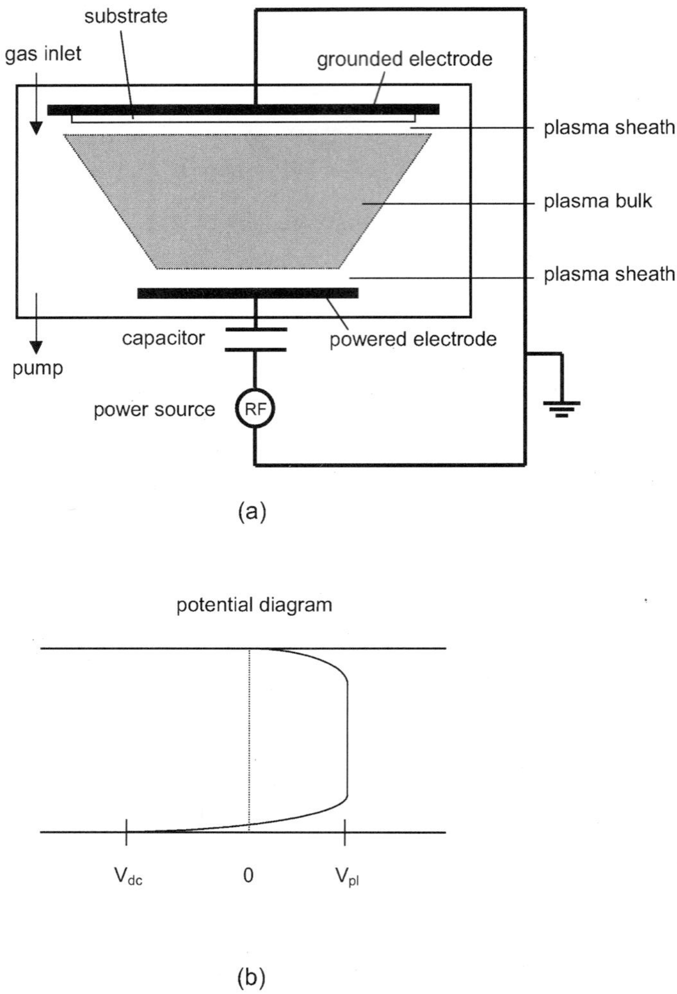

FIG. 4. Schematic representation (a) of a parallel-plate, capacitively coupled RF-discharge reactor, with unequal-size electrodes. The potential distribution (b) shows the positive plasma potential V_{pl} and the negative dc self-bias voltage V_{dc}.

phine, and arsine are extremely toxic, with TLV (threshold limited value) values in the low parts-per-million range. Therefore extreme care has to be taken in dealing with these gases. One generally installs, flow limiters (between the valve of the supply cylinder and pressure regulator, to prevent excessive gas flow in case of breakdown of the regulator), cross-purge assemblies [in order to purge the regulators and prevent release of the gas when exchanging (empty) gas cylinders], scrubbers or diluters to avoid above-TLV levels at the exhaust, detectors mounted at several critical locations, etc. More on safety issues can be found in, e.g., the proceedings of the 1988 Photovoltaic Safety Conference [132].

In the design of the reactor chamber one needs to address issues such as electrode geometry, gas flow patterns, heater design, and reactor volume (discharge and discharge-free regions). Hot walls will influence the gas temperature and the gas phase chemistry. Electrode size and electrode spacing as well as dark-space shields directly influence the discharge properties. Asymmetry introduces a bias, which may be controlled externally. Also the driving frequency is an important parameter. Over the years many reactor designs have been developed. Studies on the effects of process parameters on material quality do not always take geometric causes into account. This has prompted the development of a standard type of reactor, the GEC (Gaseous Electronics Conference) reference cell [133]. Nevertheless generalizations on the effect of process parameters on material quality have been made. For an overview see, e.g., Luft and Tsuo [6].

1.2.2. REACTOR CONFIGURATIONS

Diode reactors can be powered by an RF or dc electric field. In case of RF excitation the deposition typically is on the grounded electrode. As the combined area of grounded electrode and reactor walls usually is larger than the area of the powered electrode, a dc self-bias is developed, as shown in Figure 4b. The potential drop at the powered electrode is much larger than at the grounded electrode. The powered electrode is more negative with respect to ground, and therefore is often called the cathode. The grounded electrode then is the anode. For a dc glow discharge, the potential distribution is similar to the one shown for the RF discharge in Figure 4b. In both cases ion bombardment at the cathode is larger than at the anode. Due to this, deposition of films on the cathode or anode leads to different microstructural properties. Deposition rates at the cathode are usually higher than at the anode. Films deposited at the cathode are dense, but also stressed. Anodic films are more porous. In dc discharges one sometimes uses a mesh positioned above the cathode, which has the same potential as the cathode [134]. Ions are thus slowed down by gas phase collisions in the region between mesh and cathode, and much better material is obtained [135]. Moreover, the mesh acts as a screen for reactive radicals. The SiH_2 radical has a large sticking probability, and it will stick to the mesh easily. As a consequence, the SiH_2 radical will be filtered out, and SiH_3 will dominate the deposition. The resulting material quality has been shown in RF triode discharges to have been improved [136–139].

Planar triode RF discharges used in research generally are asymmetric; the area of the powered electrode is much smaller than the area of all grounded parts taken together (the grounded electrode may be just a small part of the grounded area). The dc self-bias therefore is large. One can reduce the asymmetry by confining the discharge with a grounded mesh [140, 141] or wall [142]. Such confinement also allows for a higher power density in the discharge, which leads to enhanced deposition rates.

An external magnetic field has also been used to confine the plasma [143]. An arrangement where electromagnets are located under the cathode is known as the *controlled plasma magnetron method* [144]. The diffusion of electrons to the walls is prevented by the magnetic field between cathode and anode. This results in an increase in electron density, and consequently in a faster decomposition of silane and a higher deposition rate. At a deposition rate of 1 nm/s, device quality material is obtained [144]. In addition, a mesh is located near the anode, and the anode can by biased externally, both in order to confine the plasma and in order to control ion bombardment.

Hot reactor walls are sometimes used as a means to increase the density of the films that are deposited on the walls. This reduces the amount of adsorbed contaminants on the walls, and leads to lower outgassing rates. A hot wall is particularly of interest for single-chamber systems without a load-lock chamber. Material quality is similar to the quality obtained with a cold reactor wall [145].

Other configurations that are used include an concentric electrode setup in a tubular reactor, where the discharge still is capacitivily coupled. Also, inductive coupling has been used, with a coil surrounding the tubular reactor [146, 147].

1.2.3. Scale-Up to Systems of Industrial Size

A plasma process that has been demonstrated to yield good quality materials in the laboratory will one day need to be scaled up to a technology that can produce the materials in larger sizes and larger quantities. Such a transfer is not straightforward, and many technological difficulties will have to be overcome before a scaled-up process is commercially viable.

A plasma process is characterized by many parameters, and their interrelations are very complex. It is of paramount importance to understand, at least to a first approximation, how the plasma parameters have to be adjusted when the geometrical dimensions of the plasma system are enlarged. Especially of use in scaling up systems are scaling laws, as formulated by Goedheer et al. [148, 149] (see also Section 1.3.2.2).

In general, the substrate temperature will remain unchanged, while pressure, power, and gas flow rates have to be adjusted so that the plasma chemistry is not affected significantly. Grill [117] conceptualizes plasma processing as two consecutive processes: the formation of reactive species, and the mass transport of these species to surfaces to be processed. If the dissociation of precursor molecules can be described by a single electron collision process, the electron impact reaction rates depend only on the ratio of electric field to pressure, E/p, because the electron temperature is determined mainly by this ratio.

The mass transport is pressure-dependent, because the product of diffusion constant and pressure, Dp, is constant. Hence, preservation of both plasma chemistry and mass transport is obtained by keeping both E and p constant during scale-up. As the area of the scaled-up reactor is larger (typically about 1 m^2) than its laboratory equivalent (1–100 cm^2), the total power or current supplied to the discharge must be enlarged in order to keep E constant. For a parallel plate reactor the power simply scales linearly with the area increase, while for other configurations this will not be a simple relation, because the plasma is not confined between the electrodes. Note that changing the area ratio of grounded to powered electrode will affect the dc self-bias voltage and the ion bombardment. The gas flow rate is the only external parameter that needs adjustment. It should be scaled up so that the average flow velocity is identical for laboratory- and industrial-scale reactors. Alternatively, one can keep the average residence time constant, which is defined as $\tau_r = pV/Q$, with V the volume of the reaction zone and Q the total mass flow rate. The residence time is a measure of the distance over which the reactive species diffuse in the reaction zone. With the pressure constant, the requirement of constant residence time scales the gas flow rate proportionally to the reaction volume.

The scale-up from a small to a large plasma reactor system requires only linear extrapolations of power and gas flow rates. However, in practice, the change in reactor geometry may result in effects on plasma chemistry or physics that were unexpected, due to a lack of precise knowledge of the process. Fine tuning, or even coarse readjustment, is needed, and is mostly done empirically.

A critical issue in scaling up a process is the uniformity in deposition rate and material quality. In general, once the deposition rate is constant within 5% over the whole substrate area, the material properties also do not vary much. After fine-tuning the power and gas flow rates, operators still may face in homogeneity issues. These can be caused by local changes in temperature, RF voltage, and gas composition, due to various causes. As an example, it has been reported that improper attachment of the substrate to the grounded electrode results in a local decrease of the deposition rate [150, 151].

Low contamination levels are readily achieved in laboratory scale UHV systems. Very high costs inhibit the use of UHV in industrial scale systems, however, so another, "local-UHV" approach has been proposed, viz. the plasma box reactor [152]. The substrate is mounted in a box, which is surrounded by a shell, which is pumped to a low pressure. The process pressure in the box is maintained by a throttle valve. As the pressure in the box is larger than the pressure in the surrounding shell, contaminants diffuse outwards and the incorporation of contaminants in the deposited layer is low.

Schropp and Zeman [11] have classified current production systems for amorphous silicon solar cells. They argue that cost-effective production of solar cells on a large scale requires that the product of the deposition time needed per square meter and the depreciation and maintenance costs of the system be small. Low deposition rates must be accompanied by low costs. The costs requirement is the main drive for ongoing investigations into the question how to increase the deposition rates while maintaining device quality material. Current production system configurations can be divided into three classes: (1) single-chamber systems, (2) multichamber systems, and (3) roll-to-roll systems [11].

In single-chamber systems there is no transport of substrates under vacuum conditions, which makes the system simple, but contamination levels may be relatively high. The addition of a load lock lowers contamination levels and requires only a linear transport mechanism. During the deposition process temperature and geometry cannot be changed easily, but changes are required for different layers in the solar cell. Substrates are loaded consecutively, and for every substrate a complete pumpdown and heating are needed. The actual deposition time is much smaller than the total processing time. A higher throughput can be achieved by loading substrates in cassettes, as is general practice in the semiconductor industry. The investment is considered low, but so is the flexibility of the system.

Industrial scale multichamber systems offer many of the advantages that are characteristic of laboratory-scale systems. Typically, the deposition of each layer in a solar cell is performed in a separate chamber, and process parameter optimizations can be done for each layer individually. However, this high flexibility comes with high investment costs. Two types of multichamber systems can be distinguished, the cluster configuration and the in-line configuration. The cluster configuration is more or less an enlarged copy of a laboratory-scale system, and offers the greatest flexibility. Transport and isolation chambers can be shared by many deposition chambers, and, depending on the actual configuration, production can be continued even while some reactors are down. A cassette of substrates (batch) can be processed completely automatically, which increases throughput of the production system.

The in-line configuration consists of deposition chambers that are separated by isolation chambers [153]. The layer sequence of a solar cell structure prescribes the actual sequence of deposition chambers. The flexibility is much less than with a cluster configuration, and costs are generally much higher, but the throughput can also be much larger. In an in-line system the substrates can move while deposition takes place, which leads to very uniformly deposited layers, as uniformity of deposition is required only in one dimension (perpendicular to the moving direction).

In the roll-to-roll configuration the substrate consists of a continuous roll, e.g. stainless steel [154], SnO_2:F-coated aluminum [155], or plastic [156–158], which is pulled through a sequence of deposition chambers.

1.2.4. ASTER, A RESEARCH SYSTEM

As a large part of the experimental and simulation results given in this chapter were obtained with or specifically calculated for the UHV multichamber deposi-

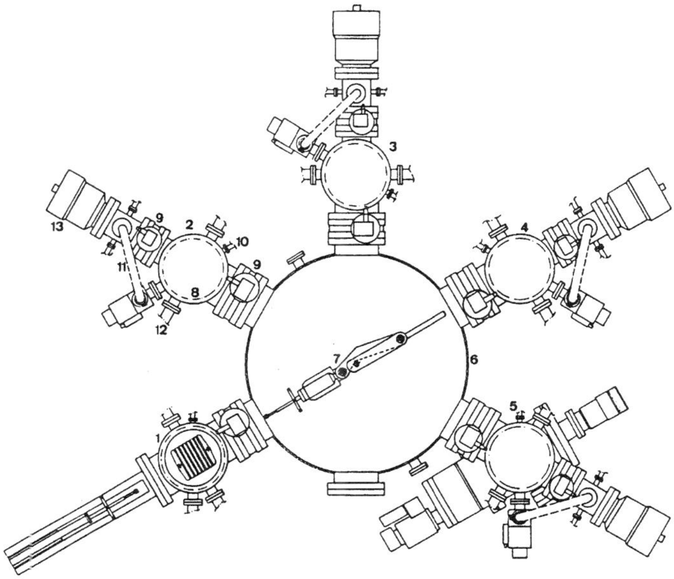

FIG. 5. Schematic representation of the ASTER deposition system. Indicated are: (1) load lock, (2) plasma reactor for intrinsic layers, (3) plasma reactor for *p*-type layers, (4) plasma reactor for *n*-type layers, (5) metal-evaporation chamber (see text), (6) central transport chamber, (7) robot arm, (8) reaction chamber, (9) gate valve, (10) gas supply, (11) bypass, (12) measuring devices, and (13) turbomolecular pump.

tion system ASTER (Amorphous Semiconductor Thin Film Experimental Reactor) at Utrecht University [14, 15], it is described in this section. It also is a more or less generic parallel plate multichamber research system. Similar commercially available multichamber equipment has been developed and described by Madan et al. [159]. It has also been installed at Utrecht University and is named Process Equipment for Amorphous Silicon Thin-Film Applications (PASTA).

The ASTER system is intended for use in research and consists of three identical plasma reactors, a spare chamber, and a load lock, all having the same outer dimensions. These chambers are connected radially via gate valves to a central transport chamber, in which a transport system or robot arm is located. All chambers are made of stainless steel, and all seals with valves, windows, gas-supply lines, and measuring devices are Conflat connections, ensuring that the vacuum will be of UHV quality. The ASTER system was completely designed and partly built in-house; the chambers and transport arm were manufactured by Leybold companies. A schematic drawing is shown in Figure 5, which illustrates the situation around 1990 [14, 15]. The spare chamber was designed first to be used as an evaporation chamber, with which it was possible to evaporate contacts without exposing the sample to air. This was stopped for practical reasons. It was used then as a chamber where samples could be parked, to enhance productivity. As research shifted in the direction of plasma analysis, this fourth chamber was retrofitted with a quadrupole mass spectrometer [160]. After having built a separate system for plasma analysis in which the mass spectrometer was integrated [161], the fourth chamber now is in use for deposition of silicon nitride.

The central transport chamber is an 80-cm-diameter stainless steel vessel, and is pumped by a 1000-l/s turbomolecular pump, which is backed by a small (50 l/s) turbomolecular pump to increase the compression ratio for hydrogen, and by a 16-m^3/h rotating-vane pump. UHV is obtained after a bake-out at temperatures above 100°C (measured with thermocouples at the outside surface) of the whole system for about a week. A pressure in the low 10^{-11}-mbar range is then obtained. With a residual gas analyzer (quadrupole mass spectrometer, QMS) the partial pressures of various gases can be measured. During use of the system, the pressure in the central chamber is in the low 10^{-10}-mbar range due to loading of samples. Water vapor then is the most abundant species in the chamber.

The robot arm can transport a substrate holder for substrate sizes up to 10×10 cm^2. The arm can both rotate and translate, and is driven mechanically by two external dc motors, one for each movement. The translation mechanism was specially designed for fast transport of the samples. The mechanism is based on an eccentrically moving hinge joint, much like a stretching human arm, and it combines high speed with a shock-free start and stop of the movement. It takes only 4 s to translate the substrate holder from the transport arm to the outermost position in the reaction chamber. A gear transmission inside the transport chamber is used to decrease the torque to be applied to the feedthroughs. WSe_2 is used as lubricant for the gear wheels. The arm position is measured by potentiometers. The position of the arm in angular direction is reproducible to within 0.03° of the preset position.

A clear advantage of the central transport chamber is that a sample can be transported to the reaction chambers in any arbitrary sequence, without breaking the vacuum in the reaction chambers. Due to the very low pressure in the central chamber, the vacuum in the reaction chambers in fact is improved when a gate valve is opened. After a complete stretch of the robot arm, the substrate holder can be taken off the arm by a lift mechanism and clamped to the upper (grounded) electrode. Experienced operators are able to manually remove a substrate holder from one chamber and deliver it to any other within 30 s, including the time needed to open and close the gate valves. Automatic control by means of a programmable logic controller, coupled to a personal computer, only slightly shortens this transfer time. Because of the large heat capacity of the 12-mm-thick titanium substrate holder and the short transfer time, the cooling of the sample is limited to a few degrees. Fast transport and a low background pressure in the central chamber ensure that clean interfaces are maintained. Transport at 10^{-10} mbar for 30 s will give rise to a surface contamination of only about 0.01 monolayer of oxygen.

New substrates are mounted on the substrate holder, which then is loaded in the load-lock chamber, which subsequently is evacuated down to 10^{-6}–10^{-7} mbar with a 150-l/s turbomolecular pump in combination with a 25-m^3/h rotating vane pump. In this chamber the substrates are preheated to the desired temperature for the deposition, and at the same time they can be cleaned by dc argon sputtering. Usually the heating time (about 1 h) is much longer than the pumpdown time (less than 10 min). The substrate is transferred to the central chamber by a sequence of operations: opening the gate valve, translating the arm, lowering the substrate holder with the lift mechanism, retracting the arm (with the substrate holder), closing the gate valve. The pumping capacity of the central chamber is such that the 10^{-10}-mbar background pressure is reestablished within 10 s after closing the gate valve.

Each plasma reactor consist of a reaction chamber, an individual pumping unit, pressure gauges, and a dedicated gas-supply system with up to four different gases, according to the layer type to be deposited. Residual gas exchange is prevented by the separation of the pumping units, as well as by the presence of the central chamber. The background pressure in the plasma reactors is lower than 10^{-9} mbar, with partial pressures of water vapor and oxygen lower than 10^{-10} mbar, as a result of the capacity of the pumping unit, which is a 360-l/s turbomolecular pump backed with a 40-m^3/h rotating-vane pump. The process pressure typically is between 0.1 and 0.5 mbar. This can be accomplished by closing the gate valve between reaction chamber and pumping unit (9 in Figure 5). The chamber then is pumped only via the bypass (11 in Figure 5), in which a butterfly valve regulates the process pressure.

A vertical cross section of the reaction chamber is shown in Figure 6. The inside diameter of the chamber amounts to 200 mm. The diameters of the grounded and powered electrodes are 180 and 148 mm, respectively, with a fixed interelectrode distance of 36.5 mm. The interelectrode distance has also been changed to 27 mm, and recently a modified powered electrode assembly has been retrofitted, with which it is possible to vary the interelectrode distance from 10 to 40 mm from the outside, i.e., without breaking the vacuum [162]. With process pressures in the range of 0.1–0.6 mbar the product of pressure and interelectrode distance, pL, may range from about 0.1 to 6 mbar cm. In practice, pL values are between 0.4 and 1.5 mbar cm, i.e., around the Paschen law minimum (see Section 1.3.2.4).

The lower electrode is coupled via a Π-type matching network to a 13.56-MHz generator. This network provides power matching between the RF power cable (50 Ω) and the plasma. Power levels are between 1 and 100 W, or between 6 and 600 mW/cm^2, using the area of the powered electrode.

The substrate holder is positioned face down (6 in Figure 6) on the upper electrode; thus deposition is upward, which prevents dust formed in the plasma from

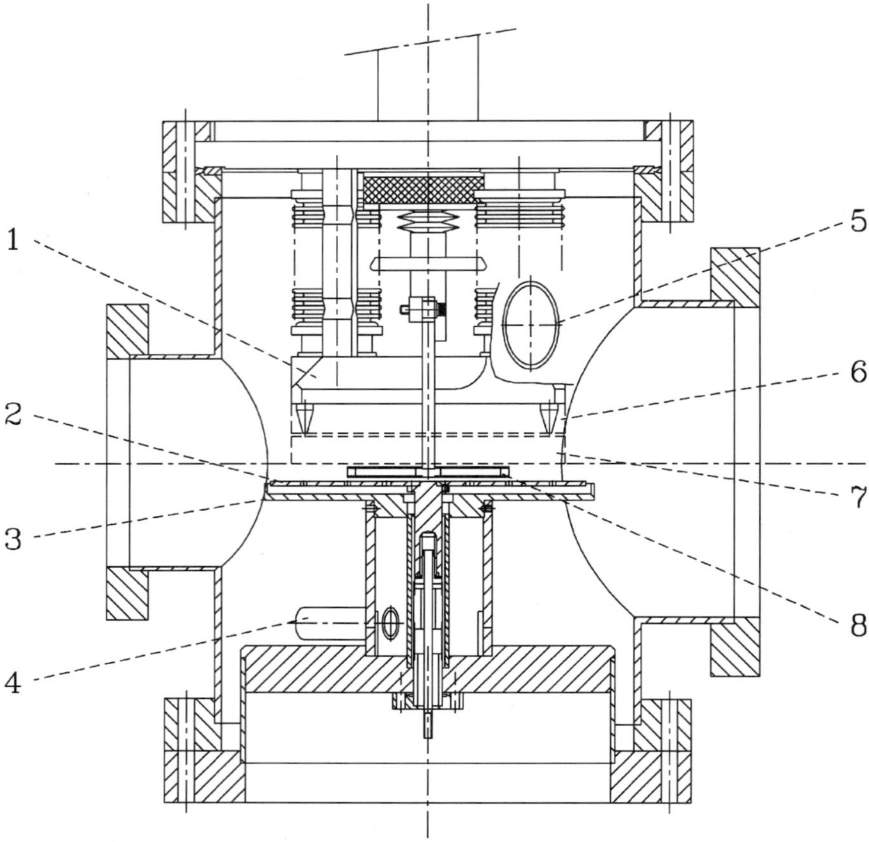

FIG. 6. Vertical cross section of the reaction chamber. Indicated are: (1) the grounded electrode, (2) the RF electrode, (3) the dark space shield, (4) the gas supply, (5) the gas exhaust, (6) the position of the sample holder during deposition, (7) the position of the sample holder when loaded, and (8) the lift mechanism.

falling onto the substrate. The plasma is confined between the powered and the grounded electrode and the reactor walls, which in addition can be water-cooled. The substrate temperature can be varied up to about 500°C by means of fire rods in the grounded electrode. The temperature is regulated by a temperature controller, which measures the temperature with a thermocouple, also mounted in the upper electrode. The actual substrate temperature deviates from the temperature set by the controller, due to heat losses in the system. Therefore the actual substrate temperature has been measured by thermocouples on the substrate over a wide range of controller temperatures, and a calibration graph is used. Throughout this chapter, the actual substrate temperature is given.

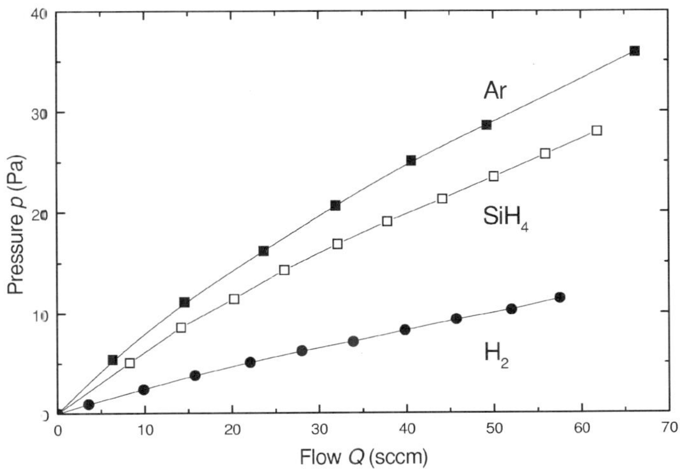

FIG. 7. The pressure p as a function of flow rate Q for different gases (Ar, SiH_4, H_2) at one and the same setting of the pressure regulating butterfly valve in the bypass (11 in Fig. 5). (Redrawn from E. A. G. Hamers, Ph.D. Thesis, Universiteit Utrecht, Utrecht, the Netherlands, 1998, with permission.)

The appropriate gas mixture can be supplied to the center of the reactor (4 in Fig. 6) via holes in the lower electrode (2 in Fig. 6), and is pumped out through the space between substrate electrode and the reactor wall to the exhaust (5 in Fig. 6). Alternatively, the gas mixture can be supplied horizontally, parallel to the electrodes, through a flange in the reactor wall, positioned between the electrodes (perpendicular to the plane of the cross section in Fig. 6, not shown). In this case, the gas is pumped out at the opposite side of the supply.

The volume of the reactor is about 10 l. At a typical process pressure of 0.2 mbar and a total gas flow rate of 60 sccm, the average residence time of molecules in the reactor amounts to about 1.3 s.

Different gases experience different conductances towards the pump. The conductance is a combined result of the manually adjustable butterfly valve in the bypass line and the pumping system. This results in gas-dependent pressures for identical gas flows, as the flow Q is related to pressure p and conductance C_{gas} as $Q = pC_{\text{gas}}$. In Figure 7 the pressures of pure argon, silane, and hydrogen are shown as a function of flow rate [163]. The setting of the butterfly valve was the same for all cases. As can be seen, the conductance for hydrogen is about a factor of 2–3 larger than for argon and silane. From the fact that the rate $dp/dQ = C^{-1}$ decreases with increasing flow, it is inferred that the flow of the gases leaving the reactor is a mixed laminar–molecular (Knudsen) flow [118].

As a consequence of gas-dependent conductances, the composition of a gas mixture may differ from the one expected on the basis of flow ratios. As an exam-

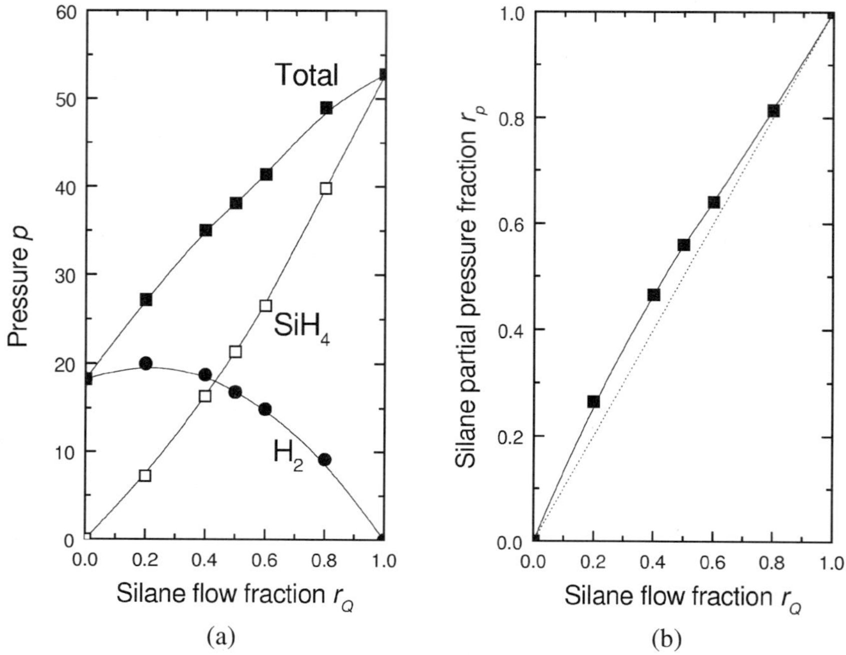

FIG. 8. (a) The total and partial pressures p and (b) the partial pressure ratio r_p of silane and hydrogen in a silane–hydrogen mixture, at different flow ratios r_Q. The total flow rate is 30 sccm. (Adapted from E. A. G. Hamers, Ph.D. Thesis, Universiteit Utrecht, Utrecht, the Netherlands, 1998, with permission.)

ple, the partial pressures (measured with a QMS) of a range of silane–hydrogen mixtures at a total flow of 30 sccm is shown in Figure 8a as a function of the silane flow fraction $r_Q = Q_{\mathrm{SiH_4}}/(Q_{\mathrm{SiH_4}} + Q_{\mathrm{H_2}})$ [163]. A flow of hydrogen yields a lower pressure (about 3 times lower) than the same flow of silane; cf. Figure 7. The actual partial pressure ratio $r_p = p_{\mathrm{SiH_4}}/(p_{\mathrm{SiH_4}} + p_{\mathrm{H_2}})$ for these mixtures is shown in Figure 8b. It is clear that equal flow rates of different gases (in this case hydrogen and silane) do not lead to equal partial pressure ratios. This is important when comparing experimental and simulation data.

Each of the reaction chambers has its own gas supply system, consisting of two channels 3–5 m in length. It also includes a nitrogen purge system. One channel is connected to a gas manifold for three types of gases; the other channel is connected to a single gas bottle. With this configuration it is possible to supply the reaction chamber with either a continuous flow of different gases or a mixture of gases, which may change rapidly during deposition (cf. chemical annealing [164, 165]). The gases are connected via pressure regulators to the supply channels. They can be switched to go via a mass flow controller to the reaction chamber or to go to a drain, which is continuously purged with argon and pumped with an 8-m^3/h rotating-vane pump. When the pressure in the drain or the exhaust lines is too high, a safety exhaust is opened with a much larger diameter than the regular exhaust line from the reaction chamber. Nitrogen is available for pressurizing the system, and also is used to dilute the exhaust lines from the reaction chambers, to ensure below-TLV values on the roof of the building. If one needs to clean a reaction chamber, a separate channel with NF_3 is available. Etching with a NF_3 discharge has been found to reduce considerably the amount of Si_xH_y compounds adsorbed on the wall [166].

In the plasma reactor dedicated for intrinsic material deposition (2 in Fig. 5), only hydrogen and silane are used, along with argon. A mixture of trimethylboron (5% TMB in H_2), SiH_4, and methane (CH_4) is used in the p-plasma reactor (3 in Fig. 5). Diborane can also be used. A mixture of phosphine [PH_3 (1% in H_2)] and SiH_4 is used in an n-plasma reactor (4 in Fig. 5). All gases are of 6.0 quality (99.9999% pure) if available from manufacturers, and otherwise as pure as possible.

The low background pressure (10^{-9} mbar) together with the purity of the gases used ensures a low concentration of contaminants. Amorphous silicon films made in the intrinsic reactor have been analyzed by using ERD, which is available in our laboratory [114]. The determined oxygen content in these films typically is lower than 3×10^{18} cm^{-3}, which is somewhat lower than the values required for obtaining *device quality* films reported by Morimoto et al. [167].

It goes without saying that safety is very important in working with such a setup. A risk analysis has been performed. Safe operation procedures have been formulated. Personnel have been trained to work safely with the setup, which includes exchanging gas bottles using pressurized masks. Interlocks are installed, as well as emergency switches. Emergency power is available in case of power failures. Gases are stored in gas cabinets that were designed and installed with assistence of gas manufacturers. All valves in the gas-supply lines are normally closed valves: if compressed air (6 bar) is not available, all valves are by definition closed. Valves for nitrogen supply are normally open valves. Gate valves are double-action valves: compressed air is needed to either open or close the valve.

1.3. Physics and Chemistry of PECVD

1.3.1. GENERAL INTRODUCTION

A plasma can be defined as a partially ionized, quasineutral gas, consisting of about equal numbers of positive and negative charges, and a different number of un-ionized neutral molecules. An external source of energy is needed to sustain the plasma for a sufficiently long time. The simplest and most widespread method that is used is the electrical discharge, dc or RF. High electric fields applied at millibar pressures yield nonequilibrium plasmas: free electrons are accelerated to 1–10 eV, while ions and neutrals have low energy (0.1 eV). These *hot* electrons initiate chemical reactions through collisions with the *cold* neutrals. Therefore, processing temperatures can be much lower than in thermal CVD, which has tremendous consequences for the applicability of PECVD. Without PECVD modern-day electronic chips would never have been possible. For general references see, e.g., Bruno et al. [8], Grill [117], Chapman [134], and Ricard [168].

This section treats the plasma physics and plasma chemistry of the typical silane–hydrogen RF discharge, with occasional examples that employ a somewhat higher excitation frequency. Electrical characterization of the discharge is followed by an analysis of the silane chemistry. An appropriate set of gas phase species is presented, which are then used in the modeling of the plasma. A comparison is made between modeling results and experimental work in ASTER. Extension to 2D modeling is presented as well.

Plasma analysis is essential in order to compare plasma parameters with simulated or calculated parameters. From the optical emission of the plasma one may infer pathways of chemical reactions in the plasma. Electrical measurements with electrostatic probes are able to verify the electrical properties of the plasma. Further, mass spectrometry on neutrals, radicals, and ions, either present in or coming out of the plasma, will elucidate even more of the chemistry involved, and will shed at least some light on the relation between plasma and material properties. Together with ellipsometry experiments, all these plasma analysis techniques provide a basis for the model of deposition.

1.3.2. Plasma Physics

1.3.2.1. Plasma Sheath

A schematic layout of a typical parallel-plate RF-discharge system is depicted in Figure 4a. The RF power is capacatively coupled to the discharge between the two electrodes. Silane is introduced between the electrodes, and reaction products and unreacted gas are pumped away from the reactor. The substrate onto which a-Si:H is deposited is mounted on the grounded (top) electrode (anode). The RF voltage $V(t) = V_{RF}\sin\omega t$ is applied to the cathode, with $\omega/2\pi = 13.56$ MHz. This particular frequency had been chosen because radiating energy at this frequency would not interfere with communications [134]. However, higher harmonics do, and one should be careful to design proper shielding.

Accelerated electrons in the applied electric field ionize gas molecules, and in these ionization processes extra electrons are created. In the steady state the loss of charged particles is balanced by their production. Due to their much lower mass, electrons move much faster than ions. As a result, charge separation creates an electric field, compensating the diffences in electron and ion velocities. The potential in the central region, or bulk plasma, is slightly positive with respect to the electrodes, due to the small surplus of positive ions.

The time-averaged potential profile is shown in Figure 4b. As ions cannot follow the oscillations in the applied electric field, it is this profile that ions experience. The bulk plasma is characterized by a constant potential, V_{pl}. In both *sheaths* (regions between plasma bulk and the electrodes), the ions experience a potential difference and are accelerated towards the electrodes. This leads to energetic ion bombardment of the electrodes. Electrons are expelled from the sheaths, so all ionization and dissociation processes must occur in the plasma bulk. Plasma light, resulting from emission from excited molecules, is emitted only from the plasma bulk; the sheaths are dark.

The electrons follow the oscillations in the electric field, and experience the time-dependent plasma potential. Due to the capacitor through which the RF power is coupled to the electrodes, no dc current flows through the plasma. The ion and electron currents towards each of the electrodes balance each other over one RF period.

In most systems the substrate electrodes are larger than the powered electrodes. This asymmetric configuration results in a negative dc self-bias voltage V_{dc} on the powered electrode. Without that, the difference in electrode areas would result in a net electron current per RF period [134, 169]. It has been shown that the ratio of the time-averaged potential drops for the sheaths at the grounded (V_{sg}) and the powered electrode (V_{sp}) are inversely proportional to a power of the ratio of the areas of the two electrodes (A_g, A_p) [134, 170–172]:

$$\frac{V_{sg}}{V_{sp}} = \left(\frac{A_p}{A_g}\right)^n \qquad \text{with} \quad 1 < n < 4 \tag{8}$$

For PECVD conditions one usually finds $n \leq 2$ [173]. The plasma potential V_{pl} equals the voltage drop at the sheath of the grounded electrode. It has been reported [170] that $V_{pl} = V_{sg} = (V_{RF} + V_{dc})/2$. In the case of an RF frequency of 13.56 MHz, typical values are 10–30 V for the plasma potential, and with $n = 2$ and with $1 < A_{sg}/A_{sp} < 2$ typical dc self-bias voltages are -50 V for low power (a few watts) to -250 V for high power (up to 100 W).

The plasma potential is the maximum value with which ions can be accelerated from the edge of the sheath towards the substrate, located at the grounded electrode. This may cause ion bombardment, which may induce ion–surface interactions such as enhancement of adatom diffusion, displacement of surface atoms, trapping or sticking of incident ions, sputtering, and implantation; see Section 1.6.2.1.

Meijer and Goedheer [174] have developed a 1D model with which, among other things, the dc self-bias and the sheath voltages can be calculated. It

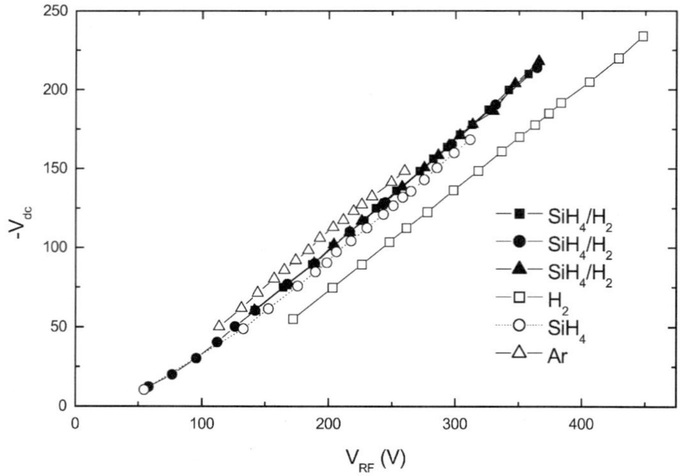

FIG. 9. Relation between dc self-bias voltage V_{dc} and applied RF voltage V_{RF} for different gases (Ar, H_2, SiH_4) and gas mixtures $[0.1 < [SiH_4]/([H_2]+[SiH_4]) < 0.9]$. Note that the slope is independent of the gas used.

is assumed that the potential variation over the *complete* discharge is harmonic:

$$V_{sp}(t) - V_{sg}(t) = V_{RF} \sin \omega t + V_{dc} \tag{9}$$

A further assumption is that the time-averaged current is zero, because of the capacitive coupling of the power supply to the electrodes. With this model unequal-size electrode systems can be modeled, by using an area ratio $\alpha = A_g/A_p$. The ratio between V_{dc} and V_{RF} has been calculated for $1 < \alpha < 10$, and is in excellent agreement [174, 175] with experimental data [176, 177].

Meiling et al. [151] have presented measured data on the relation between V_{dc} and V_{RF} in the ASTER system with an electrode area ratio $\alpha = A_g/A_p$ of 2.1 for various silane–hydrogen mixtures and different applied power levels. It was found that $V_{dc} = aV_{RF} + C$, with $a = 0.64$ and $C = 31$, independent of power and independent of silane–hydrogen mixture. The values of α and a are in excellent agreement with the calculated results from Meijer and Goedheer [174]. Calculations by Goedheer et al. [149] show that this relation also is independent of frequency. The slope a only depends on the geometry (viz. the area ratio) of the reactor system. The intercept C has a different value for pure argon or pure hydrogen plasmas, as can be inferred from Figure 9. The intercept is not predicted in the 1D model by Meijer and Goedheer [174], on account of the low dimensionality of the model. In a 2D model, the intercept is correctly calculated [178].

1.3.2.2. Scaling Laws

Goedheer et al. [148, 149] have attempted to formulate scaling laws. They varied the power P, pressure p, and frequency ω and measured the dc self-bias V_{dc}. From their experimental results (frequency range 60–100 MHz, pressure range 20–60 Pa) a scaling law was formulated [148]:

$$P^{\text{exp}} \propto \omega^2 p^{2/3} V_{\text{dc}} \tag{10}$$

Additional data (frequency range 13.56–100 MHz, pressure range 10–60 Pa) resulted in a slightly changed scaling law [149], $P^{\text{exp}} \propto \omega^2 p^{0.5} V_{\text{dc}}$, which was refined using modeling data to [149, 179]

$$P^{\text{model}} \propto \omega^{1.5} p^{0.7} V_{\text{dc}} \tag{11}$$

In addition a scaling law was derived in which also the electrode distance and radius were introduced:

$$P^{\text{model}} \propto \omega^{1.5} p^{0.7} \left(\frac{L}{R}\right)^{-0.4} R^{2.2} V_{\text{dc}} \tag{12}$$

In the ASTER system (see Section 1.2.4), experiments were performed in order to test this scaling law. To this end, a newly designed RF electrode assembly was retrofitted to a deposition chamber. With this electrode setup, it was possible to change the electrode distance from the outside, without breaking the vacuum. A large data set was taken, consisting of 420 data points [162]: at three values of the pressure ($0.1 \leq p \leq 0.45$ mbar), five of the RF power ($5 \leq P \leq 25$ W), seven of the electrode distance ($12 \leq L \leq 30$ mm), and four of the RF frequency ($13.56 \leq \omega/2\pi \leq 50$ MHz).

A selection of the data set is shown in Figure 10. The dc self-bias voltage is shown as a function of electrode distance, for different values of the RF power, at two pressures and a frequency of 40 MHz. It is clear from the experimental data that the dc self-bias voltage is not strongly dependent on electrode distance. Rauf and Kushner [180] have reported the effect of electrode distance on dc self-bias voltage for the Gaseous Electronics Conference reference cell (GECRC) [133]. They observe a gradual increase of dc self-bias versus electrode distance much like the one presented here, but only above a certain value of L. They attribute this to a change in current flows. Bringing the electrodes, which are of equal size, closer together caused a larger part of the current to flow out through the grounded electrode. Hence, the discharge became more symmetric, and as a consequence the dc self-bias decreased.

Comparing published data on the relation between the dc self-bias and RF voltage between Rauf and Kushner [180] and our group [151] shows that the proportionality constant (slope) in the GECRC is nearly twice the one observed in the ASTER reactor. In other words, the ASTER reactor is less asymmetric than the

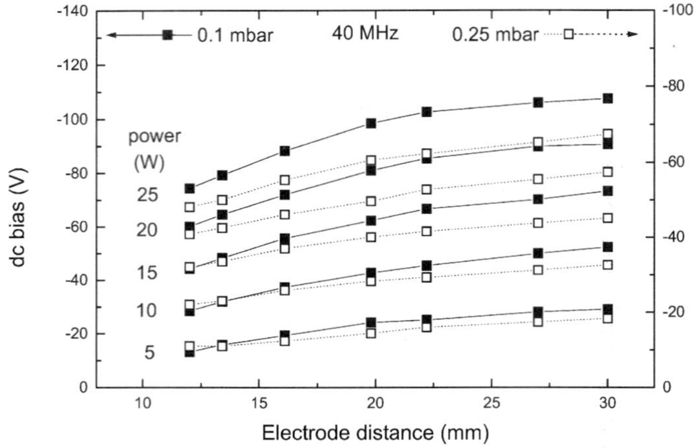

FIG. 10. Dc self-bias voltage V_{dc} as a function of electrode distance for two pressures and six power levels at 40 MHz.

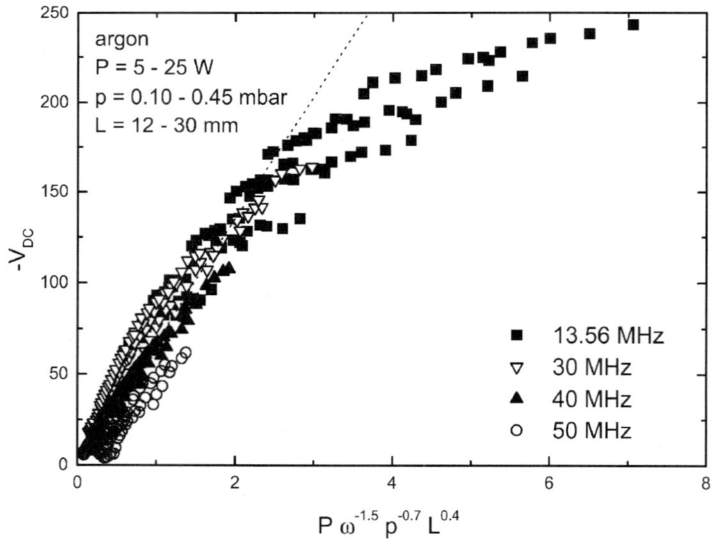

FIG. 11. Scaling law between dc self-bias voltage V_{dc} and $P\omega^{-1.5}p^{-0.7}L^{0.4}$ for an argon discharge.

GECRC. This may be an additional explanation why the V_{dc} data presented here do not show a sudden change as a function of electrode distance.

In Figure 11 the complete data set is plotted using the scaling law

$$P^{\exp} \propto \omega^{1.5} p^{0.7} L^{-0.4} V_{dc} \tag{13}$$

It is clear that the scaling law, which was deduced from modeling results, also holds for the experimental results. However, above a value for the dc self-bias of about −150 V, deviations occur. A closer look at the data reveals that the 13.56 MHz data are responsible for this. Following the arguments discussed by Rauf and Kushner [180], this may well be explained by the fact that two different matching networks were employed, viz. a Π-network for the 13.56-MHz data, and an L-network for the other frequencies. These circuits are not included in the model. Another possible explanation is the presence of higher harmonics at high power levels, and large heat dissipation in the network.

1.3.2.3. Effective Discharge Power

It is difficult to determine the true power that is dissipated in the discharge. Usually, the reading of a directional power meter in the 50-Ω line just before the matching network is reported as the RF power. The effective or true power is not simply the difference between input and reflected power, which is usually also measured with the same power meter. Godyak and Piejak have presented three methods to determine the true power: a subtractive method, a phase shift method, and an integration method [181]. All three methods give the same results. In the phase shift method the discharge power is determined from the discharge voltage and current and the phase shift ϕ between them. In the integration method the product of voltage and current waveform is integrated over one RF cycle.

The subtractive method was adapted from Horwitz [182], and is easiest in use. The principle is to measure the power delivered to the system, including the tuned matching network, in the case that the discharge is on (P_{tot}) and in the case that it is off, i.e. when the system is evacuated (P_{vac}), with the constraint that in both cases P_{tot} and P_{vac} are measured for the same electrode voltage V_{pp}. The matcher efficiency [181] or power transfer efficiency η_P [183] then is defined as

$$\eta_P = \left(\frac{P_{\text{tot}} - P_{\text{vac}}}{P_{\text{tot}}}\right)_{V_{\text{pp}}=\text{const}} = \frac{P_{\text{eff}}}{P_{\text{tot}}} \tag{14}$$

where P_{eff} is the effective power dissipated in the discharge. The power transfer efficiency η_P has been reported to vary between 40% and 90% as a function of pressure [181] and excitation frequency [183].

Both the subtractive and the phase shift method are valid as long as the RF waveform remains sinusoidal [184]. For asymmetrical discharges the integral method is to be used.

1.3.2.4. Paschen Law

In order to sustain a discharge between two parallel plate electrodes, the product of pressure p and interelectrode distance L has to satisfy the Paschen law for the

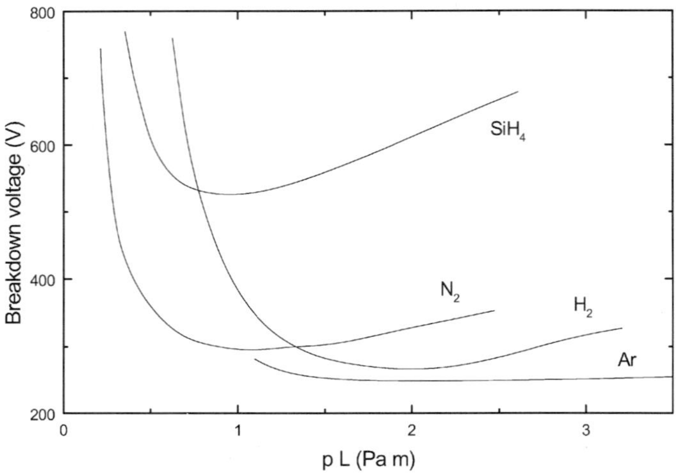

FIG. 12. Measured Paschen law for silane, hydrogen, nitrogen, and argon. (Redrawn from K. M. H. Maessen, Ph.D. Thesis, Universiteit Utrecht, Utrecht, the Netherlands, 1988, with permission.)

gas mixture used. This law relates the product pL and the breakdown voltage V_b (or minimum sustaining voltage) of the discharge:

$$V_b = \frac{C_1 pL}{C_2 + \ln pL} \tag{15}$$

with C_1 and C_2 constants that are dependent on the gas mixture. For large pL the breakdown voltage is proportional to pL. For most gases the minimum breakdown voltage is between 100 and 500 V for pL in the range of 0.1–10 Torr cm [117]. Measurements on argon, silane, hydrogen, and nitrogen [185, 186] are shown in Figure 12. Note that 1 Torr cm equals 1.33 Pa m.

At low pL the energy of the electrons is limited by collisions with the electrodes, and a large voltage is required to compensate the energy lost to the walls. At high pL electrons will collide with gas phase species, and the voltage again is large to compensate for the energy lost to collisions within the plasma, which however promotes polymerization or powder formation. Usually one employs a voltage just above the Paschen minimum to be able to deposit device quality material. Interelectrode distances typically are between 10 and 50 mm. Ishihara et al. [187] have reported only monohydride bonds to be present in the material for $pL \approx 1$ Torr cm. Dihydride bonds begin to appear above $pL = 1.5$ Torr cm.

1.3.3. Plasma Chemistry

For low-pressure plasmas containing mainly inert gases the electrons can be characterized by a Maxwellian electron energy distribution function (EEDF). However, deviations have been reported [188]. The shape of the EEDF depends on the energy lost via electron collisions and the energy gained by the time-dependent electric field in the plasma. Adding silane and/or hydrogen to the plasma results in a non-Maxwellian EEDF [189]. Electrons determine the plasma characteristics, as they are responsible for elastic collisions and inelastic collisions, which lead to rovibrational and electronic excitation, ionization, and dissociation of the neutral gas molecules. In elastic collisions the internal energies of the neutrals are unchanged, whereas the energies, momenta, and directions of the electrons undergo large changes. In inelastic collisions the energy that is transferred from the electron to the neutral species causes many processes to occur, such as dissociation, (dissociative) ionization, (dissociative) attachment, recombination, and vibrational, electronic, and dissociative excitation. As a result of vibrational excitation of the molecules in the plasma, the collisions between these excited molecules allow for redistribution of vibrational states, charge exchange between molecules, excitation of molecules, dissociation, and dissipation of vibrational energy to walls of the system. Each of these processes is characterized by its own (energy-dependent) cross section $\sigma(E)$. The rate constant k of the collisional processes is proportional to the product of EEDF $[f(E)]$ and cross section, according to $k = \int E^{1/2} f(E)\sigma(E)\,dE$. The onset of these processes is at 10 eV or higher, while the EEDF peaks around 1–2 eV. Hence, the tail of the EEDF mainly determines which processes occur.

1.3.3.1. Species Considered

In a silane–hydrogen discharge the feedstock gases SiH_4 and H_2 take part in all the processes that occur. A large number of reactions have been proposed (see e.g. Kushner [190]). Nienhuis et al. [191] have performed a sensitivity analysis in their self-consistent fluid model, from which a minimum set of reactions have been extracted for a typical low-pressure RF discharge. Tables II and III list these reactions. They will be used in the plasma models described in subsequent sections. The review articles on silane chemistry by Perrin et al. [192] and on hydrogen by Phelps [193] and Tawara et al. [194] have been used. The electron collision data are compiled in Figure 13 [189].

1.3.3.2. Electron Impact Collisions

Dissociation of the gases SiH_4 and H_2 by electron impact will create reactive species (radicals) and/or neutrals (Si_2H_6 and even higher-order silanes [195–198]). Atomic hydrogen is an important particle because it is formed in nearly all electron impact collisions, and the H-abstraction reaction [199, 200] of (di)silane is an important process, as is seen from sensitivity study. Dissociation of SiH_4 can create different SiH_x (with $x = 0, 1, 2, 3$) radicals. Only silylene (SiH_2) and

TABLE II
ELECTRON IMPACT COLLISIONS[a]

Acronym	Electron impact reaction	Threshold energy (eV)	Reference	Comment
SE1	$SiH_4 + e^- \to SiH_2^+ + 2H + 2e^-$	11.9	[206]	Lumped
S_2E1	$Si_2H_6 + e^- \to Si_2H_4^+ + 2H + 2e^-$	10.2	[206]	Lumped
SV1	$SiH_4 + e^- \to SiH_4^{(1-3)} + e^- \to SiH_4 + e^-$	0.271	[212]	Vibr. exc. 1–3
SV2	$SiH_4 + e^- \to SiH_4^{(2-4)} + e^- \to SiH_4 + e^-$	0.113	[212]	Vibr. exc. 2–4
SE2	$SiH_4 + e^- \to SiH_2 + 2H + e^-$	8.3	[199, 201, 202]	Fragment. 83%
SE3	$SiH_4 + e^- \to SiH_3 + H + e^-$	8.3	[199, 201, 202]	Fragment. 17%
S_2E2	$Si_2H_6 + e^- \to SiH_3 + SiH_2 + H + e^-$	7.0	[201]	Fragment. 91% [204]
SE4	$SiH_4 + e^- \to SiH_3^- + H$	5.7	[208]	Lumped
HE1	$H_2 + e^- \to H_2^+ + 2e^-$	15.7	[207]	Lumped
HV	$H_2^{v=0} + e^- \to H_2^{v\neq 0} + e^- \to H_2^{v=0} + e^-$	0.54	[213]	$v = 1$
		1.08		$v = 2$
		1.62		$v = 3$
HE2	$H_2 + e^- \to H + H + e^-$	8.9	[205]	$b\,^3\Sigma_u^+$ exc.

[a] Adapted from G. J. Nienhuis, Ph.D. Thesis, Universiteit Utrecht, Utrecht, the Netherlands, 1998, with permission.

silyl (SiH_3) radicals are included in the plasma chemistry, because these two radicals are considered to be the most important ones [190]. The SiH_3 radicals are assumed to be responsible for the film growth (they are *growth precursors*), and the SiH_2 radicals are assumed to be precursors of higher silanes, i.e., Si_nH_{2n+2}. Because the H-abstraction reaction is important, it is necessary to include the Si_nH_{2n+1} radical if Si_nH_{2n+2} is included. Polymers larger than Si_8H_{18} can be ignored [191]. Dissociation of Si_2H_6 produces mainly SiH_3 and Si_2H_5. The latter is also assumed to be a growth precursor.

The total dissociation cross sections of silane and disilane have been taken from Perrin et al. [201]. An uncertainty in the present knowledge of the silane chemistry is the branching ratio of the silane dissociation channels [192]. Here, the branching ratio is taken from Doyle et al. [197], who suggest using the branching ratio determined by Perkins et al. [202] for photolysis, viz., a branching of

TABLE III
CHEMICAL REACTIONS[a]

Acronym	Chemical reaction	Reaction rate ($cm^3\ s^{-1}$)	Reference	Comment
HS	$H + SiH_4 \rightarrow SiH_3 + H_2$	$2.8 \times 10^{-11} \times \exp(-1250/T_{gas})$	[192, 199]	
HS_21	$H + Si_2H_6 \rightarrow Si_2H_5 + H_2$	$1.6 \times 10^{-10} \times \exp(-1250/T_{gas})$	[192, 199, 200]	
HS_22	$H + Si_2H_6 \rightarrow SiH_3 + SiH_4$	$0.8 \times 10^{-10} \times \exp(-1250/T_{gas})$	[192, 199, 200]	
HS_n	$H + Si_nH_{2n+2} \rightarrow Si_nH_{2n+5} + H_2$	$2.4 \times 10^{-10} \times \exp(-1250/T_{gas})$	[192]	Est., $n > 2$
H_2S1	$H_2 + SiH_2 \rightarrow SiH_4$	$3.0 \times 10^{-12} \times [1 - (1 + 0.00023 p_0)^{-1}]$	[192]	p_0 (Pa)
SS1	$SiH_4 + SiH_2 \rightarrow Si_2H_6$	$2.0 \times 10^{-10} \times [1 - (1 + 0.0032 p_0)^{-1}]$	[192]	p_0 (Pa)
S_2S1	$Si_2H_6 + SiH_2 \rightarrow Si_3H_8$	$4.2 \times 10^{-10} \times [1 - (1 + 0.0033 p_0)^{-1}]$	[192]	p_0 (Pa)
S_nS	$Si_nH_{2n+2} + SiH_2 \rightarrow Si_{n+1}H_{2n+4}$	$4.2 \times 10^{-10} \times [1 - (1 + 0.0033 p_0)^{-1}]$	[192]	Est., p_0 (Pa), $n > 2$
SS2	$SiH_3 + SiH_3 \rightarrow SiH_4 + SiH_2$	1.5×10^{-10}	[192, 215]	
SS3	$SiH_3^- + SiH_2^+ \rightarrow SiH_3 + SiH_2$	1.2×10^{-7}	[214]	
S_2S2	$SiH_3^- + Si_2H_4^+ \rightarrow SiH_3 + 2\ SiH_2$	1.0×10^{-7}	[214]	
H_2S2	$SiH_3^- + H_2^+ \rightarrow SiH_3 + H_2$	4.8×10^{-7}	[214]	
S_2S_2	$Si_2H_5 + Si_2H_5 \rightarrow Si_4H_{10}$	1.5×10^{-10}	[192]	Est.

[a] Adapted from G. J. Nienhuis, Ph.D. Thesis, Universiteit Utrecht, Utrecht, the Netherlands, 1998, with permission.

83% into SiH_2 and 17% into SiH_3 (reactions SE2 and SE3 in Table II). However, Perrin et al. [192] suggest that the formation of SiH is also of importance at low electron energies (9.9 eV): it should be at least 20%. In contrast, Hertl et al. [203] have measured SiH densities in typical SiH_4–H_2 discharges, and they show that the SiH density is at least one order of magnitude lower than the SiH_2 density. Therefore, the SiH radical is not included. The fragmentation pattern of disilane dissociation (reaction S_2E2 in Table II) is taken from Doyle et al. [204]; it is 91%

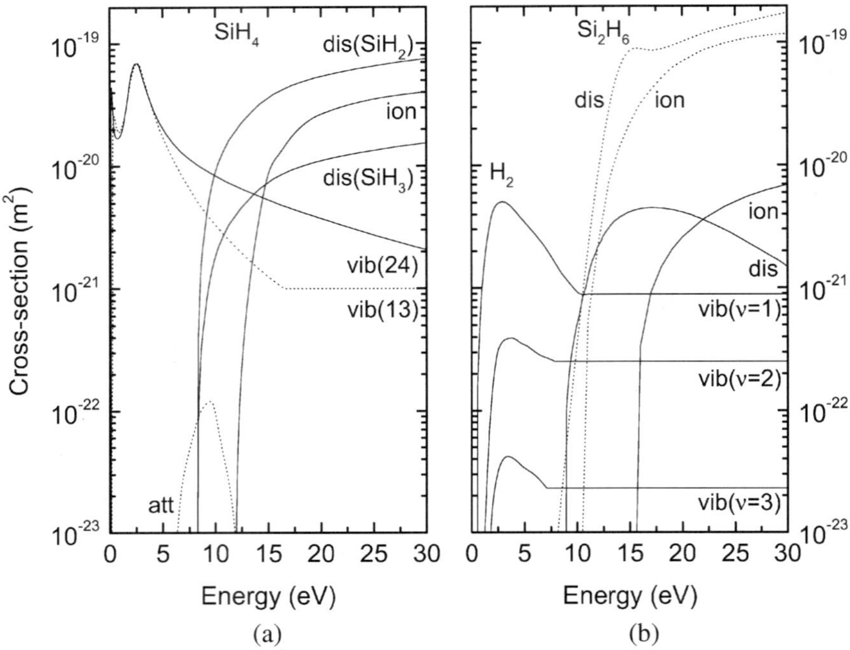

FIG. 13. Cross sections for electron collisions for SiH_4–H_2: (a) SiH_4, (b) Si_2H_6 (dotted lines) and H_2 (solid lines). Abbreviations are: ion, ionization; dis, dissociation; vib, vibrational excitation; att, attachment. See Table II for details and references. (Adapted from G. J. Nienhuis, Ph.D. Thesis, Universiteit Utrecht, Utrecht, the Netherlands, 1998, with permission.)

$SiH_3 + SiH_2 + H$ and 9% $H_3SiSiH + 2H$. As a simplification the second branching path is neglected. Dissociation of H_2 by electron impact is included [205].

Ionization of SiH_4, Si_2H_6, and H_2 can produce a large number of different positive ions [206, 207]. For simplicity all the ionization cross sections of each background neutral leading to different ions have been lumped and attached to the ion that is most likely to be formed: for SiH_4, Si_2H_6, and H_2 these ions are SiH_2^+, $Si_2H_4^+$, and H_2^+, respectively (reactions SE1, S2E1, and HE1 in Table II).

Similarly to the lumping of the positive ionization cross sections, the dissociative attachment cross sections of each background neutral leading to different anions have been lumped to one cross section (reaction SE4 in Table II). The most abundant anions of SiH_4, and H_2 are SiH_3^- and H^-, respectively [208, 209]. Since calculations have shown that dissociative attachment of H_2 has a negligible influence, it is not included. The sum of the dissociative attachment cross sections of Si_2H_6 is comparable in magnitude to that of SiH_4 [210]. However, as the partial pressure of Si_2H_6 is at most a tenth that of SiH_4, the dissociative attachment of Si_2H_6 is neglected.

Vibrational excitation by electron impact of the background neutrals is an important process, because it is a major cause of energy loss for the electrons [reactions SV1 (SiH_4 stretching mode), SV2 (SiH_4 bending mode), and HV in Table II]. Moreover, the density of the vibrationally excited molecules has been reported to be important [211]. However, information about reaction coefficients of vibrationally excited molecules is scarce [192]. Here, only the vibrational excitation of SiH_4 and H_2 is included [212, 213].

1.3.3.3. Chemical Reactions

Table III lists the set of chemical reactions, with reaction rates and references. The rate constants for mutual neutralization of a positive ion and SiH_3^- (reactions SS3, S2S2, and H2S2 in Table III) have been taken from a scaling formula due to Hickman [214], where the electron affinities are taken from Perrin et al. [192] and the ion temperature is assumed equal to the gas temperature. The end products of the mutual neutralization reaction of SiH_3^- and $Si_2H_4^+$ are taken to be SiH_3 and $2SiH_2$ (not SiH_3 and S_2H_4); hence no extra species (Si_2H_4) has to be introduced. The radical–radical reaction $SiH_3 + SiH_3$ can lead to Si_2H_6 or $SiH_2 + SiH_4$ [215]. However, Si_2H_6 will be an excited association product, and collisional deexcitation will only occur at high pressures [192]. This excited Si_2H_6 is likely to dissociate into SiH_2, and SiH_4, and therefore the only end product of $SiH_3 + SiH_3$ is assumed to be $SiH_2 + SiH_4$.

Chemical reactions of positive ions with background neutrals [193, 216], e.g., $SiH_2^+ + SiH_4 \rightarrow SiH_3^+ + SiH_3$ or $H_2^+ + H_2 \rightarrow H_3^+ + H$, have not been included in the model. The reaction coefficients of these reactions are of equal magnitude to those of some radical–neutral reactions, but the densities of the ions are much lower than the radical densities, making ion–neutral reaction of negligible influence on the partial pressure of the background neutrals and the growth rate. For a detailed analysis of the densities and fluxes of ions, the lumped sum of ion densities should be unraveled again.

1.3.3.4. Plasma–Wall Interaction

The plasma–wall interaction of the neutral particles is described by a so-called *sticking model* [136, 137]. In this model only the radicals react with the surface, while nonradical neutrals (H_2, SiH_4, and Si_nH_{2n+2}) are reflected into the discharge. The surface reaction and sticking probability of each radical must be specified. The nature (material, roughness) and the temperature of the surface will influence the surface reaction probabilities. Perrin et al. [136] and Matsuda et al. [137] have shown that the surface reaction coefficient β of SiH_3 is temperature-independent at a value of $\beta = 0.26 \pm 0.05$ at a growing *a*-Si:H surface in a temperature range from room temperature to 750 K. The surface sticking coefficient s of SiH_3 is $s = 0.09 \pm 0.02$ in a temperature range from room temperature to 575 K. Above 575 K the ratio s/β increases to 1 (saturation) at a temperature of 750 K. Note that typical deposition temperatures are below 575 K.

Information about the surface reaction coefficients of radicals Si_nH_{2n+1} where $n > 1$ is scarce. Because the structure of these radicals is similar to that of SiH_3, the same surface reaction coefficients are used. It is assumed that if Si_nH_{2n+1} radicals recombine at the surface with a hydrogen atom, a Si_nH_{2n+2} neutral is formed and is reflected into the discharge. Another possibility is the surface recombination of Si_nH_{2n+1} radicals with physisorbed Si_mH_{2m+1} radicals at the surface. Matsuda et al. [137] have shown that the probability of surface recombination of SiH_3 with physisorbed SiH_3 decreases with increasing substrate temperature. Doyle et al. [204] concluded that at a typical substrate temperature of 550 K, SiH_3 radicals mainly recombine with physisorbed H atoms.

For the SiH_2 radicals the surface reaction coefficients have been taken as $s = \beta = 0.8$ [192]. This sticking coefficient is large because there is no barrier for insertion of this species into the a-Si:H surface. Kae-Nune et al. [217] specify a surface recombination probability of about 1 for atomic hydrogen on an a-Si:H surface during deposition that results mainly in recombination of H with an H-atom bounded to the surface.

If a Si_nH_{2n+1} radical sticks to the surface, not all hydrogen will be incorporated in the layer, and therefore molecular hydrogen will be reflected into the discharge. The amount of molecular hydrogen thus reflected is taken such that the percentage of hydrogen in the deposited layer is 10%. The growth rate is calculated by summing the flux of Si-containing neutrals and ions toward the electrodes and assuming a constant density of 2.2 g/cm^3 of the deposited layer [218].

1.3.3.5. Transport Coefficients

The diffusion constant D_j of neutral particle j is calculated in two steps. First, the binary diffusion coefficient D_{ij} in each of the background gas species (SiH_4, Si_2H_6, H_2) is calculated, following Perrin et al. [192]. Then D_j is approximated using Blanc's law [219]:

$$\frac{p_{\text{tot}}}{D_j} = \sum_{i=\text{background}} \frac{p_i}{D_{ij}} \tag{16}$$

where p_{tot} is the total pressure and p_i is the partial pressure of background species i.

The ion mobility coefficients μ_j are calculated similarly. First, the ion mobility of ion j in background neutral i is calculated using the low-E-field Langevin mobility expression [219]. Then Blanc's law is used to calculate the ion mobility in the background mixture. The ion diffusion coefficient is calculated with Einstein's relation

$$D_i = \frac{k_B T_{\text{ion}}}{e} \mu_i \tag{17}$$

where T_{ion} is the ion temperature, which is assumed to be equal to the gas temperature T_{gas}.

1.4. Plasma Modeling

All properties of a-Si:H are a result of operating conditions of the discharge (pressure, RF frequency, RF power, gas mixture, geometry). Optimization of properties mostly is done empirically, due to the complex chemistry of the silane–hydrogen discharge, including plasma–wall (= substrate) interaction. Nevertheless, serious attempts have been made to model the discharge, thus providing knowledge of the chemistry and the deposition process, with the ultimate goal to find the optimum parameter space of the specific reactor system used.

Various types of models have been used to describe and study glow discharges. These models differ in the approximations that are made [e.g., the dimension, self-consistency, approach (fluid or kinetic), and number of chemical processes], which will influence the physical relevance of the modeling results. However, a higher accuracy (fewer approximations) will require a larger computational effort. Much effort has been put into the development of self-consistent models for glow discharges. Examples are fluid models [220, 221], particle-in-cell/Monte Carlo (PIC/MC) models [222–224], and hybrid models [190, 225, 226]. More extensive overviews have been presented elsewhere [195, 196, 227, 228]. Most of these self-consistent models have been used for discharges with a relatively simple chemistry and no plasma–wall interaction of the neutrals. Other models include the silane–hydrogen chemistry and the surface interaction to a relatively high degree of completeness, but these models are not self-consistent with respect to the rates of the electron impact collisions (i.e., ionization, dissociation, excitation, and attachment); see for instance Perrin [211]. Models developed for plug-flow silane–hydrogen discharges [190, 195, 196] are self-consistent with respect to the silane–hydrogen chemistry, the electron impact collisions, the plasma–wall interaction, and the transport phenomena. However, in this type of models the composition of the neutral background gas is taken equal to the gas feedstock composition, that is, the depletion of silane will be low.

1.4.1. 1D FLUID DISCHARGE MODEL

1.4.1.1. Model Description

Nienhuis et al. [189, 191] have developed a self-consistent fluid model that describes the electron kinetics, the silane–hydrogen chemistry, and the deposition process of a perfectly stirred reactor. Due to the discharge processes, the composition of the background neutrals will differ from the composition of the feedstock gases, because hydrogen and higher-order silanes are formed, while silane is consumed. Fluid models describe the discharge by a combination of balances for the particle, momentum, and energy densities of the ions, electrons, and neutrals obtained from moments of the Boltzmann transport equation, see e.g. [174, 229, 230]. A limitation of a fluid model is the requirement that the mean free path of *all* particles must be less than the characteristic dimensions of the discharge. The balance equations are coupled to the Poisson equation to calculate the electric field, where only electrostatic forces are taken into account. Solving these equations yields the behavior in space and time of the particle densities and the electric potential during one cycle of the periodic steady state RF discharge.

Rates for electron impact collisions as well as the electron transport coefficients depend on the EEDF. The collision rates, the electron transport coefficients, and the average electron energy are obtained by solving the Boltzmann equation for the EEDF. This EEDF has been expanded in two terms with respect to the velocity [231]. The EEDF is calculated as a function of the electric field for a given composition of the neutral background density. A lookup table has been constructed to obtain the collision rates and electron transport coefficients as functions of the average electron energy, which are used in the fluid model. The deposition process is modeled by the use of sticking coefficients, which actually specify the boundary conditions of the particle density balance equations.

The 1D fluid model describes a discharge created and sustained between two parallel plates, where the time-dependent discharge characteristics only vary along the z-axis, i.e., the direction normal to the plates. Plasma parameters such as the dc self-bias voltage, uniformity of the deposition, and radial transport cannot be studied. Self-consistent models in more than one dimension can model these transport phenomena more appropriately [220, 221, 232]. In reality radial dependences do exist in the discharge, due to radial transport processes. Silane is introduced and is transported from the inlet into the discharge, where it may be transformed into other species via chemical reactions. This new species may be transported outside the discharge toward the pump outlet. The model described here corrects for this by means of additional source terms. The reactor is divided into two volumes, the discharge volume (defined as the volume between the two electrodes), and the discharge-free volume; see also Figure 4a. The charged particles and the radicals are confined to the discharge volume. Radical reaction times or diffusion times are in the millisecond range, shorter than the diffusion time to the outside wall, and shorter than the residence time ($\sim$1 s) in the reactor. Hence, radial dependences are neglected, and the reactor is considered perfectly stirred. The radical densities are spatially resolved in the z-direction. while the nonradical neutral densities are assumed to be homogeneous throughout the whole reactor volume. In the remaining discharge-free volume only nonradical neutrals are considered.

The main reason for using a 1D model is the reduction of the computational effort compared to the higher-dimensional models. This reduction to one dimension is acceptable because it is possible to study the sustaining mechanisms and the chemistry in the discharge with a 1D self-consistent model.

To verify the model and to establish in which process parameter space it is valid, a comparison is made with experimental data. These experimental data are the partial pressures of silane, disilane, and hydrogen and the growth rate, obtained during deposition of amorphous silicon. Data are compared for various combinations of the total pressure in the reactor, the electrical power, and the frequency of the power source.

A sensitivity study of the influence of the elementary data (i.e., reaction coefficients, cross sections, and transport coefficients) has been performed to determine the importance of specific elementary data, so to guide further research in this area [189].

1.4.1.2. Particle Balances

For each nonradical neutral j in the discharge volume V_D, the balance equation of their total number N_j^D can be written as

$$\frac{\partial N_j^D}{\partial t} = \int_D S_{\text{reac},j}\, dV + Q_j^{F\to D} \tag{18}$$

where the source term $S_{\text{reac},j}$ represents the creation or destruction of corresponding species j by electron impact collisions and/or by chemical reactions, e.g., $\text{A} + \text{B} \to \text{C} + \text{D}$, where A, B, C, and D indicate the reaction species. The source term for the particles in such a reaction is

$$S_{\text{reac,C}} = S_{\text{reac,D}} = -S_{\text{reac,A}} = -S_{\text{reac,B}} = n_{\text{A}} n_{\text{B}} k_r \tag{19}$$

where k_r is the reaction rate. For electron collisions the reaction rate depends on the local (averaged) electron energy; see Figure 13 and Table II. Other chemical reaction rate coefficients are constant; see Table III. The term $Q_j^{F\to D}$ represents the total number of neutrals j per second that are transported from the discharge-free (F) volume into the discharge (D) volume.

For each nonradical neutral j in the discharge-free volume V_F, the balance equation of their total number N_j^F can be written as

$$\frac{\partial N_j^F}{\partial t} = Q_j^{\text{in}} + Q_j^{\text{pump}} - Q_j^{F\to D} \tag{20}$$

where the term Q_j^{in} represents the number of particles that enter the reactor volume per second as feedstock gas. The term Q_j^{pump} represents the number of pumped particles per second, and is given by

$$Q_j^{\text{pump}} = -\frac{N_j^F}{\tau} \tag{21}$$

with τ the average residence time, which is computed so that the total pressure in the discharge p_{tot}, given by the ideal gas law, equals the preset pressure.

Only the steady-state situation is considered; therefore the partial time derivatives are set equal to zero. Then, combining the above equations and solving for N_j^F yields (with $n_j = N_j^F / V_F$)

$$n_j = \frac{\tau}{V_F}\left(Q_j^{\text{in}} + \int_D S_{\text{reac},j}\, dV\right) \tag{22}$$

For each particle j in the discharge volume (electrons, ions, and radicals) the density balance can be written as

$$\frac{\partial n_j}{\partial t} + \frac{\partial \Gamma_j}{\partial z} = S_{\text{reac},j} \tag{23}$$

where n_j is the density and Γ_j is the flux of particle j.

1.4.1.3. Particle Fluxes

In the fluid model the momentum balance is replaced by the drift–diffusion approximation, where the particle flux Γ consists of a diffusion term (caused by density gradients) and a drift term (caused by the electric field E):

$$\Gamma_j = \mu_j n_j E - D_j \frac{\partial n_j}{\partial z} \tag{24}$$

where μ_j and D_j are the mobility and diffusion transport coefficients, respectively, which are charge-sign-dependent (see Section 1.3.3.5). The mobilities of electrons and negative ions are negative; that of positive ions is positive. For the neutral particles the mobility is set equal to zero. The use of the drift–diffusion approximation is allowed when two conditions are satisfied. The first condition is that the characteristic time between momentum transfer collisions (of the charged particles) is much smaller than the RF period. In other words, the drift–diffusion approximation assumes that a charged particle reacts instantaneously to a change of the electric field, because the moment equations of the fluid model are simplified by neglecting the particle inertia. The typical momentum transfer frequency for electrons is about 100 MHz. For ions the characteristic momentum transfer frequency is only a few megahertz. To use the drift–diffusion approximation for RF frequencies higher than a few megahertz the electric field is replaced by an effective electric field [233], which takes account of the fact that the ions do not respond instantaneously to the actual electric field, unless the momentum transfer frequency is much larger than the applied frequency.

The second condition for the use of the drift–diffusion approximation specifies that the mean free path of these collisions is much smaller than the characteristic gradient lengths in the discharge. In other words the motion of the particles must be collision-dominated. The mean free path of the electrons is the largest among all particles, and their description will become less accurate in the low pressure regime, i.e., at pressures lower than about 10 Pa. The cross section for momentum transfer for the electrons in silane is about 10^{-19} m^2 [212]. At 10 Pa, this gives a mean free path of about 5×10^{-3} m, which is of the order of the length scale of the electron density gradients in the discharge. For ions also this pressure limit is valid.

1.4.1.4. Poisson Equation

The electric field E and potential V are calculated using the Poisson equation,

$$\frac{\partial^2 V}{\partial z^2} = -\frac{e}{\epsilon_0}\left(\sum_{i=\text{ions}} q_i n_i - n_e\right) \qquad E = -\frac{\partial V}{\partial z} \tag{25}$$

with e the elementary charge, ϵ_0 the vacuum permitivity, n_e the electron density, n_i the ion density, and q_i the sign of the ion charge i, i.e., $+1$ or -1.

1.4.1.5. Energy Balances

The electron energy density $w_e = n_e\epsilon$ (the product of the electron density n_e and average electron energy ϵ) is calculated self-consistently from the second moment of the Boltzmann equation:

$$\frac{\partial w_e}{\partial t} + \frac{\partial \Gamma_w}{\partial z} = -e\Gamma_e E + S_w \tag{26}$$

where Γ_w is the electron energy density flux:

$$\Gamma_w = \frac{5}{3}\left(\mu_e w_e E - D_e \frac{\partial w_e}{\partial z}\right) \tag{27}$$

and μ_e and D_e are the electron mobility and electron diffusion coefficients. The term S_w in the electron energy balance equation is the loss of electron energy due to electron impact collisions. The various source terms for electron impact collisions depend on the local average electron energy ϵ, which is calculated by dividing the energy density w_e by the electron density n_e. The energy loss due to elastic collisions of electrons with neutrals [234] amounts to only about 1% of the total energy dissipation at typical discharge conditions, and this term is therefore neglected.

No energy balances for the ions and neutrals are included. The energy balance for neutrals is complex, due to the large range of possible energy exchange reactions in a typical discharge. Moreover, most of the reaction coefficients are unknown [211]. It is assumed that the ions have nearly the same energy as the neutral atoms, because of the efficient energy transfer between ions and neutrals due to their nearly equal mass. Further, the energy gain of the ions from the electric field is low because of the high inertia of the ions; the ions cannot respond to the fast oscillations of the electric field. This assumption is correct for the bulk of the discharge, where the time-averaged electric field is nearly zero. However, the time-averaged electric field in the sheaths is *not* negligible and will increase the energy of the ions far above the thermal energy of the neutrals, as is experimentally shown for silane–hydrogen plasmas [235, 236].

1.4.1.6. Boundary Conditions

In order to solve the differential equations, the boundary conditions of the potential and density profiles must be specified. The potential at the grounded electrode ($z = L$, radius R) is set equal to zero. The potential at the driven electrode ($z = 0$) is set equal to

$$V(z = 0, t) = V_{\mathrm{RF}} \sin(2\pi \nu_{\mathrm{RF}} t) \tag{28}$$

where V_{RF} is the amplitude of the potential and ν_{RF} is the frequency of the oscillating potential. The value of V_{RF} is adjusted until the dissipated power in the discharge equals the preset electrical power. The dissipated power (expressed in watts) is given by the ohmic heating of the charged particles:

$$P_{\mathrm{dis}} = \frac{e\pi R^2}{\tau_{\mathrm{RF}}} \sum_{\substack{j=\text{charged}\\ \text{particles}}} \int_0^{\tau_{\mathrm{RF}}} \int_0^L E(z,t)\Gamma_j(z,t)\, dz\, dt \tag{29}$$

where τ_{RF} is the RF period.

As the electric field always points in the direction of the electrode, the densities of the electrons and negative ions are set equal to zero at the electrode. It is assumed that the ion flux at the electrodes has only a drift component, i.e., the density gradient is set equal to zero. The conditions in the sheath, which depend on pressure, voltage drop, and sheath thickness, are generally such that secondary electrons (created at the electrodes as a result of ion impact) will ionize at most a few molecules, so no ionization avalanches will occur. Therefore, secondary electrons can be neglected.

The boundary conditions for the density balance equation of neutrals are given by the plasma–wall interaction (for instance by deposition processes), which is

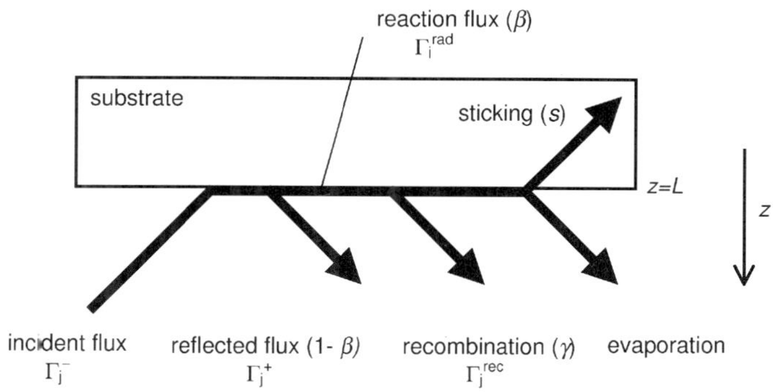

FIG. 14. Schematic representation of plasma–wall interaction for neutrals. (Adapted from G. J. Nienhuis, Ph.D. Thesis, Universiteit Utrecht, Utrecht, the Netherlands, 1998.)

described by a sticking model; see Figure 14. This surface interaction is modeled by using surface reaction coefficients β_j, which specify the probability that neutral j will react at the surface. The coefficient $R_j = 1 - \beta_j$ represents the fraction of the incident neutrals that are reflected into the gas phase keeping their initial identity. It is assumed that the nonradical neutrals do not react at the wall, whereas the radical neutrals may do so, and form a nonradical neutral or contribute to *a*-Si:H layer formation.

For the radical neutrals, boundary conditions are derived from diffusion theory [237, 238]. One-dimensional particle diffusion is considered in gas close to the surface at which radicals react (Figure 14). The particle fluxes in the two z-directions can be written as

$$\Gamma_j^- = \frac{1}{4} n_j v_{\text{th},j} + \frac{D_j}{2}\frac{\partial n_j}{\partial z} \tag{30}$$

$$\Gamma_j^+ = \frac{1}{4} n_j v_{\text{th},j} - \frac{D_j}{2}\frac{\partial n_j}{\partial z} \tag{31}$$

where $v_{\text{th},j}$ is the thermal velocity of neutral j, given by $v_{\text{th},j} = \sqrt{8k_B T_{\text{gas}}/\pi m_j}$ with m_j the mass of particle j. The first terms in Eqs. (30) and (31) are the random contributions, and the second terms are the contributions due to the gradient in the concentration of diffusing particles. The resulting net flux Γ_j is given by $\Gamma_j = \Gamma_j^+ - \Gamma_j^-$. The fraction R_j of the flux incident on the wall (Γ_j^-) is equal to the flux coming from the wall, $\Gamma_j^+ = R_j \Gamma_j^-$, when no radicals are produced at the wall. The boundary condition of radical j, which reacts at the surface of the electrode, then is given by

$$\Gamma_j^{\text{rad}} = -D_j \left.\frac{\partial n_j^{\text{rad}}}{\partial z}\right|_{z=0} = -\frac{\beta_j}{1-\beta_j/2}\frac{v_{\text{th},j}}{4} \left. n_j^{\text{rad}}\right|_{z=0} \tag{32}$$

where Γ_j^{rad} and $n_j^{\text{rad}}|_{z=0}$ are the flux of radicals j reacting at the surface, and their density at the surface, respectively.

The recombination probability γ_{ij} is defined as the probability that a neutral radical j reacts on the surface, forming a volatile stable product, i.e., a nonradical neutral i, which is reflected into the discharge (e.g. $SiH_{3(\text{gas})} + H_{(\text{surface})} \rightarrow SiH_{4(\text{gas})}$). This causes a flux into the discharge:

$$\Gamma_j^{\text{rec}} = \sum_{j \neq i} \frac{\gamma_{ij}}{\beta_j} D_j \frac{\partial n_j^{\text{rad}}}{\partial z}\bigg|_{z=0} \tag{33}$$

The sticking coefficient $s_i = \beta_i - \gamma_i$ (where $\gamma_i = \sum_{j \neq i} \gamma_{ij}$) is the probability that a neutral i sticks to the surface. In case of silane radicals Si_nH_m, this means film growth by Si incorporation. At the same time atomic hydrogen is released or evaporated by the layer. The evaporation flux is chosen such that the amount of hydrogen in the layer is 10%, i.e., the typical *device quality* hydrogen content (the possibility of a hydrogen-rich overlayer is neglected).

For each neutral particle the boundary conditions are formulated; one then obtains a coupled set of equations, which is solved to obtain the boundary conditions for all neutral density balance equations.

1.4.1.7. Electron Energy Distribution Function

In order to calculate the rates for electron impact collisions and the electron transport coefficients (mobility μ_e and diffusion coefficient D_e), the EEDF has to be known. This EEDF, $f(\vec{r}, \vec{v}, t)$, specifies the number of electrons at position $\vec{r}$ with velocity $\vec{v}$ at time t. The evolution in space and time of the EEDF in the presence of an electric field is given by the Boltzmann equation [231]:

$$\frac{\partial f}{\partial t} + \vec{\nabla}_{\vec{r}} \cdot (\vec{v} f) - \vec{\nabla}_{\vec{v}} \cdot \left(\frac{e}{m_e} \vec{E}(\vec{r}, t) f \right) = \left(\frac{\partial f}{\partial t} \right)_{\text{coll}} \tag{34}$$

where $(\partial f/\partial t)_{\text{coll}}$ is the collision term. Some approximations have to be made, as this equation is complex (details can be found in [189]).

First, it is assumed that the EEDF is spatially uniform and temporally constant, which is allowed if the energy relaxation time of the EEDF is much shorter than the RF-cycle duration, and if the relaxation length is much smaller than the typical gradient scale length. This assumption implies a spatially and temporally constant electric field. It reduces the Boltzmann equation to a problem exclusively in the velocity space.

Second, in the collision term only electron–neutral collisions are considered, because RF plasmas are weakly ionized. The inelastic collisions considered are ionization, dissociation, excitation, and attachment (see also Table II). The crude first approximation is relaxed in the fluid model by the introduction of the spatial and temporal dependence of the average electron energy. The collision rates, transport coefficients, and average electron energy are calculated in a range of electric fields E for a given composition of the background neutrals. The results are then tabulated to obtain a lookup table of the rates and transfer coefficients as a function of the averaged electron energy. In the fluid model the (time- and space-dependent) average electron energy is calculated with the energy balance equation. The validity of this assumption has been demonstrated [175, 221, 230, 234, 239]

Third, a further simplification of the Boltzmann equation is the use of the two-term spherical harmonic expansion [231] for the EEDF (also known as the Lorentz approximation), both in the calculations and in the analysis in the literature of experimental data. This two-term approximation has also been used by Kurachi and Nakamura [212] to determine the cross section for vibrational excitation of SiH_4 (see Table II). Due to the magnitude of the vibrational cross section at certain electron energies relative to the elastic cross sections and the steep dependence of the vibrational cross section, the use of this two-term approximation is of variable accuracy [240]. A Monte Carlo calculation is in principle more accurate, because in such a model the spatial and temporal behavior of the EEDF can be included. However, a Monte Carlo calculation has its own problems, such as the large computational effort needed to reduce statistical fluctuations.

The method to calculate the rates of electron impact collisions as used in this model has been studied by Meijer et al. [230] and has also been used by others [221, 234].

1.4.1.8. Calculation Scheme

The system of equations of the fluid model as well as the Boltzmann equation of the EEDF must be untangled numerically. The balance equations are spatially discretized (129 grid points in this study) using the Sharfetter–Gummel exponential scheme [241]. An implicit second-order method is used to numerically treat the time evolution of the balance equations [220]. The electron transport equation is solved simultaneously with Poisson's equation to avoid numerical instabilities. A Newton method is used to solve the resulting set of nonlinear equations. This approach limits the required number of time discretization steps (here 80 time steps per RF cycle). Doubling the number of grid points and time steps resulted in 10^{-2} and 10^{-4} for the maximum relative changes of the partial pressures (SiH_4, H_2, and Si_2H_6) and of the deposition rate, respectively. A multigrid technique is used to enhance the convergence of the fluid model [242]. Further details about the numerical techniques of the fluid model and the Boltzmann solver can be found elsewhere [175, 243].

The fluid model calculates the density profiles based on the electron impact collision rates calculated by the Boltzmann solver. This Boltzmann solver needs the densities of the background neutrals as input. Therefore, the fluid model and the Boltzmann solver are run alternately until the changes in the densities of the background neutrals are less than 10^{-4}. Convergence of the fluid model is obtained when the relative change ϵ_n of the discharge parameters (density profiles and potential profile) at the beginning of two subsequent RF periods (n and $n+1$) is less than 10^{-6}. To check whether this criterion is sufficient, some calculations were continued until ϵ_n reached a value 10^{-7}. In those cases, the partial pressures of SiH_4, H_2, and Si_2H_6 and the deposition rate differed relatively by at most 10^{-5} from the results obtained with ϵ_n equal to 10^{-6}.

1.4.1.9. Simulation Results

The 1D fluid discharge model has been applied to the ASTER deposition system (see Section 1.2.4). The deposition reactor has an inner volume of 10 l and an inner diameter of 20 cm. The upper electrode is grounded (see Fig. 4a), and the powered electrode is located 2.7 cm lower. Other typical silane–hydrogen discharge parameters are summarized in Table IV.

Using these discharge settings in the model yields the following partial pressures for the three background neutral particles considered: 7.72 Pa for SiH_4, 11.51 Pa for H_2, and 0.55 Pa for Si_2H_6. With these partial pressures, and the data on electron collisions as given in Table II, first the EEDF can be calculated, followed by the calculation of the electron transport coefficients and electron impact rates. From a comparison of a Maxwellian EEDF and the two-term Boltzmann

TABLE IV
TYPICAL SETTINGS FOR A SILANE–HYDROGEN DISCHARGE IN THE ASTER SYSTEM

Parameter	Value	Unit
Electrode spacing	0.027	m
Electrode radius	0.08	m
Reactor volume	0.010	m^3
Silane flow	30.0	sccm
Hydrogen flow	30.0	sccm
Pressure	20.0	Pa
Gas temperature	400	K
Substrate temperature	550	K
RF frequency	50	MHz
RF power	5	W

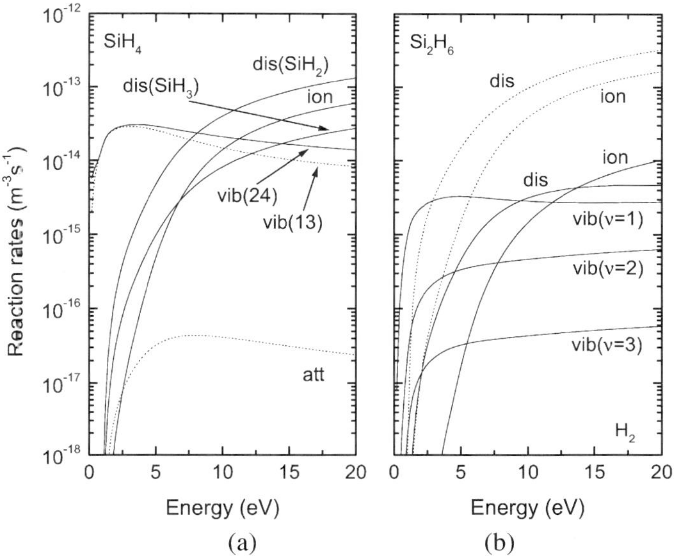

FIG. 15. Electron impact reaction rates as a function of the average electron energy. A 1 : 1 mixture of SiH_4 and H_2 was used, at a total pressure of 83 Pa. (a) Reaction rates for SiH_4, (b) reaction rates for Si_2H_6 (dotted lines) and H_2 (solid lines). Abbreviations are: ion, ionization; dis, dissociation; vib, vibrational excitation; att, attachment. See Table II for details and references. (Adapted from G. J. Nienhuis, Ph.D. Thesis, Universiteit Utrecht, Utrecht, the Netherlands, 1998.)

EEDF as calculated here, it follows that the Maxwellian EEDF underestimates the EEDF at lower energies and overestimates it at higher energies [189].

In Figure 15 the electron impact rates for the various reactions (Table II) are shown as functions of the average electron energy. In a typical silane–hydrogen discharge the average electron energy amounts to about 5 eV. As can be seen in Figure 15, at an energy of 5 eV the dissociation of silane and disilane and the vibrational excitation of silane have the highest rates.

Ohmic heating of the electrons amounts to about 80% of the plasma power. This dissipated power is mainly used for vibrational excitation of the molecules, in contrast to a inert gas discharge (e.g. Ar), where a much larger fraction of the power is used for the acceleration of the positive ions in the sheaths. Another remarkable outcome is the high density of negative ions (2.38×10^{16} m^{-3}) compared to the electron density (7.19×10^{14} m^{-3}). This is typical for an electronegative discharge, where the rate coefficient for recombination of a negative ion with positive ions is low.

Figure 16 shows the calculated time-averaged densities of charged (Fig. 16a) and neutral species (Fig. 16b). The SiH_3^- density is the highest in this case, followed by the SiH_2^+ density: quasineutrality is preserved. It can be clearly seen

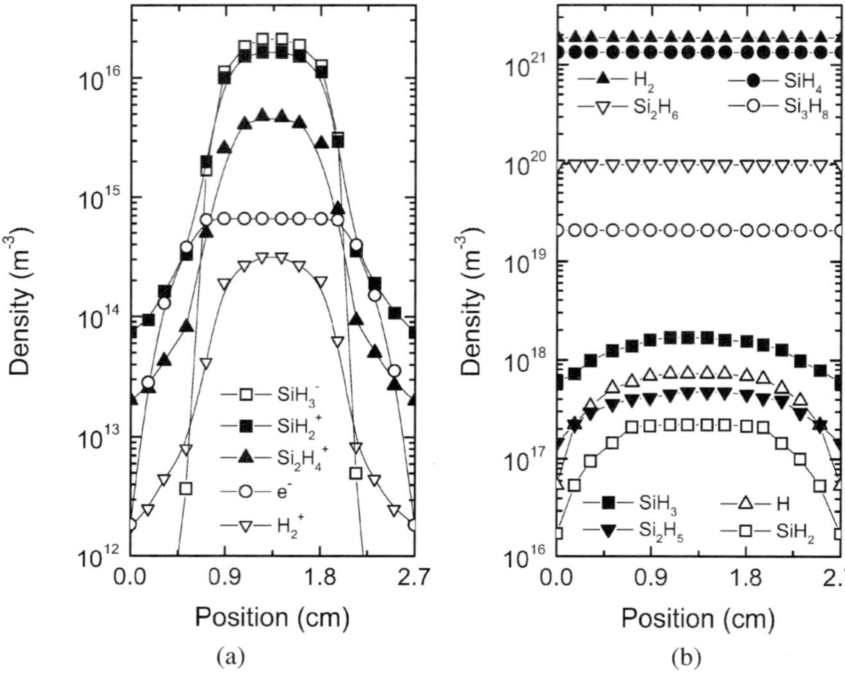

FIG. 16. Time-averaged densities of (a) charged and (b) neutral species for the discharge settings given in Table IV. Note the difference in density scales. (Compiled from G. J. Nienhuis, Ph.D. Thesis, Universiteit Utrecht, Utrecht, the Netherlands, 1998.)

that the SiH_3^- ions are confined in the discharge. The electrons are mobile and therefore much less confined. The temporal variation of the ion density profile is small, except in the sheaths. The mass of the ions is too high to respond instantaneously to the changing electric field. Ion transport is only influenced by the time-averaged electric field. The total (positive) ion flux to the electrode is 4.06×10^{18} m^{-3} s^{-1}.

From Figure 16b it is clear that the density of the nonradical neutrals (H_2, SiH_4, Si_2H_6, and Si_3H_8) is much higher than that of radicals (SiH_3, H, Si_2H_5, and SiH_2). Moreover, from the decreasing density near the electrodes, the species that react can be identified. Note that the SiH_3 density is about one order of magnitude larger than the SiH_2 density.

1.4.1.10. Comparison with Experiments

In this sub-subsection the results of the 1D model are shown together with data taken from experiments in the ASTER system. Both the partial pressures of the background neutrals (H_2, SiH_4, and Si_2H_6) and the deposition rates are accessible in experiment and model. The comparison is made as a function of one of the discharge parameters: the RF frequency, the RF power, or the total pressure in the reactor. Other process parameters have been kept constant; see Table IV.

Of course, care should be taken when comparing the experimental data and the modeling results; the model is an approximation, and the experiment has its uncertainties. The most important approximation is that the model is one-dimensional. A main uncertainty of the experiment is the relation between *source* power and *plasma* power (see Section 1.3.2.3). The source power is defined as the power delivered by the power generator. In experiments, the source power is a discharge setting (Table IV). The plasma power is the power dissipated by the plasma. As a rule, this plasma power is smaller than the source power, due to losses in the matching network, which matches the plasma impedance to the impedance of the source power generator, viz., 50 Ω [180]. From comparison of experimental data with modeling results (e.g., [220]) in a wide range of discharge settings it is inferred that the plasma power is approximately 50% of the source power. This is a crude approximation, as in fact this efficiency depends on discharge settings and the resulting discharge impedance. Agreement, therefore, is not expected between experimental and modeling results. Tendencies, however, will be clear from the results.

Another important problem to address is the regime in which the discharge is operating, whether the α (dust-free) regime or the γ'-regime, where dust plays an important role in the discharge [244] (see Section 1.6.1.6). Dust particles consist of large clusters of Si and H atoms, and are negatively charged by attachment of free electrons. They can be considered as an additional wall surface. Consequently, the free electron density decreases. In order to maintain the discharge, i.e., the amount of ionization, the average electron energy has to increase. This also leads to an enhanced dissociation of SiH_4 and thus to a larger radical production.

Generally, at low pressures and at low power levels the discharges operate in the α-regime. At a certain critical pressure, which depends more or less upon the other discharge parameters, the discharge makes a transition from the α- to the γ'-regime. The modeling does not take dust effects into account. The discharge settings used are interesting, because at increasing total pressure the discharge regime is changed from α to γ', at a critical pressure of about 30 Pa.

Pressure variation. In Figure 17 are shown the effects of total pressure on the relative pressures (i.e., the ratio of the partial pressure to the total pressure) of silane, hydrogen, and disilane (Fig. 17a) and on the deposition rate (Fig. 17b). The RF frequency is 50 MHz, and the plasma power is 5 W. The relative pressure of hydrogen slowly increases, and the relative pressure of silane slowly decreases, both in model as well as in experiment. This is caused by an increase in silane depletion at higher total pressures, which results from a higher power dissipation

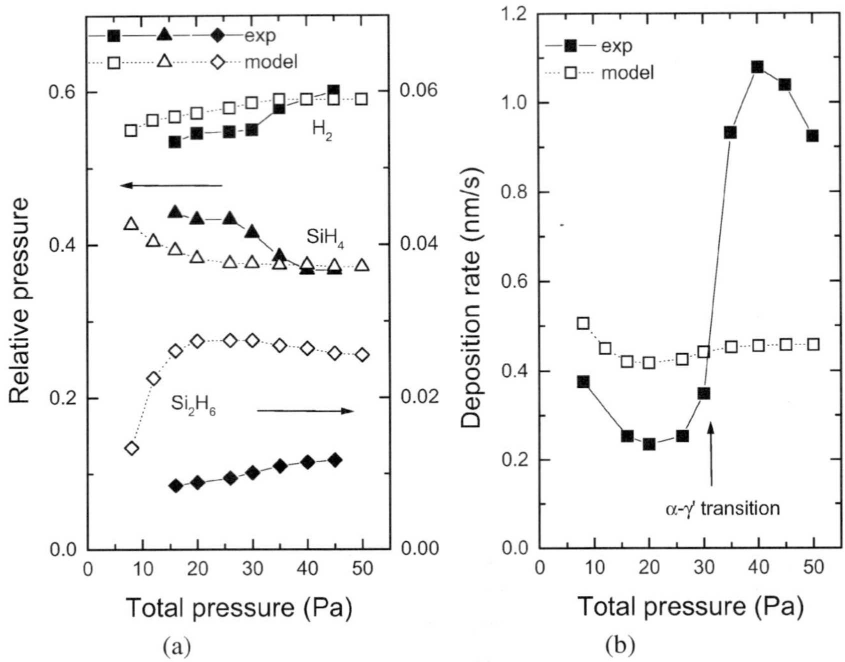

FIG. 17. (a) The relative pressures (i.e., the ratio of the partial pressure to the total pressure) of H_2, SiH_4, and Si_2H_6, and (b) the deposition rate, as a function of total pressure at an RF frequency of 50 MHz and a plasma power of 5 W. Other discharge settings are given in Table IV. Modeling results are in dotted lines and open symbols, experimental data in solid lines and filled symbols. Note the sudden increase at 30 Pa, i.e., the transition from the α- to the γ'-regime. (Compiled from G. J. Nienhuis, Ph.D. Thesis, Universiteit Utrecht, Utrecht, the Netherlands, 1998.)

by the electrons. At 16 Pa the fraction of dissipated power that is used to heat electrons is 79%, at 50 Pa this increases to 97%. This increase leads to a higher dissociation rate of silane, and explains the increased depletion.

The relative pressure of disilane increases as a function of total pressure, due to the increased production of radicals, which is a result of increased dissociation of silane, as well as to the shorter gas volume reaction times and longer diffusion times to the walls, which result from increasing the pressure.

Below 30 Pa the trends agree between model and experiment. The model slightly underestimates the silane partial pressure, by about 10%, and overestimates the hydrogen pressure by about 6%. Possible causes for this discrepancy are the simple 1D geometry of the model, the assumed relation between plasma power and source power, and the approximation for the pumping by using an average residence time. The discrepancy for the disilane data might be explained e.g. by questioning the dissociation branching ratio of silane. Layeillon et al. [195,

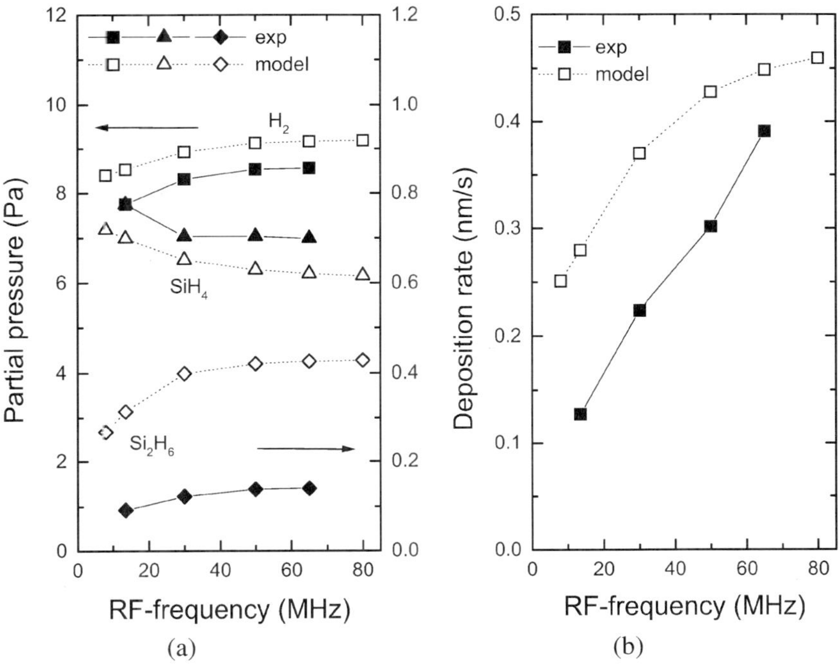

FIG. 18. (a) The partial pressures of H_2, SiH_4, and Si_2H_6 and (b) the deposition rate, as a function of RF frequency at a total pressure of 16 Pa and a plasma power of 5 W. Other discharge settings are given in Table IV. At these settings the discharge is in the α-regime. Modeling results are in dotted lines and open symbols, experimental data in solid lines and filled symbols. (Compiled from G. J. Nienhuis, Ph.D. Thesis, Universiteit Utrecht, Utrecht, the Netherlands, 1998.)

196] have discussed this, and conclude that dissociation of silane should result in a larger fraction of SiH_3.

At about 30 Pa, the experimental deposition rate (Fig. 17b) clearly shows a sudden increase, which is not revealed by the model. In addition, the partial pressure of silane decreases (and of hydrogen increases). These features have been observed in other experiments as well [245–248], and are caused by the transition from the α- to the γ'-regime.

RF frequency variation. In Figure 18 are shown the effects of RF frequency on the partial pressures of silane, hydrogen, and disilane (Fig. 18a) and on the deposition rate (Fig. 18b). The total pressure is 16 Pa, and the plasma power is 5 W. The discharge is in the α-regime.

Both model and experiment show the same tendency: the dissociation of silane increases as a function of RF frequency. This can be explained following the arguments from Heintze and Zedlitz [236]. They found that the power dissipated by the ions in the sheaths decreases with increasing RF frequency. This is confirmed by the modeling results: at 13.56 MHz about 50% of the power is used for the acceleration of the ions, whereas at 80 MHz this is reduced to about 15%. Hence, at higher RF frequencies more power is used for the heating of electrons, which in turn leads to a higher dissociation of silane. As a result, more hydrogen and silane radicals are produced, and more molecular hydrogen is formed. A saturation is observed when all power is consumed by the electrons.

The experimentally found linear increase of the deposition rate as a function of frequency is not seen in the modeling results that show saturation; see Figure 18b. The linear increase has also been measured by others [119, 120, 249], up to an RF frequency of 100 MHz. Howling et al. [250] have measured this linear relationship, while taking special care that the effective power is independent of frequency.

A possible explanation for the difference in tendencies of the deposition rate between experiment and model is that in the model the surface reaction and sticking coefficients of the radicals are taken to be independent of the discharge characteristics. In fact, these surface reaction coefficients may be influenced by the ions impinging on the surface [251]. An impinging ion may create an active site (or dangling bond) at the surface, which enhances the sticking coefficient. Recent experiments by Hamers et al. [163] corroborate this: the ion flux increases with the RF frequency. However, Sansonnens et al. [252] show that the increase of deposition rate cannot be explained by the influence of ions only.

The discrepancy may also be caused by the approximations in the calculation of the EEDF. This EEDF is obtained by solving the two-term Boltzmann equation, assuming full relaxation during one RF period. When the RF frequency becomes comparable to the energy loss frequencies of the electrons, it is not correct to use the time-independent Boltzmann equation to calculate the EEDF [253]. The saturation of the growth rate in the model is not caused by the fact that the RF frequency approaches the momentum transfer frequency ν_{me} [254]. That would lead to less effective power dissipation by the electrons at higher RF frequencies and thus to a smaller deposition rate at high frequencies than at lower frequencies.

Another possible explanation is the approximation that the vibrationally excited silane molecules are treated in the model as ground state molecules. This is based on the assumption that collisional deexcitation in the gas phase and at the walls is fast enough. The density of vibrationally excited molecules is thus neglected. The fraction of the total power used for vibrationally exciting silane molecules increases with the RF frequency, which may result in a nonnegligible excited state density. Inclusion of vibrationally excited silane molecules leads to a higher radical production and deposition rate. Vibrationally excited silane molecules have a lower threshold for electron dissociation than a ground state molecule. Also, multivibrational excitation may lead to homogeneous pyrolysis ($e^- + SiH_4 \rightarrow SiH_4^{vib} \rightarrow SiH_2 + H_2$). Perrin et al. [192, 211] have ruled out this pyrolysis channel because of the high rates of relaxation for vibrationally excited silane molecules,

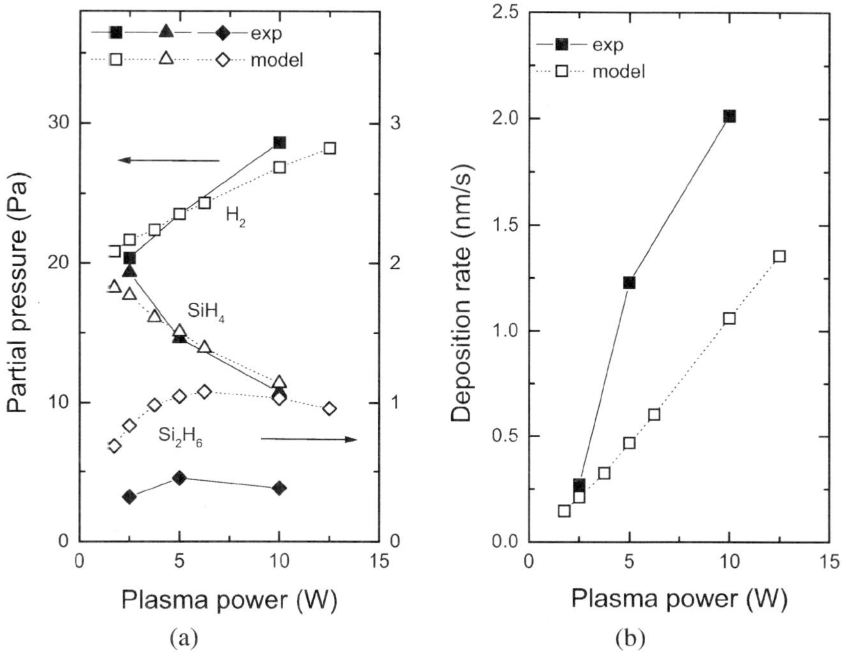

FIG. 19. (a) The partial pressures of H_2, SiH_4, and Si_2H_6 and (b) the deposition rate, as a function of plasma power at a total pressure of 40 Pa and an RF frequency of 50 MHz. Other discharge settings are given in Table IV. At these settings the discharge is in the γ'-regime. Modeling results are in dotted lines and open symbols, experimental data in solid lines and filled symbols. (Compiled from G. J. Nienhuis, Ph.D. Thesis, Universiteit Utrecht, Utrecht, the Netherlands, 1998.)

but their observation was made for an RF frequency of 13.56 MHz. In addition, vibrationally excited silane molecules have a drastically increased cross section for dissociative attachment [192, 255, 256], which increases the production of negative ions and the loss of electrons. In order to sustain the discharge, a higher average electron energy is required, which also leads to a higher dissociation.

Plasma power variation. In Figure 19 are shown the effects of RF power on the partial pressures of silane, hydrogen, and disilane (Fig. 19a) and on the deposition rate (Fig. 19b). The total pressure is 40 Pa, and the RF frequency 50 MHz. The discharge is in the γ'-regime. Results are shown as a function of the supposed effective RF power, i.e., 50% of the power set at the power source.

The decrease of the silane partial pressure and the concomitant increase of the hydrogen partial pressure as a function of plasma power can be understood in terms of the increased electron density and electron energy. Both lead to a higher dissociation of silane and hydrogen. The silane radicals and atomic hydrogen thus created diffuse to the electrodes, where deposition takes place and molecular hydrogen is formed.

The partial pressure of disilane as a function of plasma power first increases due to the higher production of silane radicals. Above a certain power the increase in disilane dissociation is higher than the increase in disilane production; hence the disilane partial pressure decreases again. Similar behavior has been obserbed by Kae-Nune et al. [217].

In Figure 19b the deposition rate is seen to increase with the plasma power. Andújar et al. [246] also measured an increase of the deposition rate with increasing total power. They found that the relation between deposition rate and electrical power depends on the pressure. At pressures lower than about 20 Pa, their measured deposition rate increases linearly with power, which is the same relation as shown by the 1D discharge model. However, at pressures higher than 20 Pa there is a saturation of the deposition rate as a function of the power, which is similar to the experiments shown here (Fig. 19b). Andújar et al. [246] conclude that at 20 Pa the discharge regime is changed from the α- to the γ'-regime. This value of 20 Pa depends on discharge parameters.

Sensitivity analysis. A possible cause for the discrepancy between experiment and model is an error in the elementary parameters (reaction coefficients, cross sections, and transport coefficients) which are obtained from the literature. With a sensitivity study it is possible to identify the most important processes [189].

To calculate the sensitivity with respect to one of the elementary parameters, its value is changed slightly and the resulting change in the plasma characteristic (i.e., the partial pressure of SiH_4, Si_2H_6, or H_2, or the deposition rate) is calculated. A measure $\Lambda = \Delta Y/\Delta X$ of the sensitivity is the relative change ΔY of the plasma characteristic Y divided by the relative change ΔX of the elementary parameter X. A sensitivity study has been made for a discharge at a pressure of 16 Pa effective power of 5 W, and RF frequency of 50 MHz. The partial pressures of SiH_4 and H_2 are not sensitive ($-0.01 < \Lambda < +0.01$) to any change in the chemical parameters. The partial pressure of Si_2H_6 is most sensitive to the reactions $SiH_4 + SiH_2$ ($\Lambda = +0.18$) and $Si_2H_6 + SiH_2$ ($\Lambda = -0.14$) and to the dissociation branching ratio SiH_3/SiH_2 of SiH_4 ($\Lambda = -0.20$). The correctness of the silane dissociation branching ratio has also been discussed by Layeillon et al. [195, 196]. They concluded that the branching ratio should result in a larger fraction of SiH_3 in order to improve modeling results. This is also observed here. The growth rate is most sensitive to the surface reaction ($\Lambda = -0.20$) and sticking probability ($\Lambda = +0.29$) of SiH_3 and to the diffusion coefficients of SiH_2 ($\Lambda = +0.12$) and H ($\Lambda = -0.12$).

Nienhuis [189] has used a fitting procedure for the seven most sensitive elementary parameters (reactions $SiH_4 + SiH_2$ and $Si_2H_6 + SiH_2$, dissociation branching ratio of SiH_4, surface reaction coefficient and sticking probability of SiH_3, and diffusion coefficients of SiH_2 and H). In order to reduce the discrepancy between model and experiment significantly, some values had to be changed by more than two orders of magnitude, which clearly is unrealistic. Uncertainties in the literature never are larger than 50%.

1.4.2. 2D Fluid Discharge Model

1.4.2.1. Model Description

In the previous section a 1D discharge model was formulated and results were compared with experimental data. Although this discharge model is sophisticated, there are some disadvantages related to its low dimensionality. Therefore, this model is extended to a 2D discharge model, the extension being only in geometry.

The self-consistent fluid model was developed by Nienhuis [189]. This model is two-dimensional (cylindrically symmetric). It is an extension, purely in geometry, of a previously published 1D model [191]. Only the important aspects of the model are summarized here. A fluid model consists of balance equations for the densities of the various species in the discharge (i.e. electrons, positive and negative ions, and neutrals) and an energy balance equation for the electrons. The heavy species (ions and neutrals) are assumed to have a uniform temperature close to the wall temperature, because of the good energy transfer both between the heavy species and between the heavy species and the wall. Heating or cooling of the electrodes is not considered. The electric field in the discharge is calculated using Poisson's equation. For the charged particle fluxes the drift-diffusion approximation is used, with the actual electric field for the electrons and an effective field for the ions, in order to allow for their inertia [233]. In the 1D model [191] the nonradical neutrals were treated with a model based on perfect mixing between the discharge volume and the discharge-free volume of the reactor vessel. Here, diffusive transport of all neutrals is taken into account, while modeling only the discharge volume.

In a fluid model the correct calculation of the source terms of electron impact collisions (e.g. ionization) is important. These source terms depend on the EEDF. In the 2D model described here, the source terms as well as the electron transport coefficients are related to the average electron energy and the composition of the gas by first calculating the EEDF for a number of values of the electric field (by solving the Boltzmann equation in the two-term approximation) and constructing a lookup table.

A sticking model is used for the plasma–wall interaction [137]. In this model each neutral particle has a certain surface reaction coefficient, which specifies the probability that the neutral reacts at the surface when hitting it. In case of a surface reaction two events may occur. The first event is sticking, which in the case of a silicon-containing neutral leads to deposition. The second event is recombination, in which the radical recombines with a hydrogen atom at the wall and is reflected back into the discharge.

The chemistry (i.e. the species and reactions included) is the same as described for the 1D model. However, here the higher-order silanes Si_nH_{2n+2} and silane radicals Si_nH_{2n+1} are limited to $n < 4$ to reduce the computational effort. Thus, Si_3H_7 and Si_3H_8 are representative for all silanes with $n > 2$. The formation of powder (large silane clusters) is not taken into account in this model. The discharge settings for the calculations shown here are a total pressure of 20 Pa, a power input of 250 W m^{-2}, an RF frequency of 50 MHz, and an inlet flow of 30 sccm of SiH_4 and 30 sccm of H_2. This parameter set is chosen because it results in a situation where most of the silane is consumed in a large reactor. This situation is required for economic reasons in industrial applications.

1.4.2.2. Comparison with Experiments: Small Reactor

The 2D model has been validated and compared with the 1D model and with the experimental data shown in the previous section. Details can be found in [189]. It was found that the partial pressures (averaged in the case of the 2D model) differed by at most 9%. On the symmetry axis in the 2D model the partial pressures of SiH_4, H_2, and Si_2H_6 are 7.77, 11.23, and 0.64 Pa, respectively. For the 1D model the values are 7.72, 11.51, and 0.55 Pa, respectively. Averaged over the whole discharge, the 2D model calculates the partial pressures of SiH_4, H_2, and Si_2H_6 to be 8.05, 11.16, and 0.56 Pa, respectively, which is closer to the experimental data (see Fig. 17a). A partial-pressure profile of SiH_4 as calculated with the 2D model for both the volume between the electrodes (discharge volume) and the volume next to the electrodes (discharge-free volume) is shown in Figure 20. The dimensions are chosen to mimic the experimental reactor, with an electrode distance L of 2.7 cm (axial coordinate z) and electrode radius R of 7.6 cm (radial coordinate r). The inlet of SiH_4 is a (cylindrically symmetric) ring inlet, shown at the top right of the figure, where the partial pressure is at its highest value. The pump outlet is at the bottom right. It can be seen that the silane partial pressure varies only 10% as a function of the axial position, and it decreases slightly towards the axis of the discharge. This reflects the consumption of silane. Most of the silane flows from the inlet to the outlet; only a small amount diffuses to the discharge region. Similar profiles for the partial pressures of hydrogen and disilane are found.

Other discharge characteristics calculated with the 2D model differ only slightly from the ones calculated with the 1D model. The calculated growth rate is 0.419 nm/s, versus 0.426 nm/s for the 1D model. Electron and ion densities are lower for the 2D model: the electron density is 5.78×10^{14} m^{-3} (2D) and 7.19×10^{14} m^{-3} (1D); the negative-ion density is 2.00×10^{16} m^{-3} (2D) and 2.38×10^{16} m^{-3} (1D). The ion fluxes also are lower: the ion flux to the grounded electrode is 3.38×10^{18} m^{-2} s^{-1} (2D) and 4.06×10^{18} m^{-2} s^{-1} (1D); the ion flux to the powered electrode is 3.82×10^{18} m^{-2} s^{-1} (2D) and 4.06×10^{18} m^{-2} s^{-1}

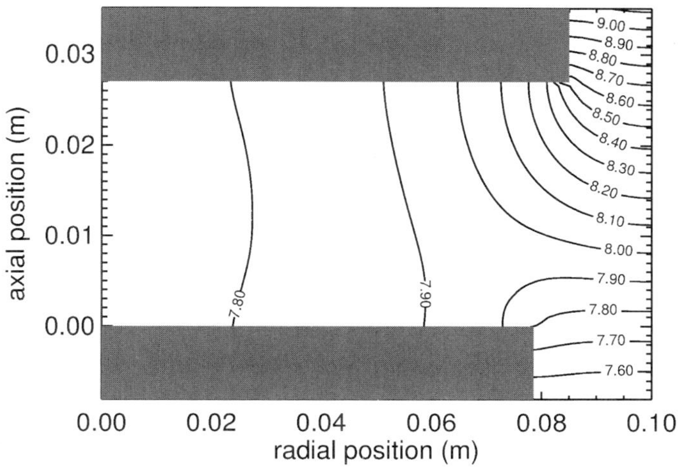

FIG. 20. The partial-pressure profile of SiH_4 calculated with the 2D model. Discharge settings are a total pressure of 20 Pa, a power of 5 W, and an RF frequency of 50 MHz. (From G. J. Nienhuis, Ph.D. Thesis, Universiteit Utrecht, Utrecht, the Netherlands, 1998, with permission.)

(1D). This is due to the radial losses in the 2D model. The difference in ion fluxes to the grounded and powered electrode results in a dc self-bias voltage of −13.96 V, whereas the experimental V_{dc} is −22 V.

From the profile of the average electron energy it is seen that a relatively high average electron energy exists near the edge of the powered electrode ($z = 2$–5 mm, $r = 7$ cm), which is independent of the phase in the RF cycle [189]. This is caused by the high electric field between the outside of the powered electrode and the grounded shield around it (known as the dark space shield).

The partial pressure profile of SiH_3 shows a maximum between the electrodes, much like the 1D profile (see Fig. 16). The SiH_2 partial-pressure profile shows a maximum towards the edge of the discharge volume ($z \approx 8$ mm, $r \approx 6.5$ cm). Both the SiH_3 and SiH_2 source term also peak in this region. The high average electron energy in this region of the discharge is responsible for the formation of these radicals by electron dissociation of silane.

The total deposition rate at the grounded electrode and the contribution of each individual species to the total deposition rate is shown in Figure 21 as a function of radial position. The deposition rate is related to the flux into the surface of silicon-containing species, SiH_3, SiH_2, Si_2H_5, Si_3H_7, SiH_m^+, and $Si_2H_m^+$. For the discharge conditions used here it follows that SiH_3 is the radical with the highest contribution to the deposition rate, 37%. Other radicals contribute 16% (SiH_2), 17% (Si_2H_5), and 13% (Si_3H_7). The total contribution of ions to the deposition rate is 17%: 13% for SiH_m^+, and 4% for $Si_2H_m^+$.

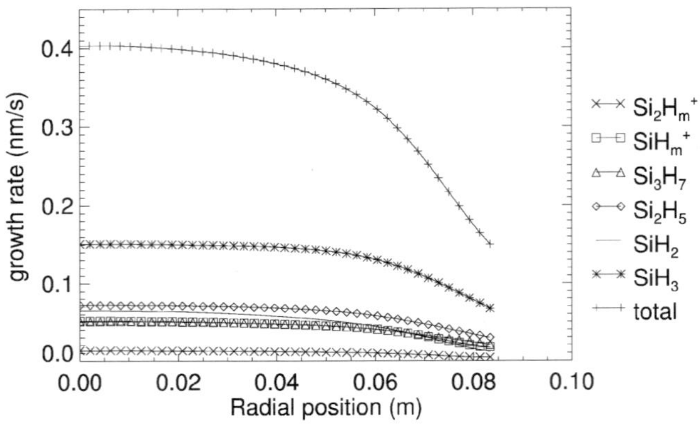

FIG. 21. Deposition rate at the grounded electrode, with the contribution of each silicon-containing species. Discharge settings are a total pressure of 20 Pa, a power of 5 W, and an RF frequency of 50 MHz. (From G. J. Nienhuis, Ph.D. Thesis, Universiteit Utrecht, Utrecht, the Netherlands, 1998, with permission.)

The deposition rate is not uniform over the grounded electrode. A common requirement for large-scale reactors is a 5% difference in deposition rate over a certain electrode area, preferably the complete electrode. Using this, the 2D model would predict a useful electrode area of about 30%, i.e., an electrode radius of about 4 cm.

In the ASTER reactor deposition experiments were performed in order to compare with the 2D model results. Normalized deposition rates are plotted in Figure 22 as a function of radial position for data taken at 25 and 18 Pa. The deposition takes place on a square glass plate. For each pressure two profile measurements were performed, each profile perpendicular to the other (a and b in Fig. 22). A clear discrepancy is present. The use of the simplified deposition model is an explanation for this. Another recent 2D fluid model also shows discrepancies between the measured and calculated deposition rate [257], which are attributed to the relative simplicity of the deposition model.

Another explanation for the observed discrepancy might be the influence of the electrode material. In all model calculations it is assumed that the potential at the grounded electrode is zero. In reality, as one generally uses insulating substrates such as glass, other boundary conditions are needed. Nienhuis [189] has reformulated boundary conditions taking into account the actual square $10 \times 10\text{-cm}^2$ glass substrate mounted on the cylindrically symmetric grounded electrode. It was found that the deposition rate at the center of the grounded electrode (covered with the glass) was reduced by about 15%. Going towards the edges of the glass, the deposition rate increases up to the level that is calculated for the full metal electrode. For larger radial positions the deposition rate equals the one for the full metal electrode. In addition, calculation of the contribution of individual

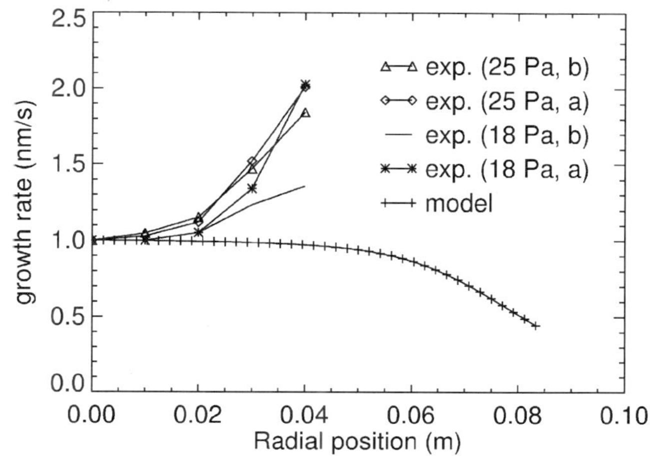

FIG. 22. Normalized deposition rate at the grounded electrode of the 2D model compared with experiments. Discharge settings for the model are a total pressure of 20 Pa, a power of 5 W, and an RF frequency of 50 MHz. The experiments were performed at 18 and 25 Pa. The measurements were done for two perpendicular directions, a and b. (From G. J. Nienhuis, Ph.D. Thesis, Universiteit Utrecht, Utrecht, the Netherlands, 1998, with permission.)

species to the deposition rate show that the contributions of SiH_3 and SiH_2 increase around the glass–metal transition. It is concluded that the effect of a mixed boundary condition may at least partially explain the observed increase of deposition rate towards the edges of the glass.

1.4.2.3. Comparison with Experiments: Large Reactor

In a small reactor the depletion of silane is usually low, and consequently the background gas composition can be considered homogeneous. In a large reactor the composition varies throughout the discharge volume, as a result of the much higher consumption of silane. For commercial reasons the flow of silane is kept as low as possible, so as to obtain full consumption; this, however, may lead to loss of homogeneity. Large reactor systems have recently been modeled using the 2D model by Nienhuis and Goedheer [232]. A reactor is modeled with the same electrode spacing as above (2.7 cm), but with electrode radius of 50 cm and reactor radius 60 cm, which are typical dimensions used in reactors for industrial applications [252]. Two types of gas inlet are modeled: a showerhead, i.e., a powered electrode with small pores through which the gas is introduced [258], and a ring inlet, i.e., an inlet through the ring-shaped area between the grounded electrode and the reactor wall. In both cases, pumping occurs through the area between the powered electrode and the wall.

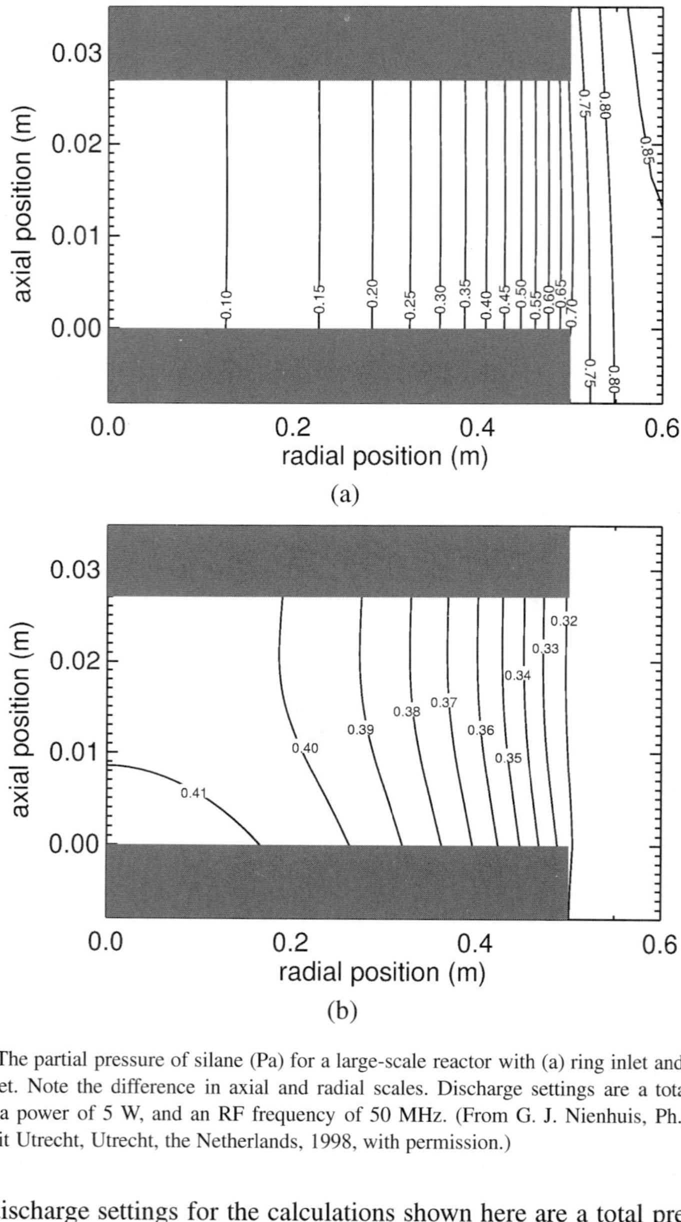

FIG. 23. The partial pressure of silane (Pa) for a large-scale reactor with (a) ring inlet and (b) showerhead inlet. Note the difference in axial and radial scales. Discharge settings are a total pressure of 20 Pa, a power of 5 W, and an RF frequency of 50 MHz. (From G. J. Nienhuis, Ph.D. Thesis, Universiteit Utrecht, Utrecht, the Netherlands, 1998, with permission.)

The discharge settings for the calculations shown here are a total pressure of 20 Pa, a power input of 250 $W\,m^{-2}$, an RF frequency of 50 MHz, and an inlet flow of 30 sccm of SiH_4 and 30 sccm of H_2. The resulting silane pressure distribution is shown in Figure 23 as a function of radial position for both the ring inlet (Fig. 23a) and the showerhead inlet (Fig. 23b). In case of the ring inlet the

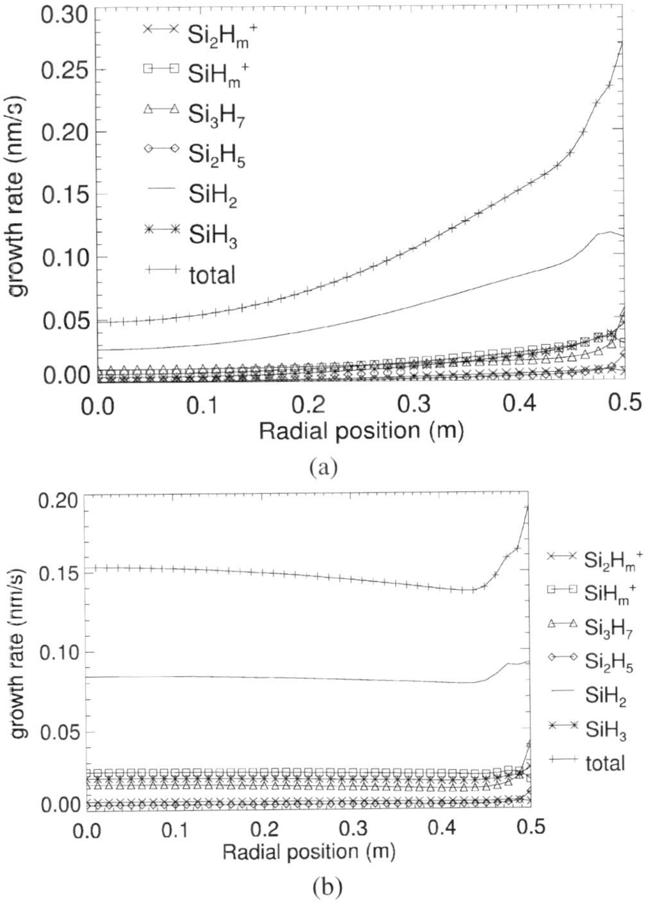

FIG. 24. The deposition rate at the grounded electrode with the contribution of each silicon-containing species, for a large scale reactor with (a) ring inlet and (b) showerhead inlet. Note the difference in uniformity. Discharge settings are a total pressure of 20 Pa, a power of 5 W, and an RF frequency of 50 MHz. (From G. J. Nienhuis, Ph.D. Thesis, Universiteit Utrecht, Utrecht, the Netherlands, 1998, with permission.)

silane pressure distribution is very inhomogeneous; for the showerhead inlet it is relatively homogeneous. A strong depletion of silane is seen for the ring inlet on going from the outside to the center of the discharge. In fact, the discharge consists of an almost pure hydrogen discharge at the center and a mixture of silane and hydrogen towards the edges. For the showerhead inlet the situation is reversed, and the effect is much smaller.

The depletion of silane has large consequences for the uniformity of the deposition, as is shown in Figure 24. In the case of the showerhead the deposition is very homogeneous, and decreases slowly from the center to the edge of the discharge, which can be related to the small decrease in silane partial pressure as seen in Figure 23b. The ring inlet shows a large increase in the deposition rate from the electrode center to the outside. This reflects the high silane consumption in the opposite direction. The sharp increase of the deposition rate at the very edge of the electrode has also been observed experimentally [259].

The large depletion of silane has consequences for the chemistry. In the electron impact dissociation of silane the two radicals SiH_2 and SiH_3 are formed with branching ratio 83%:17% [197]; the other product is H (two atoms or one, respectively). When there is enough silane available, the SiH_2 radical is lost by the insertion reaction $SiH_2 + SiH_4 \rightarrow Si_2H_6$, and via the H-abstraction reaction $H + SiH_4 \rightarrow SiH_3 + H_2$ the radical SiH_3 is produced. Deposition of SiH_2 and SiH_3 is the loss process at the walls.

The loss and production depend on silane partial pressure. For a high partial pressure of silane most SiH_2 radicals react with silane before reaching the wall. As the production of SiH_3 radical is also high, the SiH_3 radical is the main precursor for deposition. This is the case in the small reactor (see Fig. 21). For a low partial pressure, the reaction time of SiH_2 with SiH_4 is longer than the diffusion time to the wall. Together with the lower production of SiH_3, this causes SiH_2 to be the main precursor for deposition (see Fig. 23). The main precursor for deposition thus depends on the silane partial pressure. This will have serious consequences for the quality of the deposited material. The SiH_2 radical has a much larger sticking probability than the SiH_3 radical. The structure of a layer deposited with SiH_2 will be different from that of layer deposited with SiH_3. SiH_2 leads to columnar deposition, which is unwanted. *Device quality* material is obtained when deposition occurs via the SiH_3 radical. Scaling up of a small to a large reactor therefore is not straightforward.

1.4.3. Particle-in-Cell Discharge Models

1.4.3.1. Particle-in-Cell Model Principles

A complete model for the description of plasma deposition of *a*-Si:H should include the kinetic properties of ion, electron, and neutral fluxes towards the substrate and walls. The particle-in-cell/Monte Carlo (PIC/MC) model is known to provide a suitable way to study the electron and ion kinetics. Essentially, the method consists in the simulation of a (limited) number of computer particles, each of which represents a large number of physical particles (ions and electrons). The movement of the particles is simply calculated from Newton's laws of motion. Within the PIC method the movement of the particles and the evolution of the electric field are followed in finite time steps. In each calculation cycle, first the forces on each particle due to the electric field are determined. Then the equations of motion for electrons and ions are integrated. Subsequently collisional processes are treated by the Monte Carlo method. The chance of a collision for each particle during one time step is calculated using the cross sections for the various processes, and random number generation controls the actual occurrence of the collision. Finally the particle distribution is interpreted into electron and ion densities, and the Poisson equation for the electric field is solved. Thus the ions and electrons move in an electric field that is itself directly determined by the particle densities (via the Poisson equation). The result is referred to as the *consistent* solution for the electric field.

Despite the simplifications and the incomplete information with respect to the cross sections, the PIC method gives remarkably good results in RF modeling [260, 261]. The advantage of a fully kinetic description is accompanied by the disadvantage of very time-consuming computations. The PIC method allows only for a fully explicit treatment of the particle transport. Numerical instabilities due to large particle velocities and the coupling between particle densities and the electric field can only be avoided by reducing the time step in the simulation and so increasing the computational effort. Large numerical fluctuations in the calculated quantities as a result of the small number of computer particles (small with respect to the real physical numbers of ions and electrons) can also not be avoided in PIC simulations.

Some authors have made their code for PIC calculations in RF discharges available for a larger public. In particular, Birdsall et al. [261–263] developed the XPDP1 code, which is an implementation of the PIC method for RF discharges, written in C for use in UNIX/X-window environments. This XPDP1 code is developed for simulations in 1D RF discharges with plasmas consisting of one type of ion (argon, hydrogen), which have, as collisional processes, elastic collisions of electrons and ions with background-gas neutrals, excitation, and ionization. The code is built up in a modular way, which makes it easy to modify or add parts of the code, e.g., to make it suitable for use with other gases with more collisional processes. The code is intended to run interactively on workstations or PCs using X-windows, and communicates with the outside world via windows and dialog boxes.

Treatment of the SiH_4–H_2 discharge requires the addition of extra processes (ionization, dissociation, and attachment for the electrons, recombination for the ions) in the collision component. This resulted in major changes in this part of the code [264, 265]. Especially the incorporation of collision processes with target particles, which do not form a constant background density, but have a space- and time-varying density, needed some significant modifications. They were necessary to handle ion–ion recombination, but may also be used in future, e.g. for ion–electron collisions and for electron–radical collisions. In the original code partial collision probabilities were determined via cross sections; in the modified one, inverse free paths had to be applied. The cross sections for the electrons were obtained from the literature [266]. The cross sections for ion–neutral scattering had to be estimated. For the energy dependence of the scattering cross section an ion-induced dipole interaction has been assumed [267]. The recombination cross section is directly estimated from rate coefficients [190], and results in a total recombination sink (integrated over velocity space) identical to the fluid one.

1.4.3.2. Hybrid PIC/MC–Fluid Model

As a first attempt to modify the code to be able to run simulations on SiH_4–H_2 discharges, a hybrid PIC/MC–fluid code was developed [264, 265]. It turned out in the simulations of the silane–hydrogen discharge that the PIC/MC method is computationally too expensive to allow for extensive parameter scans. The hybrid code combines the PIC/MC method and the fluid method. The electrons in the discharge were handled by the fluid method, and the ions by the PIC/MC method. In this way a large gain in computational effort is achieved, whereas kinetic information of the ions is still obtained.

The fluid model is a description of the RF discharge in terms of averaged quantities [268, 269]. Balance equations for particle, momentum, and/or energy density are solved consistently with the Poisson equation for the electric field. Fluxes described by drift and diffusion terms may replace the momentum balance. In most cases, for the electrons both the particle density and the energy are incorporated, whereas for the ions only the densities are calculated. If the balance equation for the averaged electron energy is incorporated, the electron transport coefficients and the ionization, attachment, and excitation rates can be handled as functions of the electron temperature instead of the local electric field.

Especially for the electrons, the fluid model has the advantage of a lower computational effort than the PIC/MC method. Their low mass (high values of the transport coefficients) and consequent high velocities give rise to small time steps in the numerical simulation ($v\Delta t < \Delta x$) if a so-called explicit method is used. This restriction is easily eliminated within the fluid model by use of an implicit method. Also, the electron density is strongly coupled with the electric field, which results in numerical instabilities. This requires a simultaneous implicit solution of the Poisson equation for the electric field and the transport equation for the electron density. This solution can be deployed within the fluid model and gives a considerable reduction of computational effort as compared to a nonsimultaneous solution procedure [179]. Within the PIC method, only fully explicit methods can be applied.

The disadvantage of the fluid model is that no kinetic information is obtained. Also, transport (diffusion, mobility) and rate coefficients (ionization, attachment) are needed, which can only be obtained from experiments or from kinetic calculations in simpler settings (e.g. Townsend discharges). Experimental data on silane–hydrogen Townsend discharges are hardly available; therefore PIC/MC simulations have been performed.

The quasi-Townsend discharge is an infinitely extended plasma, in which the electrons move in a constant, spatially homogeneous electric field E. The electrons collide with neutrals in a background gas of infinite extent. The constant electric field replaces the solution of the Poisson equation and makes the PIC method applicable for the electrons within reasonable computational limits. During the particle simulation in the quasi-Townsend discharge, averaged values, over all electron particles i, of the energy ϵ_i, diffusion constant D_i, mobility μ_i, ionization rate $k_{\text{ion},i}$, and attachment rate $k_{\text{att},i}$ are calculated. This results in values for the transport and rate coefficients: the electron transport coefficients D_e and μ_e, the ionization and attachment rates k_{ion} and k_{att} and the averaged electron energy $\langle\epsilon\rangle = 3k_B T_e/2$. All parameters are obtained for several values of the electric field E.

Subsequently, because the electron temperature is known as a function of the electric field E, the temperature T_e can be used instead of E as a parameter for coefficients and rates, by elimination of E. Thus, the coefficients are available both as a function of E and of T_e. Both the local electric field [225, 269] and the electron temperature [239, 268] have been used as parameters in fluid modeling.

A Townsend PIC simulation was performed for several gas mixtures ($SiH_4 : H_2 = 100:0, 50:50, 20:80$, and $0:100$) [264, 265]. It was found that the dependence of the electron temperature on the electric field strongly depends on the mixture ratio. As a result, the rate coefficient parametrized by the electric field, $k = k(E)$, also strongly depends on the mixture ratio. In contrast, it was found that the rate as a function of the electron energy, $k = k(T_e)$, shows only minor changes ($\leq 2\%$) for different mixing ratios. In Figure 25 the ionization (Fig. 25a) and attachment (Fig. 25b) rates are given. Here, two often used coefficients in RF fluid modeling are also shown for reference: an argon ionization rate derived from Townsend experiments (Fig. 25a), and a CF_4 attachment coefficient [179] (Fig. 25b).

For the following reasons the choice of the temperature T_e as the experimental parameter is more appropriate [268, 270]: (a) the high-energy tail of the energy distribution function of the electrons determines ionization versus attachment; (b) both in pure silane and hydrogen and in mixtures, T_e is a good parameter for the distribution function; (c) T_e as a function of the electric field varies strongly with the mixture ratio SiH_4/H_2.

Methods that compensate for nonequilibrium effects in the situation of E-parametrized coefficients are very complicated, and are sometimes not firmly grounded. Because the electron temperature also gives reasonable results without correction methods, the rate and transport coefficients were implemented as a function of the electron energy, as obtained from the PIC calculations presented in Figure 25.

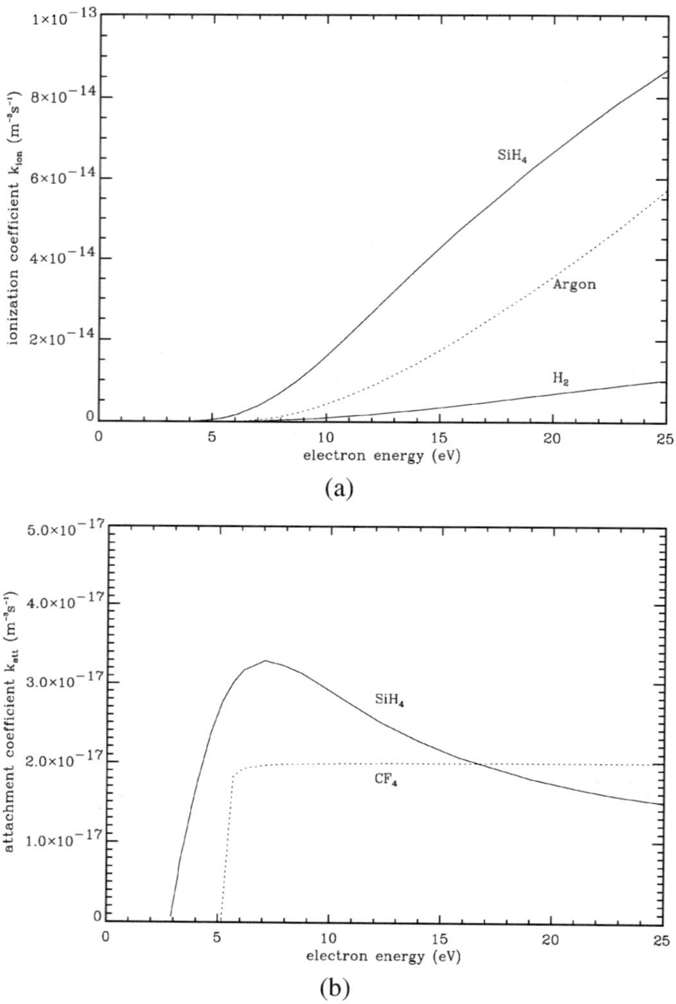

FIG. 25. Simulation results for quasi-Townsend discharges: (a) ionization coefficients for SiH_4, H_2, and Ar; (b) attachment coefficients for SiH_4 and CF_4.

1.4.3.3. Simulation of RF Discharges in Silane–Hydrogen Mixtures with the Hybrid Model

In the deposition of amorphous silicon the RF frequency plays an important role. Therefore, in the simulations the frequency dependence was emphasized. Calculations were performed for frequencies $\nu_{RF} = 10$, 40, and 80 MHz.

The $SiH_4 : H_2$ ratio was 20% : 80% in all simulations. As most experiments in the ASTER deposition system are performed at a constant total power, the

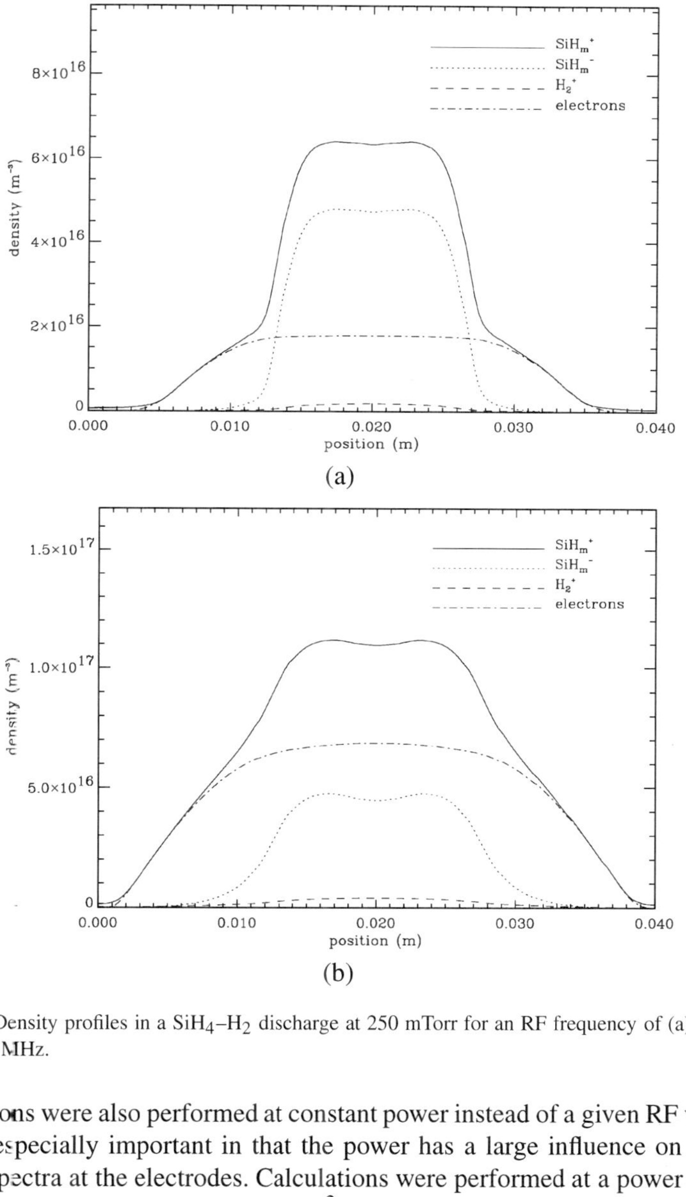

FIG. 26. Density profiles in a SiH_4–H_2 discharge at 250 mTorr for an RF frequency of (a) 10 MHz and (b) 80 MHz.

calculations were also performed at constant power instead of a given RF voltage. This is especially important in that the power has a large influence on the ion energy spectra at the electrodes. Calculations were performed at a power of $P =$ 12 W, with an electrode area of 0.010 m^2 and an electrode distance of 0.04 m. The pressure was $p = 250$ mTorr. Also a calculation was performed at $p = 150$ mTorr for the standard RF frequency of 10 MHz.

In Figure 26 density profiles for two frequencies, 10 MHz (Fig. 26a) and 80 MHz (Fig. 26b), are shown. The densities of the species (electrons, positive

TABLE V
SIMULATION RESULTS FOR THE HYBRID PIC/MC–FLUID MODEL FOR A SiH_4–H_2 DISCHARGE DRIVEN WITH A POWER OF 12 W

p (mTorr)	ν_{RF} (MHz)	Density (10^{16} m^{-3})				Power (W)				E_{max} (10^4 V/m)	Outflux (10^{18} m^{-2} s^{-1})	
		e^-	SiH_m^+	SiH_m^-	H_2^+	e^-	SiH_m^+	SiH_m^-	H_2^+		SiH_m^+	H_2^+
250	10	1.8	6.3	4.7	0.18	1.3	7.2	0.0	3.5	11	15	1.1
250	40	4.8	9.4	4.9	3.2	2.7	5.9	0.0	3.4	8.5	28	2.8
250	80	6.9	10.1	4.5	3.9	3.7	5.0	0.0	3.3	7.1	36	3.9
150	10	1.2	6.9	5.7	0.048	1.1	7.5	0.0	3.4	9.4	14	0.9

ions, negative ions) and power dissipated in these species are presented in Table V, as well as the maximum electric field over the sheath and the outflux of positive ions.

It can be clearly observed that the electron density increases with frequency. The increase of frequency at a constant power results in lower RF voltages and, consequently, lower electric fields in the sheath. The maximum electric field E_{max} in the sheath diminishes with increasing frequency. The field decrease is too small to prevent the displacement current, $\epsilon_0 \partial E/\partial t = \epsilon_0 \nu_{RF} \partial E/\partial \tau$, with $\tau = t/\tau_{RF}$, from increasing. As a result, the electron conduction current in presheath and bulk increases, and the heating of electrons becomes stronger, whereas the lower potential drop across the sheath causes a lower power dissipation into the ions. The increase of the electron heating results in a higher ionization of silane (SiH_m^+) and hydrogen (H_2^+). This production is partly compensated for by higher transport losses (outfluxes) The net result is an increasing electron and positive ion density with increasing frequency. On the other hand the negative ion density decreases due to the enhanced recombination between positive ions and SiH_m^-. As a result of the different curves of $k_{ion}(\epsilon)$ for H_2 and SiH_4, the ionization of H_2 is enhanced relatively more by the increasing electron heating than that of SiH_4. The increased ion fluxes lead to a shrinking of the sheath.

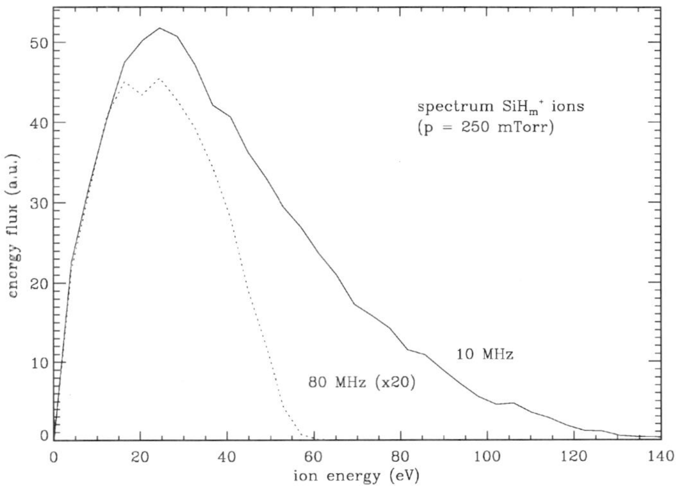

FIG. 27. Ion energy distribution of SiH_m^+ ions in a SiH_4–H_2 discharge at 250 mTorr for an RF frequency of 10 MHz (solid line) and 80 MHz (dotted line).

A further result of the increase of power dissipation in the electrons has consequences for the plasma chemistry. Besides the increased ion densities, also the production of radicals will be increased, which may lead to higher deposition rates.

It is observed that a decrease of the pressure (from $p = 250$ to 150 mTorr) mainly results in a decrease of the densities due to higher transport losses, and in an extension of the sheath due to a higher ion mobility. The electric field and the electron heating diminish slightly for lower pressure. The electron and H_2^+ density, and consequently the H_2^+ outflux, are much more influenced by the pressure decrease than the SiH_m^+ and SiH_m^- ion densities and the SiH_m^+ outflux.

The ion energy distribution of SiH_m^+ is presented in Figure 27 for 10 and 80 MHz. Because the total power deposition was kept constant at 12 W, they give insight into the positive ions' behavior in deposition experiments. It is seen that for silane ions (Fig. 27)—and also for hydrogen ions (not shown)—the ion energy distribution shifts to lower energies for increasing frequency, due to the decreasing electric field and sheath thickness. The saddle structure due to the RF frequency, which can be seen at very low pressures (<10 mTorr) [271], has totally disappeared.

For one specific set of discharge parameters, in a comparison between the hybrid approach and a full PIC/MC method, the spectra and the ion densities of the hybrid model showed some deviations from those of the full particle simulation. Nevertheless, due to its computational advantages, the hybrid model is appropriate for gaining insight into plasma behavior under varying parameters, especially with respect to the ion energy distribution at the deposition layer.

1.4.3.4. Simulation of RF Discharges in Silane–Hydrogen Mixtures with a Full PIC/MC Model

The full PIC/MC method has not been used frequently to study SiH_4–H_2 discharges [223], because of the large amount of computer time required. Yan and Goedheer [224] have developed a 2D PIC/MC code with some procedures to speed up the calculation. In the code, electrons and positive and negative ions are described in a kinetic way. The capacitively coupled RF discharge in a mixture of SiH_4 and H_2 has been simulated for pressures below 300 mTorr and frequencies from 13.56 to 65 MHz. Yan and Goedheer [224] studied the effects of frequency on (1) the EEDF in the discharge area, (2) the ion energy distribution (IED) at the substrate, (3) the ion flux to the substrate, and (4) the power dissipation at constant power density.

The PIC/MC method is based on a kinetic description of the particle motion in velocity–position space. Positive and negative *superparticles* move in the self-consistent electric field they generate. The electric field is obtained by solving the Poisson equation. A Monte Carlo formalism is used to describe the effect of collisions. The displacement between collisions is ruled by Newton's law. There are four kinds of charged particles in the 2D model that is used by Yan and Goedheer [224]: electrons, positive ions (SiH_3^+, H_2^+), and negative ions (SiH_3^-). The negative hydrogen ion (H^-) is not considered, because its density is much lower than the other ion densities. For a computationally stable simulation, the distance between adjacent points of the grid on which the Poisson equation is solved is taken to be less than several Debye lengths, and the product of the time step and the electron plasma frequency is taken to be much smaller than one. The number of superparticles for each species is around 10^4. The superparticle size for species with a low density (electrons and H_2^+) is between 10^6 and 10^7; for species with a high density (SiH_3^+ and SiH_3^-) it is between 10^8 and 10^9.

The simulation of an electronegative gas discharge converges much more slowly than that of an electropositive discharge. This is mainly caused by the slow evolution of the negative-ion density, which depends only on attachment (to create negative ions) and ion–ion recombination (to annihilate negative ions), both processes with a very small cross section. In addition to the common procedures adopted in the literature [222, 223, 272, 273], such as the null collision method, and different superparticle sizes and time steps for different types of particle, two other procedures were used to speed up the calculation [224].

The first procedure is to use the rate equilibrium equations for electrons and ions at the quasi-steady state for a new guess of the particle densities. In the final result, quasi-steady balances of the positive ions and negative ions integrated over the whole discharge region will be reached, i.e., the balance for the creation of positive ions (SiH_3^+) by ionization and their loss to the reactor boundary and by recombination, and the balance for the creation of negative ions by attachment and their loss by recombination. These balances are used in the following way. After a number of cycles, the average ionization, recombination, and attachment rates are known, as well as the average loss of positive ions to the wall. Also the deviation from the balances is known. The new densities of electrons and positive and negative ions can be guessed by combining charge neutrality with the balance equations in a global model. These densities will be different from the values obtained in the previous simulation, and the number of simulation particles thus has to be adapted. The easiest way to implement the changes is through a change of the size of the superparticles. To do this, only the interior part of the discharge is used, where the negative ions are present. The changes are spatially distributed according to the profile of the negative ions. After the correction the standard PIC/MC calculation is restarted. This procedure is repeated several times during a full simulation and reduces the computational effort considerably.

The second procedure is to use a small number of large particles in the beginning of the simulation. The results of that are used as the initial conditions for a simulation with a large number of small particles. Thus, in the final results, the statistical fluctuations are reduced. Also, results from a previous simulation can be used as initial conditions for a new simulation with changed parameters, which saves a large amount of time during parameter scans.

Most of the electron and ion impact reactions that are included in this code are listed in Table II. Added to this are the elastic collisions $SiH_4 + e^-$ [274] and $H_2 + e^-$ [275], the recombination $SiH_3^- + SiH_3^+ \rightarrow 2SiH_3$ [214], and the charge exchange $H_2^{+*} + H_2 \rightarrow H_2^+ + H_2^*$ [193]. Note that disilane is not included. The cross sections are taken from the references in Table II. Ion–neutral elastic collision cross sections are obtained according to the Langevin formula [192]. The total number of ion–ion recombinations in the discharge is obtained from the integral of the product of the recombination rate coefficient k_{rec} and the ion density profiles. The recombination event is attributed to the negative ion; it recombines with the nearest positive ion. In this model the focus is on the energy dissipation, EEDF, IED, and ion flux to the wall, but not on the details of chemical reactions. Chemical reactions are not included in our code. The experiments described above showed that the background pressure and the gas composition can be regarded as being constant for the dimensions used. In the simulation, the pressure and the gas composition are assumed to be constant.

The frequency effects are studied in the cylindrical geometry ($R = 0.08$ m, $L = 0.027$ m) at a constant power of 25 W, which corresponds to a volume power density 45 mW cm^{-3}. The pressure is 120 mTorr with 45% SiH_4 and 55% H_2. It is found that the RF voltage at this power scales with the frequency as $V_{RF}\nu_{RF}^{1.15} = C$, with C a constant. Because the induced displacement current increases with the frequency, a lower RF voltage is needed at high frequencies to keep the same power density. This relation can explain the increase of the deposition rate with frequency at a constant RF voltage [276], because in that case the dissipated power increases strongly.

The simulations show further that the average electron energy decreases slightly as a function of frequency, from about 5.8 to 5 eV on going from 13.56 to 65 MHz. The low voltage applied at high frequencies leads to a weaker sheath electric field. Consequently positive ions are pulled out less efficiently, so their density increases. Due to the charge neutrality in the bulk, the electron density also increases, from 1.8×10^{15} m^{-3} at 13.56 MHz to 3.2×10^{15} m^{-3} at 65 MHz. On the other hand, the current density $j \approx e\mu_e E$ in the bulk is constant because of the constant power. Therefore a higher electron density leads to a weaker electric field in the bulk, which is responsible for heating electrons. This causes the decrease of the average energy. The explanation of the increase of the deposition rate with frequency as observed by Curtins et al. [277] is that there is a larger relative population of high-energy electrons. These high-energy electrons would yield higher ionization and dissociation rates. The simulation described here shows that the density of low-energy electrons increases strongly while the density of high-energy electrons even goes down as a result of the combination of high-frequency effects on the EEDF and on the electron density. Stochastic heating [278] does not seem to lead to an increase of the electron energy, probably because its effect is canceled by the decreasing voltage and by the efficient energy loss to vibrational excitation.

The flux of SiH_3^+ to the substrate and the average ion energy in this flux at different frequencies are shown in Figure 28a. The ion flux and ion energy are sampled over a disk one-quarter of the electrode radius around the center of the grounded electrode, which avoids the influence of the edge. The average ion energy decreases quickly with increasing frequency due to the decrease of the RF voltage. The ion flux to the wall remains approximately the same (1.7×10^{19} m^{-2} s^{-1}) when the frequency is varied. This result is quite different from the results presented by Heintze and Zedlitz [279], where the ion flux showed a significant increase with frequency. Their results, however, are not in agreement with the Bohm criterion, because the electron density did not change much while the ion flux changed one order of magnitude [224]. This may be caused by dust formation. Another inconsistency is that the dissociation rate measured by the same authors [280] only increased by a factor of 1.5 when the frequency was increased from 50 to 250 MHz. The dissociation cross section is larger than the ionization cross section and with lower threshold; see also Figure 13. This implies that dissociation has higher probability than ionization. The obvious increase in ion flux should therefore be accompanied by at least the same amount of dissociation.

The IED of SiH_3^+ on the substrate at different frequencies is shown in Figure 28b. With increasing frequency, the energy distribution profile becomes more

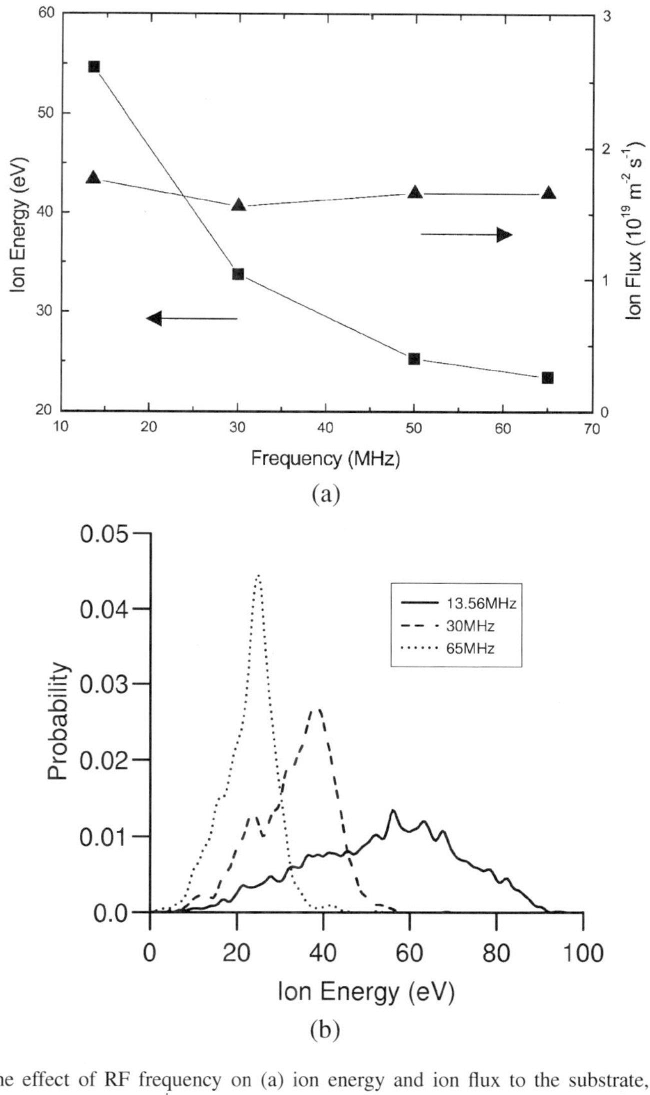

FIG. 28. The effect of RF frequency on (a) ion energy and ion flux to the substrate, and (b) the ion energy distribution for SiH_3^+. The power is 25 W, and the pressure is 120 mTorr. [From M. Yan and W. J. Goedheer, *Plasma Sources Sci. Technol.* 8, 349 (1999), © 1999, Institute of Physics, with permission.]

peaked. That is attributed to a decrease in the number of collisions experienced by ions, because at higher frequencies the lower applied RF voltage leads to a narrower sheath. The same reason obviously causes the maximum ion energy to decrease with increasing frequency. This result is consistent with the evolution of the estimated maximum ion energy by Dutta et al. [281]. But in other experiments by Hamers [163] it was found that the maximum ion energy did not change much.

The power dissipated at two different frequencies has been calculated for all reactions and compared with the energy loss to the walls. It is shown that at 65 MHz the fraction of power lost to the boundary decreases by a large amount compared to the situation at 13.56 MHz [224]. In contrast, the power dissipated by electron impact collision increases from nearly 47% to more than 71%, of which vibrational excitation increases by a factor of 2, dissociation increases by 45%, and ionization stays approximately the same, in agreement with the product of the ionization probability per electron, the electron density, and the ion flux, as shown before. The vibrational excitation energy thresholds (0.11 and 0.27 eV) are much smaller than the dissociation (8.3 eV) and ionization (13 eV) ones, and the vibrational excitation cross sections are large too. The reaction rate of processes with a low energy threshold therefore increases more than those with a high threshold.

The simulated rate of vibrational excitation at 30 MHz is already twice that at 13.56 MHz. At 65 MHz, this rate is three times the rate at 13.56 MHz. The dissociation and ionization rates do not change much. The increase of the vibrational excitation leads to an increase of the density of excited SiH_4 molecules. This agrees with an N_2 discharge experiment where the vibrational temperature of N_2 was observed to rise with the applied frequency [282].

The excited SiH_4 can participate in various reactions. For example, one can speculate that the excited SiH_4 molecules could enhance pyrolysis ($SiH_4 \rightarrow SiH_2 + H_2$) [283] and consequently contribute to the deposition rate. SiH_2 can further dissociate to SiH, and it was observed that the emission of SiH shows a similar increase to the deposition rate [250]. This explanation of an increased deposition via excited SiH_4 molecules, however, is not supported by the results obtained when the pressure is varied. Simulations were performed at 13.56 MHz and a power of 16 W, at pressures ranging from 150 down to 90 mTorr. Although the behavior of the vibrational excitation, dissociation, and ionization are quite similar to the behavior as a result of frequency variation, the deposition rate in the experiments remains the same at low pressures [163, 245]. At high pressures the discharge will reach the dusty (γ') regime. This implies that the vibrational excitation does not influence the deposition directly.

The IED reacts quite differently to pressure changes than to frequency changes. Although the IEDs are somewhat different due to the larger number of collisions in the sheath at higher pressures, they cover approximately the same energy range. It has been reported by Dutta et al. [284] that the quality of the deposited material is better at high frequency than at lower frequency, because the intrinsic stress goes down when the frequency is increased. Since the displacement energy of atoms in the solid is about 20 eV [285], there most likely is an energy range of ions that effectively assists the radical deposition on the surface by influencing the reactivity of the films. At high frequencies more ions are in the effective (lower) energy range than that at a lower frequency (Fig. 28b). This could well cause the higher deposition rate. At different pressures, the energy spectrum of the ions is almost the same, and hence the deposition rate is the same. Chemical reactions are not included here, so the influence of changes in the radical fluxes on the deposition process is not studied. The simulation and experimental results [191, 247], however, indicate that the composition of the SiH_4–H_2 mixture does not change dramatically.

1.5. Plasma Analysis

In order to relate material properties with plasma properties, several plasma diagnostic techniques are used. The main techniques for the characterization of silane–hydrogen deposition plasmas are optical spectroscopy, electrostatic probes, mass spectrometry, and ellipsometry [117, 286]. Optical emission spectroscopy (OES) is a noninvasive technique and has been developed for identification of Si, SiH, Si^+, and H^+ species in the plasma. Active spectroscopy, such as laser induced fluorescence (LIF), also allows for the detection of radicals in the plasma. Mass spectrometry enables the study of ion and radical chemistry in the discharge, either *ex situ* or *in situ*. The Langmuir probe technique is simple and very suitable for measuring plasma characteristics in nonreactive plasmas. In case of silane plasma it can be used, but it is difficult. Ellipsometry is used to follow the deposition process *in situ*.

1.5.1. Optical Emission

A plasma emits light, as a result of excited species that undergo a radiative transition from a high to a lower level state. By measuring the wavelenghts and intensities of the emitted spectral lines, it is possible to identify and quantify neutrals and ions in the plasma. This spectral fingerprint allows for the study of physical and chemical processes that occur in the plasma. OES is a nonintrusive technique, which can provide spatial and temporal information. By its nature, OES is limited to the detection of light-emitting species.

Optical emision spectra nowadays are simply measured using a fiber optic cable that directs the plasma light to a monochromator, which is coupled to a photodetector. By rotating the prism in the monochromator a wavelength scan of the emitted light can be obtained. Alternatively, an optical multichannel analyzer can be used to record (parts of) an emission spectrum simultaneously, allowing for much faster acquisition. A spectrometer resolution of about 0.1 nm is needed to identify species.

Quantitative analysis of emission spectra is difficult. As a first assumption the emission intensity from a species is proportional to its concentration, but the proportionality constant depends among other things on the electron density and on the quantum efficiency of the excitation process. As the electron density depends on the RF power and the gas mixture, and the quantum efficiency on the total pressure and the gas mixture, the proportionality "constant" never really is constant [117].

Optical emission is a result of electron impact excitation or dissociation, or ion impact. As an example, the SiH radical is formed by electron impact on silane, which yields an excited or superexcited silane molecule ($e^- + SiH_4 \rightarrow SiH_4^* + e^-$). The excess energy in SiH_4^* is released into the fragments: $SiH_4^* \rightarrow SiH^* + H_2 + H$. The excited SiH* fragments spontaneously release their excess energy by emitting a photon at a wavelength around 414 nm, the bluish color of the silane discharge. In addition, the emission lines from Si*, H*, and H_2^* have also been observed at 288, 656, and 602 nm, respectively.

Matsuda and Hata [287] have argued that the species that are detectable using OES only form a very small part (<0.1%) of the total amount of species present in typical silane deposition conditions. From the emission intensities of Si* and SiH* the number density of these excited states was estimated to be between 10^6 and 10^7 cm^{-3}, on the basis of their optical transition probabilities. These values are much lower than radical densities, 10^{12} cm^{-3}. Hence, these species are not considered to partake in the deposition. However, a clear correlation between the emission intensity of Si* and SiH* and the deposition rate has been observed [288]. From this it can be concluded that the emission intensity of Si* and SiH* is proportional to the concentration of deposition precursors. As the Si* and SiH* excited species are generated via a one-electron impact process, the deposition precursors are also generated via that process [123]. Hence, for the characterization of deposition, discharge information from OES experiments can be used when these common generation mechanisms exist [286].

LIF has been used to determine spatial profiles of SiH and Si radicals at various deposition conditions. These radicals are excited by a tuned pulsed laser, and the fluorescent intensities are a measure of the density of the particles studied. Spatial profiles can be obtained by translating incident laser beam and detection optics. LIF is highly selective and highly sensitive, but it is a complex technique. Mataras et al. [289, 290] have compared OES and LIF profiles of SiH* and SiH. From the differences observed it is concluded that the generation processes of these radicals are different. The SiH* concentration is related to the magnitude of both cathode and anode sheath potentials, whereas the SiH concentration is related to the density of energetic electrons in the plasma bulk. As the generation of SiH_2 and of SiH_3 radicals follow similar pathways, it is inferred that the LIF profile of SiH is a good measure of the generation profile of SiH_2 and SiH_3 too [286]. This seems to be confirmed by data obtained by infrared laser absorption spectroscopy (IRLAS) [215, 291, 292], where the SiH_3 profile is measured to be similar to the LIF SiH profile.

Profiles of SiH measured by OES as a function of RF voltage have been reported by Böhm and Perrin [184]. In their nearly symmetrical discharge a clear relation exists between SiH emission intensity and RF voltage, at low pressure (55 mTorr). Further, the SiH profiles are M-shaped, i.e., they peak around the time-averaged sheath boundary and are lower at the center of the plasma, reflecting the time-averaged electron density profile. Going to high pressures (180 mTorr) the so-called α–γ' transition occurs, and the SiH emission intensity is highest at the center of the discharge. LIF experiments at these pressures are impossible due to the presence of powder, which causes strong light scattering.

1.5.2. ELECTROSTATIC PROBES

An electrostatic, or Langmuir, probe is a metallic electrode, insulated except at the tip, which is introduced into the plasma. The metal mostly used is tungsten, because of its high melting point. The insulator of the probe should be chemically inert. The probe is inserted into a discharge chamber via an electrically insulating vacuum seal, within which the probe is either fixed or movable, usually in one direction, via a bellows. This *in situ* technique locally perturbs the plasma by changing the electric field. The magnitude of the disturbance depends on the dimensions of the probe relative to the Debye length λ_D, which in turn depends on the plasma properties. Usually λ_D is in the micrometer range, and probe diameters in the millimeter range, so disturbances are very small.

Employing a Langmuir probe, it is possible to determine the plasma densities (n_i and n_e), electron temperature T_e, plasma potential V_{pl}, and floating potential V_{fl} [117, 134, 293, 294]. It is often used as a complementary technique to optical and/or mass spectrometry. A Langmuir probe is operated by applying an external variable potential (usually between -100 and $+100$ V) and measuring the current (in the milliampere range) through the probe. The sweep rate usually is 1–10 V/μs, and multiple *I*–*V* characteristics are acquired and averaged. From such an averaged *I*–*V* characteristic all plasma parameters mentioned above can in principle be determined.

When a probe is inserted into a plasma, it will experience electrons and ions colliding with its tip. Due to the high mean speed of electrons, the flow of electrons is higher than the flow of ions. Consequently, the tip will charge up negatively until the electrons are repelled, and the net current then is zero. The probe potential then is the floating potential, V_{fl}. The electron current density J_e then balances the ion current density J_i. At potentials lower than V_{fl} the ion current cannot increase further—in fact, only ions are collected from the plasma—and the ion saturation current I_{is} is measured. The plasma potential V_{pl} is defined as the potential at which all electrons arriving near the probe are collected and the probe current equals the electron current. Note that the plasma assumes the plasma potential in the absence of a probe; hence probe perturbation at V_{pl} is minimal. At potentials higher than V_{pl} the electron saturation current I_{es} is measured.

The *I–V* characteristic thus is divided into three regions: (I) $V < V_{\text{fl}}$, (II) $V_{\text{fl}} < V < V_{\text{pl}}$, and (III) $V > V_{\text{pl}}$. In region I the ion saturation current dominates; in region III, the electron saturation current. They are derived to be [117, 294]:

$$I_{is} = 0.4 n_i e A_p \sqrt{\frac{2k_B T_e}{m_i}} \tag{35}$$

$$I_{es} = n_e e A_p \sqrt{\frac{8k_B T_e}{m_e}} \sqrt{1 + \frac{eV}{k_B T_e}} \tag{36}$$

with A_p the probe area. The expression for I_{es} is derived for a cylindrical probe tip.

In the intermediate region II, assuming a Maxwellian EEDF, the electron retarding current is given by

$$I_{er} = n_e e A_p \sqrt{\frac{k_B T_e}{2\pi m_e}} \exp\left(\frac{eV}{k_B T_e}\right) \tag{37}$$

The electron temperature is derived from the *I–V* characteristic by first subtracting the measured ion current from the measured current in region II. From the slope of the logarithm of I_{er} versus V, T_e is determined [Eq. (37)]. Plotting the square of the probe current versus voltage in region III yields a straight line [Eq. (36)], and the electron density is derived from its slope (and the known T_e). The ion density then is easily calculated with Eq. (35). The plasma potential is determined from the intersection between the extrapolated currents from regions II and III. The floating potential is the potential at which the current is zero. Other methods for the extraction of plasma properties from *I–V* data, not assuming a Maxwellian EEDF, are described by Awakowicz [294].

In RF discharges one has to compensate for the RF voltage component across the probe sheath. This is done with a pickup element, situated close to the probe tip, which should be in the plasma bulk as well. Compensating electronics within the probe is usually designed to work at a prescribed RF frequency, and cannot be used at other frequencies.

In the ASTER system a data series is measured for an argon and a hydrogen plasma running at 13.56 MHz, in which the power (5–30 W) and pressure (5–50 Pa) are varied [265, 295]. The probe tip is positioned exactly between powered and grounded electrode, at the center of the discharge.

The plasma potential determined from *I–V* data for argon increases linearly from about 24 to 27 V with increasing power at a pressure of 0.05 mbar. At the highest pressure of 0.35 mbar these values have shifted downwards by about 2 V only. For the hydrogen discharge similar behavior is observed, with an increase

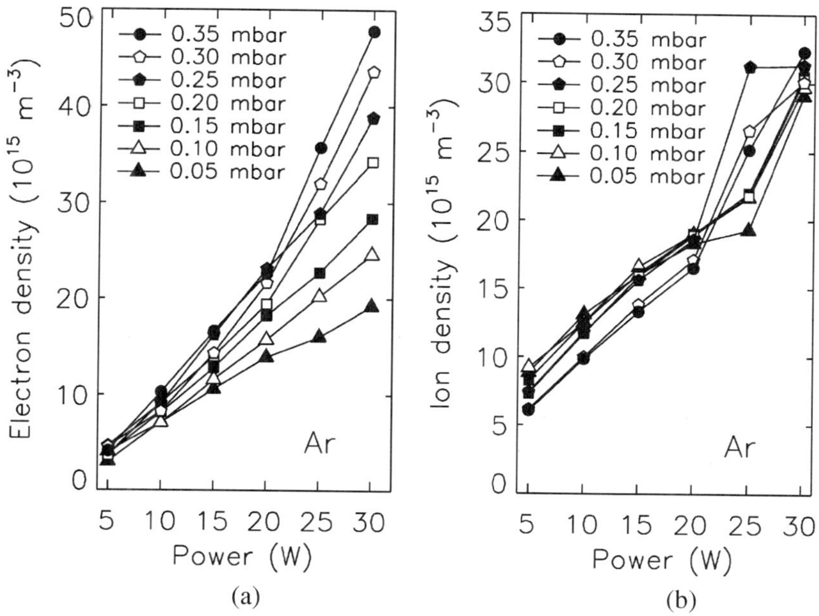

FIG. 29. Electron (a) and ion (b) density as a function of power for different pressures in the case of an argon discharge.

from 25 to 32 V with power, at 0.10 mbar. These values shift by about 3 V on going to a pressure of 0.5 mbar. The variation of the plasma potential in the case of hydrogen is larger than in the case of argon. This difference in behavior is also seen in the electron temperature and the electron and ion density. In the case of argon, T_e varies between 1.8 and 2.8 eV. The effect of pressure variation is larger than the effect of power variation. At low pressure (0.05 mbar) T_e is around 2.5 eV, whereas at high pressure (0.35 mbar) T_e is around 2 eV. In the case of hydrogen, T_e varies between 1.5 and 4.2 eV. At low pressure (0.10 mbar) T_e varies slightly around 4 eV, whereas at high pressure (0.5 mbar) T_e changes from 4 eV at low power (5 W) to 1.5 eV at high power (30 W).

The electron and ion densities for argon and hydrogen are shown in Figure 29 and Figure 30, respectively. The absolute values of n_i for argon (Fig. 29b) are about 5–8 times larger than the ones for hydrogen (Fig. 30b). Further, the increase of n_i as a function of applied power is only slightly pressure-dependent in the case of argon. The dependence is much clearer in the case of hydrogen. Similar observations can be made for the electron density.

Employing a extendable bellows, the probe tip can be moved towards the edge of the discharge over a distance of about 5 cm. Measurements performed in an argon discharge at 0.15 mbar at three different power levels show that V_{pl}, T_e, n_e,

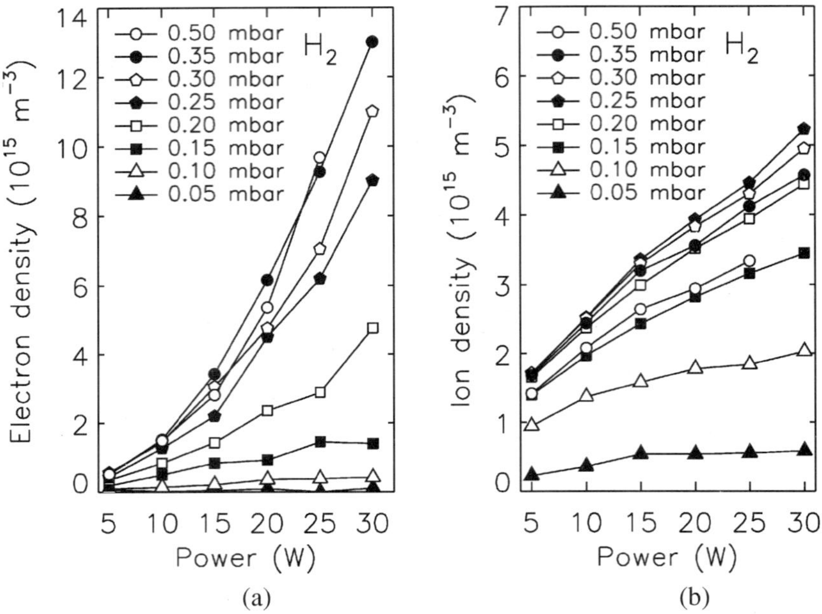

FIG. 30. Electron (a) and ion (b) density as a function of power for different pressures in the case of a hydrogen discharge.

and n_i do not vary more than 5–10%. This shows that the discharge is homogeneous.

The Langmuir probe method is easily employed in discharges; however, especially in deposition plasmas, care has to be taken in interpreting *I–V* data. The presence of negative ions in silane plasmas complicates the analysis. Their contribution should be taken into account. Further, (thin) insulating layers may form on the probe, which can result in short-circuiting of the probe. These contamination effects distort the probe characteristics, especially around V_{pl}. The probe surface may be cleaned *in situ* by ion bombardment, i.e., by biasing the probe to a large negative potential in a *non*depositing plasma such as argon. Melting the deposited contaminations off the probe should be possible by passing a high electron current through the probe.

Probe measurements in silane discharges have been reported [296, 297]. Apparently, no difficulties were experienced, as the deposited amorphous silicon layer on the tip was sufficiently photoconductive. For typical silane discharge conditions values for T_e are found to be between 2 and 2.5 eV. Electron densities are around 1×10^9 cm^{-3} [296]. Probe measurement in the ASTER system failed due to strong distortions of the probe current, even after following cleaning procedures.

1.5.3. MASS SPECTROMETRY

Mass spectrometry is used to obtain information on the neutral and ionic composition of the deposition discharge. Three modes of sampling are commonly used: analysis of discharge products downstream, line-of-sight sampling of neutrals, and direct sampling of ions from the discharge. Mass spectrometry is a simple method to measure and quantify neutral species in a silane discharge, although calibrations are required. For ionic species, mass spectrometry is the only method available. A quadrupole mass filter is a proven instrument that is essential in mass spectrometry of low-pressure discharges.

1.5.3.1. Analysis of Neutrals

The analysis of the neutral gas composition in a discharge yields useful information on the mechanisms and kinetics of silane dissociation. However, it should be borne in mind that with mass-spectrometric analysis one only detects the final products of a possibly long chain of reactions.

The partial pressures of the stable neutral molecules in the discharge (silane, hydrogen, disilane, trisilane) can be measured by a quadrupole mass spectrometer (QMS). The QMS usually is mounted in a differentially pumped chamber, which is connected to the reactor via a small extraction port [286]. In the ASTER system a QMS is mounted on the reactor that is used for intrinsic material deposition. The QMS background pressure (after proper bake-out) is between 10^{-12} and 10^{-13} mbar. The controllable diameter in the extraction port is adjusted so that during discharge operation the background pressure never exceeds 10^{-11} mbar.

The gas that enters the QMS is ionized by electron impact at a factory-preset electron energy of 70 eV (or 90 eV), and subsequently mass-analyzed. The ion currents at different mass-to-charge ratios (m/e) need to be converted to partial pressures by careful calibrations, as reported by Hamers [163]. Gas X is admitted to the reactor and the corresponding pressure p is measured, as well as the ion current I at a specific ratio $m/e = \mu_X$ of the gas, e.g., for argon $m/e = 40$ amu/e. The background signal at the same $m/e = \mu_X$ that results from residual gases in the QMS chamber is subtracted from the measured ion current. The calibration factor γ_X is defined as the ratio between the corrected ion current and the reactor pressure. Calibrations are performed typically at three pressures before and after measurements on the discharge, as the sensitivity of the channeltron in the QMS may rapidly change. The calibrations are performed with hydrogen, argon, silane, and disilane, with the main contributions at $m/e = 2, 40, 30$, and 60 amu/e, respectively.

In the deposition of a-Si:H the dissociation of the process gas silane leads to the formation of hydrogen, disilane, trisilane, higher-order silanes, and a solid film. Silane is depleted. From the consumption of silane, one may estimate the deposition rate of the solid film. The difference in the silane partial pressure between

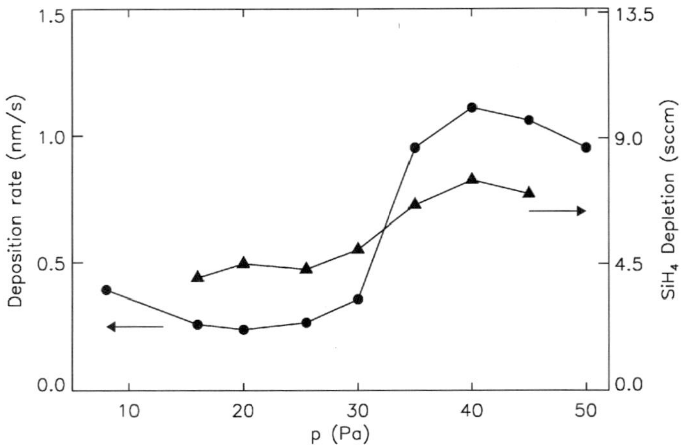

FIG. 31. The deposition rate and the corresponding silane depletion as a function of the total process pressure. Other conditions are 50 MHz, 30 : 30 $SiH_4 : H_2$. (From E. A. G. Hamers, Ph.D. Thesis, Universiteit Utrecht, Utrecht, the Netherlands, 1998.)

the discharge-on and the discharge-off state is often used as an indication of the amount of silane that is depleted [298–300]. However, the relative change in pressure in general is not equal to the relative change in flow, due to gas-dependent conductances [163, 301] (see also Section 1.2.4). For example, in the ASTER deposition system equal partial pressures of silane and hydrogen are observed for a flow ratio of silane to hydrogen of about 1.25. Hence, another way of determining the silane depletion is used [163, 301]. First, the partial pressures of silane and hydrogen are measured in the discharge-on state. Subsequently, the discharge is switched off, and the silane and hydrogen flow rates are adjusted so that the same partial pressures as in the discharge-on state are reached. These flow rates then are the flow rates of silane and hydrogen that leave the reactor in the discharge-on state. The silane depletion $\Delta Q_{\mathrm{SiH}_4}$ is the difference between the admitted silane flow rate and the flow rate of silane that leaves the reactor when in the discharge-on state. Additionally, an estimate can be made of the deposition rate, under the assumptions that all depleted silane is used for the deposition (no large amount of higher-order silanes or dust is formed). Also the deposition is assumed to occur homogeneously over the surface of the reactor. The estimated deposition rate is $r_d^{\mathrm{est}} = \Delta Q_{\mathrm{SiH}_4}/k_B T_0 \rho A_r$, with ρ the atomic density of the amorphous silicon network (5×10^{22} cm^{-3}), T_0 the standard temperature (300 K), and A_r the reactor surface (0.08 m^2). This yields an estimated deposition rate of 0.11 nm/s per sccm SiH_4 depletion [163]. In Figure 31 the deposition rate is compared with the corresponding silane depletion as a function of process pressure for a silane–hydrogen discharge at 50 MHz and 10 W. For these conditions the α–γ' transition occurs

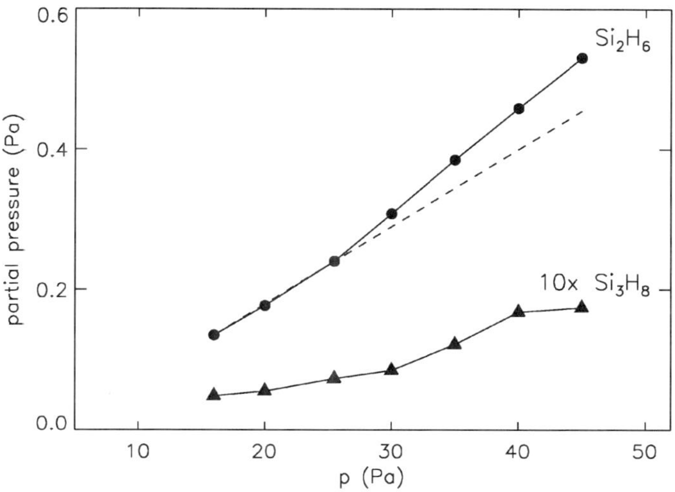

FIG. 32. The partial pressures of disilane and trisilane. The dashed line is an extrapolation of the disilane partial pressure in the α-regime. (From E. A. G. Hamers, Ph.D. Thesis, Universiteit Utrecht, Utrecht, the Netherlands, 1998.)

at 30 Pa: the deposition rate is increased by a factor of 4, whereas the depletion is increased only by a factor of 2. The right-hand axis is scaled using the value 0.11 nm/s·sccm. In the α-regime the deposition rate is underestimated, whereas in the γ'-regime it is overestimated. This is in part due to the assumed uniformity of deposition. In the α-regime the deposition rate at the center of the substrate is lower than at the edges; in the γ'-regime this is reversed. Moreover, from the low dc bias voltages observed in the γ'-regime it is inferred that the discharge is more confined.

The partial pressures of disilane and trisilane are shown in Figure 32, for the same process conditions as in Figure 31. Both partial pressures increase as a function of pressure. Around the α–γ' transition at 30 Pa the disilane partial pressure increases faster with increasing pressure, as can be seen from the deviation from the extrapolated dashed line. The disilane partial pressure amounts to about 1% of the total pressure, and the trisilane partial pressure is more than an order of magnitude lower. Apparently, in the γ'-regime the production of di- and trisilane is enhanced.

Not only the silane depletion, but also the hydrogen production can be used to obtain information on the reaction products of the decomposition process. In Figure 33 the silane depletion and the corresponding hydrogen production are shown for a number of experiments, with process parameters such as to cover both the α- and the γ'-regime [163, 301]. A clear correlation exists. In addition,

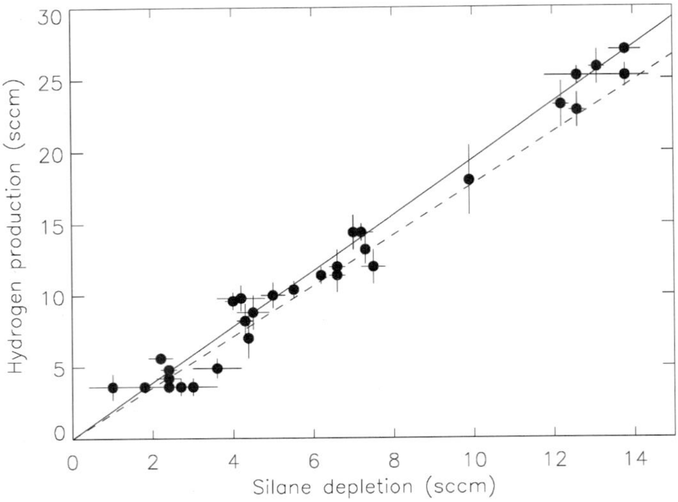

FIG. 33. The depletion of silane and the corresponding production of hydrogen for several process conditions, covering both the α- and the γ'-regime. The solid line represents the case where all the consumed silane is converted into a-Si:$H_{0.1}$ and $1.95H_2$. The dashed line represents the case where 30% of the consumed silane is converted into disilane instead of being deposited. (From E. A. G. Hamers, Ph.D. Thesis, Universiteit Utrecht, Utrecht, the Netherlands, 1998, with permission.)

the solid line in Figure 33 relates the hydrogen production and silane depletion in the case that only a-Si:H is formed with 10 at.% H, i.e., a-Si:$H_{0.1}$ and H_2. The dashed line represents the case where 30% of the silicon from the consumed amount of silane would leave the reactor as disilane, instead of being deposited as a-Si:$H_{0.1}$. The data can well be explained by these two cases. In fact, they are an indication that part of the consumed silane leaves the reactor with a larger amount of H per Si atom than a-Si:H, which may be caused by the formation of higher silanes and/or powders.

Mass-spectrometric research on silane decomposition kinetics has been performed for flowing [298, 302–306] and static discharges [197, 307]. In a dc discharge of silane it is found that the reaction rate for the depletion of silane is a linear function of the dc current in the discharge, which allows one to determine a first-order reaction mechanism in electron density and temperature [302, 304]. For an RF discharge, similar results are found [303, 305]. Also, the depletion and production rates were found to be temperature-dependent [306]. Further, the depletion of silane and the production of disilane and trisilane are found to depend on the dwell time in the reactor [298]. The increase of di- and trisilane concentration at short dwell times (<0.5 s) corresponds to the decrease of silane concentration. At long dwell times, the decomposition of di- and trisilane produces a much greater decrease in silane concentration. Moreover, a strong correlation is observed between deposition rate and di- and trisilane concentration. This has led to a proposal [283, 298, 308] of the main electron impact dissociation channel of silane, viz., $e^- + SiH_4 \rightarrow SiH_2 + H_2 + e^-$. The higher silanes are partially decomposed into radicals, which are responsible for deposition [308].

In a static discharge, first silane is introduced and subsequently all valves are closed. Then the discharge is ignited, and the mass-spectrometer signals for silane ($m/e = 30$ amu/e), disilane ($m/e = 62$ amu/e), trisilane ($m/e = 92$ amu/e), and tetrasilane ($m/e = 120$–128 amu/e) are followed as a function of time [197]. Rate equations can be formulated that explain the decrease in silane concentration and the increase in di-, tri-, and tetrasilane. From these data another proposal [197] of the main electron impact dissociation channel of silane is formulated: $e^- + SiH_4 \rightarrow SiH_2 + 2H + e^-$, in agreement with others [201, 309, 310]. The abstraction reaction $H + SiH_4 \rightarrow SiH_3 + SiH_3$ explains the large production of SiH_3 radicals in this case. The discrepancy as reflected in the hydrogen production (either H or H_2) cannot be resolved solely by mass-spectrometric data.

1.5.3.2. *Analysis of Radicals*

It is possible but difficult to detect directly, by mass spectrometry, the neutral radicals produced in a silane discharge. To discrimate between the radicals and the much more abundant neutrals, ionization of the species by low-energy electrons can be used [307], and the term *threshold ionization mass spectrometry* (TIMS) is used to denote this technique. This method is based on the principle of ionization of neutrals by electrons that have a well-defined energy. These ionized neutrals then are detected by means of a quadrupole mass spectrometer. In general, the radical A can be detected when the electron impact ionization process leads to the formation of an ion A^+. In order for this to occur, the energy of the electrons should be larger than the ionization potential E_i. Also, dissociative ionization of a molecule containing A may lead to A^+, following $AB + e^- \rightarrow A^+ + B + 2e^-$. Here, the energy of the electrons should be larger than the appearance potential E_a. It holds that $E_a > E_i$, and $E_a \approx E_i + E_{A-B}$, where E_{A-B} is the energy of the A$-$B bond [311]. When the electron energy is larger than E_a, the dissociative ionization process dominates, because the density of the molecule AB is much larger than that of the radical A. When the electron energy has a value between E_i and E_a, then only the radicals are ionized, and one is able to determine the radical density.

In silane discharges, one observes the following: when the discharge is off, the mass spectrometric signal at $m/e = 31$ amu/e (SiH_3^+) as a function of electron energy is due to dissociative ionization of SiH_4 in the ionizer of the QMS, with an ionization potential of 12.2 eV [312]. The signal with the discharge on is due to ionization of the radical SiH_3 plus the contribution from dissociative ionization of silane (the ions from the discharge are repelled by applying a positive voltage to the extraction optics of the QMS). The appearance potential of the SiH_3 radical is 8.4 eV [312], and therefore a clear difference between discharge on and off is observed. The corresponding threshold energies for SiH_2 are 11.9 and 9.7 eV (these four numbers are also given by Kae-Nune et al. [311], as 12.0, 8.0, 11.5, and 9.0 eV, respectively). The net radical contribution to the ion signal is given by [311]

$$\Delta I(\mathrm{A}^+) = (I_{\mathrm{on}} - I_{\mathrm{bg}} - I_{\mathrm{ions}}) - (I_{\mathrm{off}} - I_{\mathrm{bg}}) \frac{n_{\mathrm{SiH_4,on}}}{n_{\mathrm{SiH_4,off}}} \tag{38}$$

where I_{on} (I_{off}) is the signal when the discharge is on (off), I_{bg} the background signal, and I_{ions} the signal due to ions (which should be zero). The factor $n_{\mathrm{SiH_4,on}}/n_{\mathrm{SiH_4,off}}$ takes account of the depletion of the density of SiH_4, which can be obtained from the ratio $I_{\mathrm{on}}/I_{\mathrm{off}}$ at high electron energies. The concentration of radical SiH_x can be expressed as [311]

$$\frac{n(\mathrm{SiH}_x)}{n(\mathrm{SiH_4})} = C \frac{Q(\mathrm{SiH_4} \to \mathrm{SiH}_x^+, E_2 > E_a)}{I(\mathrm{SiH}_x^+, E_2 > E_a)} \times \frac{1}{E_1'' - E_1'} \int_{E_i - E_1'}^{E_1'' < E_a} \frac{\Delta I(\mathrm{SiH}_x^+, E)}{Q(\mathrm{SiH}_x \to \mathrm{SiH}_x^+, E)} \, dE \tag{39}$$

where $I(\mathrm{SiH}_x^+, E_2 > E_a)$ is the signal due to dissociative ionization of SiH_4, $\Delta I(\mathrm{SiH}_x^+, E)$ the net radical signal [Eq. (38)], $Q(\mathrm{SiH_4} \to \mathrm{SiH}_x^+, E_2 > E_a)$ the electron impact cross section for the dissociative ionization of SiH_4, $Q(\mathrm{SiH}_x \to \mathrm{SiH}_x^+, E)$ the electron impact cross section of the ionization of the radical SiH_x, and C a geometrical correction factor. Clearly, the knowledge of the electron impact cross sections is essential for the exact determination of the radical densities. Kae-Nune et al. [311] have used data for deuterated methane radicals (CD_3 and CD_2 [313]) to determine $Q(\mathrm{SiH_3} \to \mathrm{SiH_3^+}, E)$ and $Q(\mathrm{SiH_2} \to \mathrm{SiH_2^+}, E)$ by scaling. The value of $Q(\mathrm{SiH} \to \mathrm{SiH^+}, E)$ is determined by taking the average of $Q(\mathrm{SiH_2} \to \mathrm{SiH_2^+}, E)$ and the measured $Q(\mathrm{Si} \to \mathrm{Si^+}, E)$ [314]. In Figure 34 the measured SiH_3^+ signal is compared with fitted curves for radical and dissociative ionization [311].

Nowadays, a commercially available QMS system (Hiden EQP300) is used by many groups, with which detection of radicals, and also of ions (see Section 1.5.3.3), is possible [315]. Kae-Nune et al. [198, 217, 311] have mounted this QMS system in the grounded electrode of their parallel plate PECVD reactor. Neutral (and ionic) species are sampled from the discharge via a 0.3-mm-diameter hole in the grounded electrode. Two-stage differential pumping ensures a low background pressure ($<10^{-6}$ Torr) in the ionization chamber of the QMS.

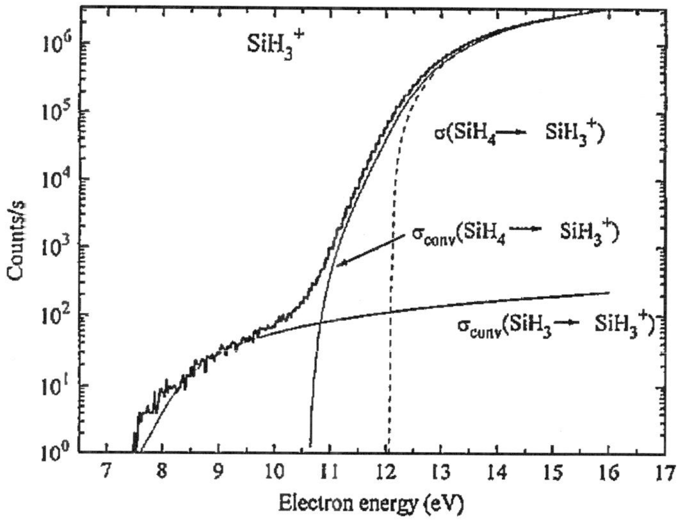

FIG. 34. Measured and fitted curves for the ionization of the SiH_3 radical. [From P. Kae-Nune, J. Perrin, J. Guillon, and J. Jolly, *Plasma Sources Sci. Technol.* 4, 250 (1995), © 1995, Institute of Physics, with permission.]

With this setup a series of measurements was taken in which the power was varied from 5 to 30 W. The pressure was low, 0.06 Torr, while the temperature was 250°C. The SiH_x $(0 < x < 3)$ radical densities near the surface were measured as a function of effective power delivered to the discharge. It was found that the SiH_3 radical is the most abundant one, with a concentration varying between 1.8×10^{11} and 3.2×10^{11} cm^{-3}. The SiH_2 radical concentration was about a factor of 20 lower. Both radical concentrations increase with increasing power.

From the radical concentrations it is possible to derive the radical flux to the surface:

$$\Phi_{SiH_x} = n_{SiH_x} \frac{v_{SiH_x}}{4} \frac{\beta_{SiH_x}}{1 - \beta_{SiH_x}/2} \tag{40}$$

where v_{SiH_x} is the thermal velocity $(= \sqrt{8k_B T/\pi m})$ of the radical SiH_x, β_{SiH_x} the surface reaction probability (see also Fig. 14 in Section 1.4.1.6). The quantity β is the adsorbed fraction of radicals, which either stick (sticking probability s_{SiH_x}) or recombine as SiH_4 or Si_2H_6 and go to the gas phase (recombination probability γ_{SiH_x}). By definition, $\beta = s + \gamma$. An amount $r = 1 - \beta$ is reflected from the surface directly.

The contribution of radical SiH_x to the total deposition rate follows from

$$R_{d,SiH_x} = \Phi_{SiH_x} \frac{s_{SiH_x}}{\beta_{SiH_x}} \frac{M_{Si}}{N_A \rho_{a\text{-}Si:H}} \tag{41}$$

where N_A is Avogadro's number, M_{Si} the molar mass of silicon, and $\rho_{a\text{-Si:H}}$ the mass density of a-Si:H (2.2 g/cm^3). For SiH_3 Matsuda et al. have determined β and s [137]: $\beta_{SiH_3} = 0.26$ and $s_{SiH_3} = 0.09$. For SiH_2, SiH, and Si large β and $s \approx \beta$ are assumed, viz., $\beta_{SiH_2} = s_{SiH_2} = 0.8$ [311], $\beta_{SiH} = s_{SiH} = 0.95$ [316], and $\beta_{Si} = s_{Si} = 1$ [311]. Using these values for β and s, Kae-Nune et al. [311] have determined the contribution of radicals to the deposition rate. At low power levels, the deposition rate can fully be accounted for by the SiH_3 and SiH_2 radicals in a 60 : 40 ratio. At high power levels the other radicals also become important. At their highest power level of about 50 mW/cm^2, the sum of all radical contributions to the deposition rate is only 65%, indicating that dimer and trimer radicals, as well as ions, also contribute to the deposition.

Using the same threshold ionization mass spectrometry setup, Perrin et al. [317] have measured the temporal decay of radical densities in a discharge afterglow. From these experiments the coefficient β for the radical SiH_3 has been determined to be 0.28, which is in agreement with already known results from other (indirect) experimental approaches [136, 137, 318]. For the Si_2H_5 radical β is determined to be between 0.1 and 0.3. The coefficient β for atomic hydrogen on a-Si:H lies between 0.4 and 1, and is thought to represent mainly surface recombination to H_2.

1.5.3.3. Analysis of Ions

Detection of ions from a discharge is done by direct sampling through an orifice. In order to extract the ions collisionlessly the dimensions of the sampling orifice should be smaller than the sheath thickness; they are typically of the order of 100 μm. Moreover, the detected ions and their energies are representative of the plasma bulk situation only when the sheath is collisionless, i.e., at low pressures [286]. One generally is interested in the interaction of ions with a growing surface; hence normal operating pressures are used.

In RF discharges of silane, SiH_3^+ usually is the most abundant ion, but others (SiH_2^+, and $Si_nH_m^+$ with $1 < n < 9$) also are present [319]. The relative abundance of SiH_3^+ ions increases with increasing pressure, while that of SiH_2^+ decreases [319]. The ionization cross section of SiH_2^+ is higher than that of SiH_3^+ [320], but SiH_2^+ is lost via the reaction $SiH_2^+ + SiH_4 \rightarrow SiH_3^+ + SiH_3$ [305]. At pressures lower than 0.1 Torr SiH_2^+ becomes the dominant ion.

The ion clusters ($Si_2H_m^+$, $Si_3H_n^+$, ...) are also present, and larger clusters (positive and neutral) can be formed through reactions with silane molecules. Negative ions have been detected [321], which are thought responsible for the powders in the discharge [322].

Ion energy distributions (IEDs) are measured by several groups [323–326]. The reliability of IEDs depends strongly on the knowledge of the transmission function of the instruments, which most likely is energy-dependent. Improper ad-

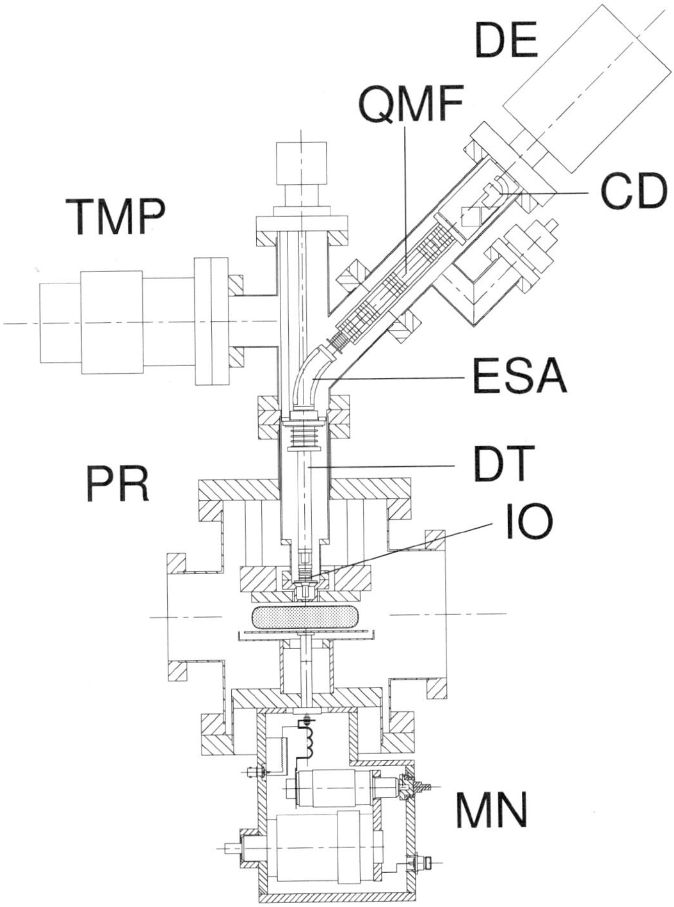

FIG. 35. Vertical cross section of the reaction chamber equipped with the mass spectrometer system. Indicated are QMF, the quadrupole mass filter; ESA, the electrostatic analyzer; CD, the channeltron detector; DE, the detector electronics; DT, the drift tube; IO, the ion optics; TMP, the turbomolecular pump; PR, the plasma reactor; and MN, the matching network.

justment of the various potential levels throughout the instruments can result in "ghost" structures in the IED, and they reflect the physics of the instrument rather than of the discharge. Hamers et al. [161, 163] have presented a method to determine the proper transmission function of an electrostatic lens system of a commercially available ion energy and mass spectrometer, the Hiden EQP [315]. This system is used by many groups to study plasmas [160, 198, 311, 321, 327, 328]. It is very versatile with respect to the control of voltages on the many electrostatic lenses.

The EQP is mounted in a plasma reactor identical to the ones in ASTER; see Figure 35. The plasma is generated between the two parallel electrodes. In the grounded electrode, where normally the substrate is placed for the deposition of a-Si:H, a small orifice is located to sample particles that arrive from the discharge. The orifice is made in a 20-μm-thick stainless steel foil, and has a diameter of 30 μm. The foil is integrated in a stainless steel flange. The ratio of the thickness of the foil to the diameter of the orifice results in a physical acceptance angle of 55° with respect to the normal to the electrode. The small size of the orifice compared to the sheath thickness and the mean free path of the particles in the plasma ensures that the orifice does not influence the discharge [323]. Using this orifice in deposition plasmas leads to deposition of a-Si:H on the surface of the orifice, but also on its inner sides, thereby narrowing the sampling diameter. This can be observed by monitoring the pressure rise in the mass spectrometer during experiments. Typically, a 50% reduction in orifice area takes 3 h. The a-Si:H film on the orifice then needs to be etched away, which is done *ex situ* by a KOH etch.

The design of the instrument, together with the pumping capacity, ensures a low background pressure ($<10^{-9}$ mbar). Under process conditions the pressure directly behind the orifice is about a factor of 10^5 lower than the process pressure; in the mass filter, even a factor of 10^6. The mean free path of particles that have entered the EQP therefore is several meters.

The ion optics (IO in Figure 35) consists of a number of electrostatic lenses, which direct the sampled ions through the drift tube (DT) to the electrostatic analyzer (ESA). The ESA transmits only those ions with a specific energy-to-charge ratio ϵ_{pass}. An ESA has a constant relative energy resolution $\Delta\epsilon/\epsilon_{\text{pass}}$. Because this ESA operates at a constant $\Delta\epsilon$, the pass energy ϵ_{pass} is constant. Here, $\epsilon_{\text{pass}} = 40$ eV, and $\Delta\epsilon = 1.5$ eV (FWHM). The measurement of IEDs requires that sampled ions be decelerated or accelerated to this pass energy. This is also done in the ion optics part.

The other very important function of the ion optics is to shape the ion beam. Voltages on the various lenses should be set to avoid chromatic aberration, which causes energy-dependent transmission of ions in the instrument, and as a result erroneous IEDs [161, 163]. The correct lens settings have been found by simulations of ion trajectories in the EQP using the simulation program SIMION [329]. In addition, an experimental method to find the correct settings has been presented [161, 163].

The acceptance angle depends on the lens settings and on the ion energy. Low-energy ions are deflected more towards the optical axis than high-energy ions, which results in a larger acceptance angle. The acceptance angle varies between about 6° and 1° for ion energies between 1 and 100 eV [161, 163]. For lower energies the acceptance angle can reach values above 20°, which causes drastic increases of the low-energy part of the measured IED.

The measured ion energy distributions are affected by the processes that ions experience during their passage through the plasma sheath. In a collisionless situation, the IED is solely determined by the RF modulation of the plasma potential. The extent to which an ion will follow these modulations depends on its mass. Electrons are able to follow the electric field variations instantaneously, while ions experience a time-averaged electric field. The energy of an ion that arrives at the electrode depends on the value of the plasma potential at the time that the ion entered the sheath. Hence, the ion energy is modulated in much the same way that the plasma potential is modulated. Many ions arrive at the electrode with energies close to the extreme values of the harmonically varying voltage, and this results in a saddle structure in the IED, in which the two peaks reflect the minimum and maximum plasma potential [134]. The heavier the ion, the more it experiences a time-averaged plasma potential, and the narrower the saddle structure is.

Elastic collisions and chemical reactions in the sheath lead to a broad angular distribution of the ion velocity at the electrode. Examples for SiH_3^+ and SiH_2^+ are the H^- abstraction reactions $SiH_3^+ + SiH_4 \rightarrow SiH_4 + SiH_3^+$ [330] and $SiH_2^+ + SiH_4 \rightarrow SiH_3^+ + SiH_3$ [331]. Due to the small acceptance angle of the EQP, these processes are only in part reflected in the IED [163, 332].

Charge exchange. Much more important for the shape of the IED are charge exchange processes. In a charge exchange process an electron is exchanged between a neutral and an ion. In the plasma sheath a charge exchange between a neutral of thermal energy and a fast ion leads to the formation of a thermal ion and a fast neutral. These newly created thermal ions are accelerated towards the electrode, and the kinetic energy that they gain depends on the phase in the RF period and their position at the time of creation. This leads to distinct peaks in the IED, and the origin of these peaks can be explained as follows [325]. Electrons respond to the RF modulation and move to and from the electrodes. The movement of the electron front is a function of the phase ϕ in the RF period. When an ion is created (by a charge exchange process) behind the electron front, i.e., in the plasma bulk, it will not experience an electric field. Once the electron front moves inward to the plasma bulk, the ion will be accelerated to the electrode. Depending on the place of creation x, the ion will be able to reach the electrode before the electron front is at the position of the ion again. Once in the plasma bulk, the ion will not be accelerated until the electron front moves inward again. This may be repeated several times. Wild and Koidl [325] have presented a description of the origin of the maxima in IEDs. They argue that these maxima only occur if the ion energy ϵ at the electrode is independent of the phase ϕ and the position x of creation of the newly formed ion. Two conditions are to be met: $d\epsilon/d\phi = 0$ and $d\epsilon/dx = 0$. As shown by Wild and Koidl [325], the energy ϵ varies between ϵ_{min} and ϵ_{max}, and for every x, the values of ϵ_{min}, ϵ_{max}, and $\epsilon_{\phi=0}$ fulfill the condition $d\epsilon/d\phi = 0$. Further, $\epsilon_{\phi=0}$ exhibits a series of extrema as a function of x, and at these extrema is alternately equal to $\epsilon_{\min}$ and $\epsilon_{\max}$. At these extrema both conditions are met, and a saddle structure is observed in the IED.

In addition, ions that arrive with the mean energy of the saddle structure have reached the electrode after an integer number of periods. Thus, ions in successive charge exchange saddle structures have needed an increased number of periods to reach the electrode, on average. Hamers has very nicely illustrated this by performing Monte Carlo simulations of charge exchange processes [163], using a time- and position-dependent electric field, which is calculated from the sheath model as described by Snijkers [333]. About 300,000 SiH_2^+ ions were followed. The ion density at the electrode was 3×10^8 cm^{-3}, the plasma potential 25 V, and the excitation frequency 13.56 MHz. The main saddle structure is found to be around 25 V (23–27 V). In the IED four charge exchange peaks are clearly distinguished, at about 2, 7.5, 14, and 19 eV. From a comparison with the transit time, i.e. the time an ion needs to cross the sheath, it is clear that the peaks in the IED correspond to ions that needed an integer number of periods to cross the sheath. The first peak at 2 eV corresponds to ions that needed one period, and so on. In general, ions in charge exchange peak p needed p periods to cross the sheath.

The energy position ϵ_p of the peak and the corresponding number p of periods of time T are used by Hamers et al. [163, 332] to reconstruct the time-averaged potential profile $V(x)$ in the sheath. An ion that arrives with energy ϵ_p at the electrode has a velocity that follows from $\frac{1}{2}mv^2 = \epsilon_p - eV(x)$. With $v = dx/dt$ one derives

$$\int_0^{x_p} \sqrt{\frac{m/2}{\epsilon_p - eV(x)}}\, dx = \int_0^{pT} dt = pT \tag{42}$$

where x_p is the mean position of creation of the ions that arrive at the electrode with energy ϵ_p. The unknown $V(x)$ now is found by further assuming that $V(x)$ can be represented by a parabolic function [328, 333]. A parabolic potential profile corresponds to a constant (net) charge carrier density n as a function of distance x in the sheath, according to $-\vec{\nabla}^2 V = \vec{\nabla} \cdot \vec{E} = en/\epsilon_0$. The electric field $E(x)$ in the sheath then is a linear function of x:

$$E(x) = -E_0 + \frac{enx}{\epsilon_0} \tag{43}$$

with E_0 the electric field at the electrode ($x = 0$). Integrating $E(x)$ from the electrode ($V = 0$) over a distance x gives the potential

$$V(x) = E_0 x + \frac{en}{2\epsilon_0} x^2 \tag{44}$$

The sheath thickness d_s is found from the position where the electric field vanishes: $d_s = \epsilon_0 E_0/(en)$. Here the potential is equal to the time-averaged plasma potential V_{pl}. We then find, with Eq. (44), for the electric field at the electrode

$$E_0 = \sqrt{\frac{2enV_{\mathrm{pl}}}{\epsilon_0}} \tag{45}$$

and for the sheath thickness

$$d_s = \sqrt{\frac{2\epsilon_0 V_{\mathrm{pl}}}{en}} \tag{46}$$

Rewriting Eq. (44) yields

$$V(x) = V_{\mathrm{pl}}\left[\frac{2x}{d_s} - \left(\frac{x}{d_s}\right)^2\right] \tag{47}$$

Substitution of $V(x)$ in Eq. (42) and solving for ϵ_p then yields

$$\epsilon_p = eV_{\mathrm{pl}}\left\{1 - \left[\cosh\left(pT\sqrt{\frac{e^2n}{m\epsilon_0}}\right)\right]^{-2}\right\} \tag{48}$$

The energy position ϵ_p of peak p in the IED of an ion with mass m is seen to be dependent on the plasma potential V_{pl}, the RF period T, and the ion plasma frequency $\omega_i = \sqrt{e^2n/(m\epsilon_0)}$. Equation (48) can be used to determine the (net) charge carrier density in the sheath and the time-averaged potential V_{pl} from measured IEDs. The mean position x_p follows from combining Eq. (47) and Eq. (48):

$$x_p = d_s\left\{1 - \left[\cosh\left(pT\sqrt{\frac{e^2n}{m\epsilon_0}}\right)\right]^{-1}\right\} \tag{49}$$

In silane discharges several ions are observed to be involved in a charge exchange process, and therefore maxima in their ion energy distribution at distinct energies are observed. The charge carrier density and the plasma potential that result from the fit of the IED allow for the quantification of the related parameters sheath thickness and ion flux. This method has been be used to relate the material quality of a-Si:H to the ion bombardment [301, 332]; see also Section 1.6.2.3.

In the following, IEDs measured in silane–argon and silane–hydrogen discharges are shown, and V_{pl} and n are determined from fitting the data using Eq. (48). In fact, V_{pl} is determined from an IED that is not affected by collisions in the sheath; this is the case for $Si_2H_4^+$, where only one peak is observed in the IED. It should be noted that in general the width of the saddle structure is smaller than the energy resolution (1.5 eV) of the instrument, and therefore cannot be distinguished.

Silane–argon discharges. Silane–argon discharges are very illustrative, for no less than four different ions have charge exchange maxima in their ion energy distribution. In Figure 36 IEDs are shown that were measured in a plasma that

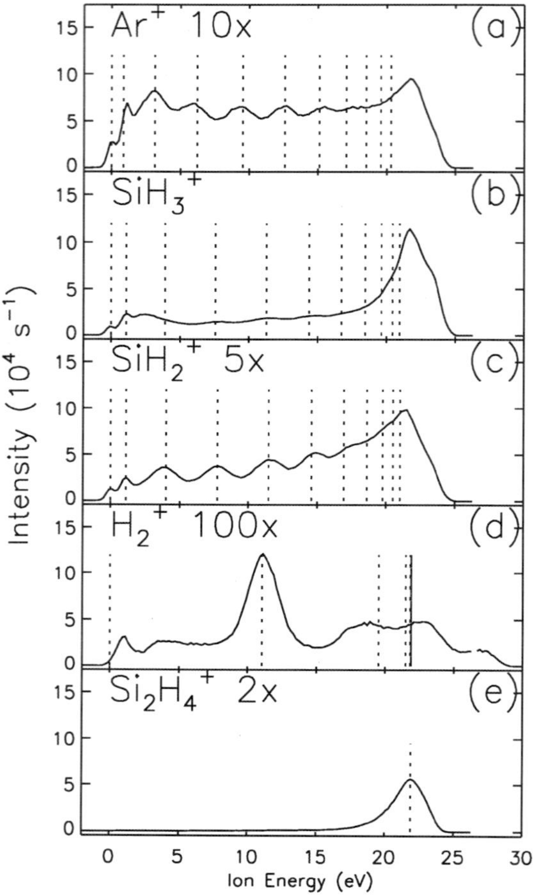

FIG. 36. The measured ion energy distributions of (a) Ar^+, (b) SiH_3^+, (c) SiH_2^+, (d) H_2^+, and (e) $Si_2H_4^+$ from a SiH_4/Ar plasma. The conditions are 13 sccm SiH_4, 13 sccm Ar, 13.56 MHz, 10 W, 0.2 mbar, 250°C, $V_{dc} = -135$ V. The dotted lines in (a), (b), (c), and (d) indicate the position of the first 10 charge exchange peaks based on a plasma potential of 21.8 V and a charge carrier density n of 1.66×10^8 cm^{-3}. The dotted line in the IED of $Si_2H_4^+$ (e) indicates the value of the plasma potential. Note the different scaling factors of the various IEDs. (From E. A. G. Hamers, Ph.D. Thesis, Universiteit Utrecht, Utrecht, the Netherlands, 1998, with permission.)

was created in a mixture of 13 sccm Ar and 13 sccm SiH_4, at a pressure of 0.1 mbar. The applied RF frequency was 13.56 MHz, at a power of 10 W. The dc self-bias voltage that developed was -135 V. The substrate temperature was 250°C. At least the first six peaks in the IED of Ar^+ and the first five peaks in the IED of SiH_2^+ can clearly be discerned. From the IED of $Si_2H_4^+$ (Fig. 36e) a time-average plasma potential V_{pl} of 21.8 ± 0.3 V is derived. Using now only *one* fitting parameter, i.e., the charge carrier density n, all distinct peaks in the IEDs of Ar^+, SiH_3^+, SiH_2^+, and H_2^+ can be fitted with Eq. (48). Performing the fit yields $n = (1.66 \pm 0.05) \times 10^8$ cm^{-3}. The energy positions are fitted well, as can be seen by comparing the data in Figure 36 with the dotted vertical lines. The sheath thickness as calculated from Eq. (46) is 3.81±0.05 mm. The first charge exchange peak in the Ar^+ IED originates at a distance of 100 μm from the electrode, as calculated from Eq. (49). The sixth peak, which is still visible, originates at a distance of about half the sheath thickness from the electrode.

It is observed that the SiH_3^+ ion is the most abundant ion, see Figure 36b. The peak around 23 V is the main saddle structure and is asymmetric due to collision processes. SiH_2^+ (see Figure 36c) exhibits a 5 times lower peak intensity. The first five charge exchange maxima in the IED of SiH_2^+ (Fig. 36c) are clearly visible. The IED intensity is about 5 times lower than for SiH_3^+. Although the SiH_2^+ ion is the main product of dissociative ionization of SiH_4 [320], the measured IED intensity of SiH_3^+ is always larger. The H^- transfer reaction causes SiH_3^+ to be more abundant than SiH_2^+. Also, in the IED of Ar^+ (Fig. 36a) about six of the charge exchange maxima are clearly visible. The broad overall IED shows that many ions underwent a charge exchange reaction in the sheath. Even H_2^+ (Fig. 36d) IEDs can be measured. The low intensity is due to the low partial pressure of hydrogen (it is created by silane dissociation only), and to the fact that the ionization potential of hydrogen is, like that of argon, higher than that of silane. Therefore the number of created hydrogen ions will be small.

The maximum in the IED at 11 eV is the first charge exchange maximum of H_2^+. The energy of this maximum is always somewhat lower than the energy of the fourth charge exchange maximum in the SiH_2^+ IED. This is due to the almost four times higher plasma frequency of the H_2^+ ion ($m/e = 2$ amu/e) than that of the SiH_2^+ ion ($m/e = 30$ amu/e).

The H_2^+ saddle structure is broader than the saddle structures of the other ions, due to the larger ion plasma frequency of H_2^+. The ions with an energy larger than 25 eV are part of the high-energy side of the main saddle structure.

$Si_2H_4^+$ is an ion that is created in the plasma by polymerization reactions. Several pathways may lead to this ion. The first pathway is the dissociative ionization of Si_2H_6 that is formed in a radical–neutral reaction. The second pathway is the direct formation in the ion–molecule reaction [192] $SiH_2^+ + SiH_4 \rightarrow Si_2H_4^+ + H_2$.

Silane–hydrogen discharges. The IEDs of several ions in a typical silane–hydrogen plasma are shown in Figure 37. The plasma was created in a mixture of 26 sccm SiH_4 and 24.5 sccm H_2, at a pressure of 0.2 mbar. The applied RF frequency is 13.56 MHz, at a power of 10 W. The self-bias voltage that developed was −118 V. The substrate temperature was 250°C. The discharge is in the α-regime. The IEDs of H_2^+ and SiH_2^+ show distinct peaks. The positions of the

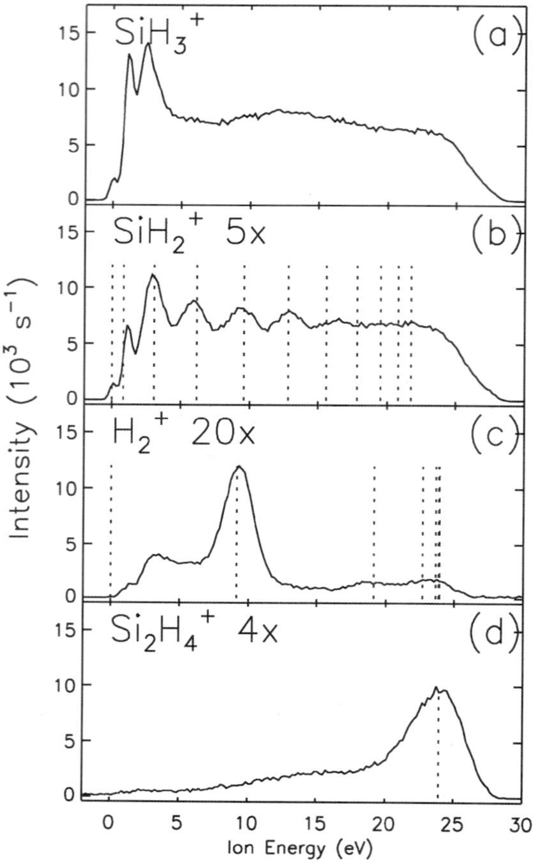

FIG. 37. The ion energy distributions of (a) SiH_3^+, (b) SiH_2^+, (c) H_2^+, and (d) $Si_2H_4^+$ from a SiH_4–H_2 plasma. The conditions are 26 sccm SiH_4, 24.5 sccm H_2, 13.56 MHz, 10 W, 0.2 mbar, 250°C, $V_{dc} = -118$ V. The dotted lines in (b) and (c) indicate the position of the first 10 charge exchange peaks based on a plasma potential of 23.8 V and a charge carrier density n of 1.11×10^8 cm^{-3}. The dotted line in the IED of $Si_2H_4^+$ (d) indicates the value of the plasma potential. Note the different scaling factors of the various IEDs. (From E. A. G. Hamers, Ph.D. Thesis, Universiteit Utrecht, Utrecht, the Netherlands, 1998, with permission.)

peaks are again described with the plasma potential, as deduced from the $Si_2H_4^+$ IED, and the charge carrier density n. The value for V_{pl} is 23.8 ± 0.2 V, and the value for n is $(1.11 \pm 0.04) \times 10^8$ cm^{-3}. The sheath thickness is 4.88 ± 0.10 mm. Under these and under most other conditions, the SiH_3^+ IED shows no distinguishable charge exchange maxima.

The measured intensity in the SiH_3^+ IED again is higher than that in the SiH_2^+ IED. There are two distinct peaks at low energies in the SiH_3^+ IED, and such peaks are observed in most SiH_3^+ IEDs. Their energy positions cannot be explained by charge exchange processes. The intensity in the H_2^+ IED is higher than in the silane–argon plasma, simply because the partial pressure of hydrogen in this silane–hydrogen mixture is higher.

Charge exchange maxima are in general pronounced in the SiH_2^+ IED in discharges at 13.56 MHz. The successive charge exchange peaks tend to overlap each other at higher frequencies. The IED of H_2^+ exhibits charge exchange peaks that are still well separated at higher frequencies due to the almost four times higher ion plasma frequency of H_2^+ than that of SiH_2^+.

Very different IEDs can be observed in discharges that are operated in the γ'-regime and that contain powder. The discharge, in this case, is created in a mixture of 30 sccm SiH_4 and 30 sccm H_2 at a pressure of 0.60 mbar. The RF frequency is 13.56 MHz, and the applied power is 33.7 W. The dc self-bias voltage is -32.5 V. The substrate temperature is 250°C. The measured IEDs are shown in Figure 38. The plasma potential is determined from the IED of $Si_2H_4^+$, and is 49.2 ± 0.5 V, which is about twice the value in discharges operated in the α-regime. The charge carrier density is $(9.9 \pm 0.3) \times 10^8$ cm^{-3}. The sheath thickness is 2.34 ± 0.05 mm.

The peaks in both the SiH_2^+ (Fig. 38a) and the SiH_3^+ (Fig. 38b) IED are well separated, and the intensity in the region between the charge exchange peaks is relatively low. The IED of $Si_2H_4^+$ also shows charge exchange maxima on the low-energy side of the IED, as well as an asymmetrical main saddle structure around 50 eV.

The sheath characteristics of the argon–silane and the two silane–hydrogen discharges of which the IEDs were shown above are summarized in Table VI. The sheath characteristics of the 10-W silane–argon discharge and the 10-W silane–hydrogen discharge in the α-regime are rather similar. The high-power, high-pressure discharge (γ'-regime) has a twice larger plasma potential, a twice smaller sheath distance, a four times larger electric field E_0, and a much larger ion flux and ion energy flux (see Section 1.6.2.2) than the discharge in the α-regime.

Charge exchange reactions. The charge exchange reactions between Ar and Ar^+ and between H_2 and H_2^+ are well known [325, 328]:

$$Ar^{+(f)} + Ar^{(t)} \rightarrow Ar^{(f)} + Ar^{+(t)} \tag{50}$$

$$H_2^{+(f)} + H_2^{(t)} \rightarrow H_2^{(f)} + H_2^{+(t)} \tag{51}$$

where (f) and (t) denote a fast ion or neutral and an ion or neutral with thermal energy. These reactions are called symmetrical resonant charge exchange reactions, as they occur between an ion and the corresponding neutral.

The ionization energies of hydrogen (15.4 eV) and argon (15.8 eV) are higher than those of silane (11.6 eV) and disilane (9.9 eV). Therefore, ion–molecule reactions of H_2^+ or Ar^+ with silane or disilane will result in electron transfer from

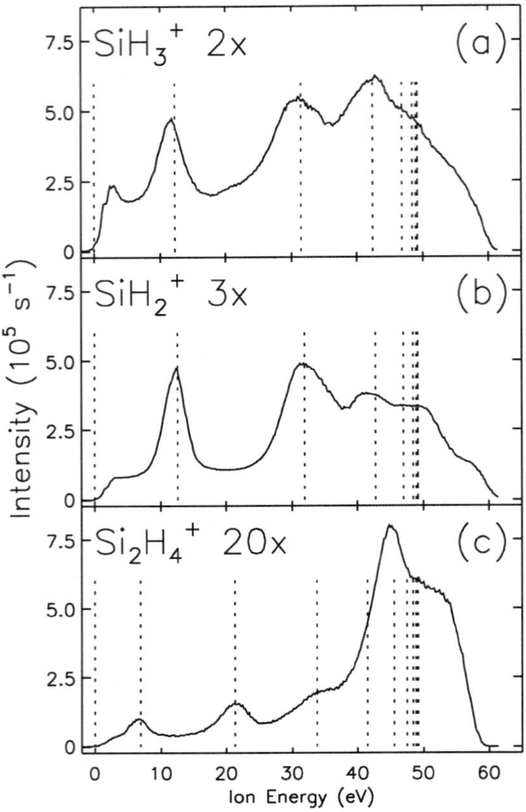

FIG. 38. The ion energy distributions of (a) SiH_3^+, (b) SiH_2^+, and (c) $Si_2H_4^+$ from a SiH_4–H_2 plasma in the γ'-regime. The dotted lines in (a), (b), and (c) indicate the position of the first 10 charge exchange peaks based on a plasma potential of 49.2 V and a charge carrier density n of 9.9×10^8 cm^{-3}. Note the different scaling factors of the various IEDs. (From E. A. G. Hamers, Ph.D. Thesis, Universiteit Utrecht, Utrecht, the Netherlands, 1998, with permission.)

silane to the ion [190], which leads to dissociative ionization of silane. These known asymmetric charge exchange reactions result in SiH_x^+ $(0 < x < 3)$ ions with thermal energy in the sheath; see Table VII [334, 335]. The charge exchange of H_2^+ with SiH_4 is a major loss process for the hydrogen ions in typical silane plasmas.

The cross sections of these asymmetric charge exchange reactions are in general lower than the cross sections of resonant charge exchange processes [134]. For comparison, the cross section of the Ar^+–Ar charge exchange reaction is 4×10^{-15} cm^2, about one order of magnitude larger than the reactions listed in Table VII. The total cross section of charge exchange reactions between Ar^+ and

TABLE VI

SHEATH PROPERTIES OF AN ARGON–SILANE AND TWO SILANE–HYDROGEN DISCHARGES[a]

Property	Ar/SiH_4	H_2/SiH_4 (α)	H_2/SiH_4 (γ')	Unit
V_{pl}	21.8 ± 0.3	23.8 ± 0.2	49.2 ± 0.5	V
n	1.66 ± 0.05	1.11 ± 0.04	9.9 ± 0.3	10^8 cm^{-3}
d_s	3.81 ± 0.05	4.88 ± 0.10	2.34 ± 0.05	mm
E_0	11.4 ± 0.2	9.8 ± 0.2	42.0 ± 0.8	kV/m
Γ_{max}	1.93 ± 0.05	1.35 ± 0.04	17.3 ± 0.7	10^{18} m^{-2} s^{-1}
$(\epsilon\Gamma)_{max}$	6.7 ± 0.2	5.2 ± 0.2	136 ± 4	W m^{-2}

[a] α and γ' regimes; plasma potential V_{pl}, charge carrier density n, sheath thickness d_s, electric field at the electrode E_0, ion flux Γ_{max}, and ion energy flux $(\epsilon\Gamma)_{max}$. Compiled from E. A. G. Hamers, Ph.D. Thesis, Universiteit Utrecht, Utrecht, the Netherlands, 1998.

TABLE VII

CHARGE EXCHANGE REACTIONS OF Ar^+ AND H_2^+ WITH SiH_4[a]

Reaction	σ (10^{-16} cm^2)
$Ar^+ + SiH_4 \rightarrow Ar + Si^+ + 2H_2$	0.2
$Ar^+ + SiH_4 \rightarrow Ar + SiH^+ + H_2 + H$	0.3
$Ar^+ + SiH_4 \rightarrow Ar + SiH_2^+ + H_2$	0.4
$Ar^+ + SiH_4 \rightarrow Ar + SiH_3^+ + H$	2
$H_2^+ + SiH_4 \rightarrow H_2 + Si^+ + 2H_2$	2
$H_2^+ + SiH_4 \rightarrow H_2 + SiH^+ + H_2 + H$	2
$H_2^+ + SiH_4 \rightarrow H_2 + SiH_2^+ + H_2$	11
$H_2^+ + SiH_4 \rightarrow H_2 + SiH_3^+ + H$	34

[a] The cross sections σ for reactions with Ar^+ are determined at thermal energies of the reactants [334]; the cross sections for reactions with H_2^+ are determined at a kinetic energy of the reactants of 1 eV [335]. (Compiled from E. A. G. Hamers, Ph.D. Thesis, Universiteit Utrecht, Utrecht, the Netherlands, 1998.)

SiH_4 is much smaller than the total cross section of charge exchange reactions between H_2^+ and SiH_4.

The discharge in which the charge exchange peaks of $Si_2H_4^+$ are observed is in the γ'-regime. A considerable amount of silane will be depleted and gas phase polymerizations occur, which means that Si_2H_6 is likely to be present in large quantities. Therefore, Hamers et al. [235] have suggested that Si_2H_6 is involved in the charge exchange reaction where $Si_2H_4^+$ is created:

$$H_2^+ + Si_2H_6 \rightarrow H_2 + Si_2H_4^+ + H_2 \quad (52)$$

1.5.4. ELLIPSOMETRY

Another *in situ* technique that is used to study deposition in real time is ellipsometry. This optical technique is noninvasive and does not perturb the discharge. In ellipsometry one measures the amplitude and phase of the reflected light from a surface. Spectroscopic ellipsometry (SE) allows for the wavelength-dependent measurement of the dielectric function of the material, at submonolayer sensitivity [336]. Spectroscopic phase-modulated ellipsometry (SPME) gives the possibility of fast data recording, and therefore is exploited for kinetic studies of the deposition of *a*-Si:H [337]. Spectroscopic ellipsometry is a very sensitive tool to study surface and interface morphology in the UV–visible range. The use of IR light allows for the identification of Si−H bonding configurations [338].

Polarization modulation can be done by rotating the polarizer, as is done in rotating-element ellipsometers [339], or by using photoelastic devices [337]. Data acquisition nowadays is fast, which makes real-time measurements of film and interface formation possible: a full spectrum ranging from 1.5 to 5 eV can be measured in less than a second [340–342].

In a typical ellipsometry experiment a sample is irradiated with polarized light, which subsequently is reflected from the sample surface and detected after passing an analyzer. The ratio ρ of complex reflectances for perpendicularly (s) and parallelly (p) polarized light usually is represented as follows:

$$\rho = \frac{r_{\mathrm{p}}}{r_{\mathrm{s}}} = \tan\Psi \exp(i\Delta) \tag{53}$$

with r_{p} and r_{s} the complex reflectances for p- and s-polarized light, and Ψ and Δ two convenient angles. This ratio can be related directly to the complex dielectric function $\epsilon = \epsilon_1 + i\epsilon_2$ in the special case of light incident at an angle ϕ on the surface of a semiinfinite medium [336]:

$$\epsilon = \sin^2\phi \left(1 + \tan^2\phi \left[\frac{1-\rho}{1+\rho}\right]^2\right) \tag{54}$$

In case of multilayer structures much more complex expressions result.

In the IR one usually presents spectroellipsometric measurements using the optical density $D = \ln(\overline{\rho}/\rho)$, where $\overline{\rho}$ refers to the substrate before deposition of a film [343]. The presence of a vibrational mode (rocking, bending, stretching) of a silicon–hydrogen bond, for instance, is revealed by a peak at the corresponding energy of that mode in the real part of D versus energy (in cm^{-1}). The contribution of a specific vibration mode to the dielectric function of the film is estimated from the Lorentz harmonic-oscillator expression [344]:

$$\epsilon = \epsilon_\infty + \frac{F}{\omega_0^2 - \omega^2 - i\Gamma\omega} \tag{55}$$

with ω_0 the frequency of the mode, F its oscillator strength, and Γ a damping constant. The oscillator strength can be expressed as $F = \omega_0^2(\epsilon_0 - \epsilon_\infty)$, where ϵ_0 and ϵ_∞ are the high- and low-frequency dielectric constants, respectively. The vibration mode of a chemical bond will show up as a maximum or minimum in the real part of D versus energy, depending on the dielectric function of the substrate.

The interpretation of ellipsometric data is based on the description of the optical properties of the material under study. Effective-medium theories (EMTs) are widely used for this purpose, as they can be applied when materials are heterogeneous and when the size of inhomogeneities is small compared to the wavelength of the light, so that scattering of light can be neglected. In a-Si:H the presence of voids will have an effect on the dielectric function. Also, during deposition the surface will roughen more or less, depending on deposition conditions, which will have a perturbing effect on the polarization of the incident light. An optical model of a substrate with an a-Si:H film could consist of a substrate, an a-Si:H film with voids, and a surface layer of a-Si:H with a large number of voids, which reflects the surface roughness.

The general form that is used in EMTs to describe the dielectric function ϵ of an a-Si:H layer with a certain fraction of voids is given by [345]

$$\epsilon = \frac{\epsilon_a \epsilon_v + \kappa \epsilon_h (f_a \epsilon_a + f_v \epsilon_v)}{\kappa \epsilon_h + (f_a \epsilon_v + f_v \epsilon_a)} \tag{56}$$

where κ is the screening parameter [$\kappa = (1/q) - 1$ and $0 \leq q \leq 1$], ϵ_h is the host dielectric function, ϵ_a and f_a are the dielectric function and volume fraction of the a-Si:H component, and ϵ_v and f_v are the dielectric function and volume fraction of the void component. Several approximations can be made. At the end of the seventies Aspnes et al. [346] found that the Bruggeman approximation provided the best description for the data available. Recently Fujiwara et al. [345], with new spectroellipsometric data, confirmed this fact. The Bruggeman approximation [61] defines $\epsilon_h = \epsilon$. For spherical inclusions $q = 1/3$, i.e., $\kappa = 2$, and the dielectric function is isotropic. For the effect of voids present in the a-Si:H layer we further simplify Eq. (56) by taking $\epsilon_v = 1$, and substituting $f_a = 1 - f_v$.

In a real-time spectroellipsometric measurement in which the kinetics of a-Si:H deposition is studied, trajectories are recorded in the Δ–Ψ plane at various photon energies between 2 and 4 eV. These trajectories can be simulated and fitted to models that represent the growing a-Si:H layer. Canillas et al. [347] have made a detailed study of the deposition of the first few layers of a-Si:H on a NiCr/glass substrate. Similar results are obtained for a c-Si substrate. They have proposed several models to explain the data. One possible model is the *hemispherical* nucleation model, which describes a hexagonal network of spherical a-Si:H nuclei located at an average distance d between them. The growth is represented by an increase in the radii of the nuclei until the nuclei make contact. This results in an a-Si:H layer with a large void fraction, of $f_v = 0.39$ [286]. From the start of the deposition ($f_v = 1$) to the point where the nuclei make contact ($f_v = 0.39$), the dielectric function is calculated using the Bruggeman EMT, and a Δ–Ψ trajectory can be simulated. In another model, the *columnar* nucleation model, columns of a-Si:H material start to grow from the initial hexagonal network, where the growth is represented by an increase in column radius and height. Again, f_v is calculated, and the dielectric function can be deduced from that. A further refinement is that the columns coalesce, and form a dense layer, with a layer of a certain thickness on top, which consists of free standing columns. This extra layer, typically 5 nm in thickness, represents the roughness, and increases with film thickness for typical deposition conditions. A recent study supports the concept of coalescing 3D islands of a-Si:H [340]: directly from the start of the deposition, the thickness of the surface roughness layer increases to 2 nm. At that thickness the islands start to combine into a continuous layer, and the thickness of the surface roughness layer decreases to remain constant at 1.5 nm.

Andújar et al. [246] have used SE to study the effects of pressure and power on the properties of a-Si:H. In their system the α–γ' transition occurs just above 0.2 mbar at a power of 10 W. They found that the density of the material is decreased on going from the α- to the γ'-regime, as is deduced from the decrease in the imaginary part of the dielectric function ϵ_2.

Collins et al. [342] have used SE to study the effects of hydrogen dilution. On native oxide c-Si substrates they have found that the thickness of the roughness layer is nearly constant at 1.5–2 nm in the case of deposition with pure silane, for a total deposited layer thickness up to 1000 nm. At a dilution factor $R = [H_2]/[SiH_4]$ of 10, the thickness of the roughness layer decreases from 1.5 to just below 1 nm. In case of $R = 20$, a similar decrease is observed (with a minimum of 0.3 nm at a bulk layer thickness of 20 nm), but followed by a steep increase to 4.5 nm at a bulk layer thickness of 100 nm, which is consistent with the formation of a microcrystalline film. At values above $R = 25$, the deposited layer is microcrystalline. These deposition experiments have also been carried out starting on a 300-nm-thick a-Si:H film, prepared with $R = 0$. Here, the amorphous-to-microcrystalline transition, as evidenced by the steep increase of the thickness of the roughness layer, occurs at 200 nm for $R = 15$ and at 60 nm for $R = 30$. Based on these observations, a phase diagram was constructed, showing the amorphous-to-microcrystalline transition as a function of the bulk layer thickness and the dilution factor R. The importance of this phase diagram lies in the fact that a deposition process that is close to the amorphous-to-microcrystalline transition yields material that has excellent optoelectronic properties [348–352].

Spectroscopic ellipsometry has also been applied in the characterization of compositionally graded a-SiC:H alloys [353], where the flow ratio $z =$

$[CH_4]/([CH_4] + [SiH_4])$ was varied during deposition. The analysis of the SE data on the graded layer prompted the development of a new four-medium model, which consists of the ambient, a surface roughness layer, an outer layer, and the pseudo-substrate. A virtual interface approximation is applied at the interface between outer layer and pseudo-substrate. This model gives the near-surface carbon content and the instantaneous deposition rate, as well as the thickness of the roughness layer. Besides *a*-Si:H–*a*-SiC:H interfaces, also *p*–*i* interfaces have been characterized, which is possible because the dielectric function changes at the interface [354, 355].

The use of IR in SE allows for the investigation of hydrogen incorporation in *a*-Si:H. The identification of vibrational modes in nanometer thin films is difficult, due to the weak signals. Nevertheless, Blayo and Drévillon have shown that monohydride and dihydride bonding configurations can be discerned in 0.5- to 1-nm-thick *a*-Si:H layers [338], using IR phase-modulated ellipsometry (IRPME). In *in situ* IRPME studies the real part of D is presented in the stretching mode region of the IR, i.e., 1900–2200 cm^{-1}. Three stretching modes can be revealed in this range, typically around 2000, 2100, and 2160 cm^{-1}, corresponding to SiH, SiH_2, and SiH_3 bonding configurations. From the evolution of the SiH– and the SiH_2 stretching mode with increasing deposited thickness, it was found [356] that the deposited layer is built up from two layers. The bottom layer contains SiH bonds, and linearly increases in thickness with deposition time. The top layer has a nearly constant thickness of 1–2 nm, and contains SiH_2 bonds. At the start of the deposition a layer consisting of SiH_2 is formed, and its thickness increases with time. At a thickness of about 1 nm, another layer between substrate and the SiH_2-containing layer is formed, which contains SiH. The deposition proceeds as the thickness of this latter layer increases with time. Attenuated total reflection IR spectroscopy has revealed that SiH_2 and SiH_3 surface modes are present even before the SiH_2-containing layer is formed [340].

The position of the SiH and SiH_2 stretching modes varies with the thickness of the deposited layer. In the first few monolayers the SiH peak shifts upwards by about 20 cm^{-1}, as a result of the very high hydrogen content [340]. At the end of the coalescence phase, the SiH peak is at 1995 cm^{-1}, and its position does not change any more. The substrate used was *c*-Si with a native oxide layer. In contrast, another study shows an increase in the SiH peak position from 1960 to 1995 cm^{-1}, together with an increase in the SiH_2 peak position from 2060 to 2095 cm^{-1}, for thicknesses of the deposited layer between 0 and 500 nm [344]. Moreover, these shifts depend on the deposition temperature and on the substrate. These findings were related to the change in disorder as a function of deposited thickness. The disorder, characterized by the width of the Raman TO peak, decreased with increasing thickness.

1.6. Relation between Plasma Parameters and Material Properties

1.6.1. External Parameters

Luft and Tsuo have presented a qualitative summary of the effects of various plasma parameters on the properties of the deposited *a*-Si:H [6]. These generalized trends are very useful in designing deposition systems. It should be borne in mind, however, that for each individual deposition system the optimum conditions for obtaining device quality material have to be determined by empirical fine tuning. The most important external controls that are available for tuning the deposition processs are the power (or power density), the total pressure, the gas flow(s), and the substrate temperature. In the following the effects of each parameter on material properties will be discussed.

1.6.1.1. Plasma Power

When the plasma power (density) is increased, the deposition rate is increased monotonically up to the point where the gas flow rate becomes the limiting factor [81, 357] (see Figure 39). This deposition rate increase comes with a number of disadvantages, including poor film quality and powder formation. At low power levels a large part of the SiH_4 remains undissociated [303], and the corresponding films contain only silicon monohydrides, inferred from FTIR measurements. At high power levels the SiH_4 is strongly dissociated, and the films contain a large amount of SiH_2 [358]. The microstructure parameter R^* is high. The high power levels also lead to columnar microstructure, which is accompanied by high spin densities [359, 360]. Experimental results obtained in ASTER show similar trends, as is shown in Section 1.6.2.3.

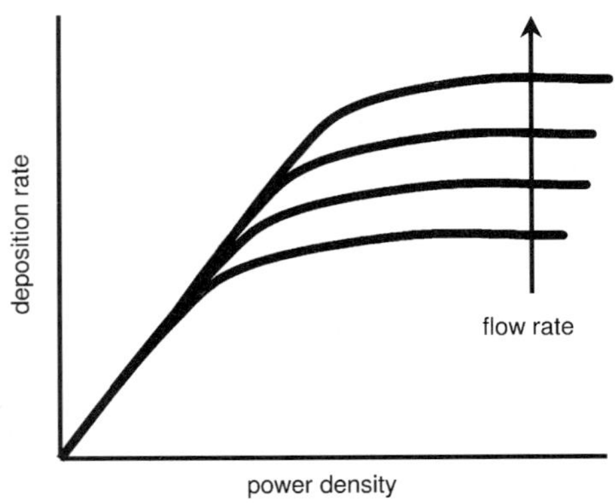

FIG. 39. Schematic representation of the influence of power density and flow rate on the deposition rate. [After A. Matsuda, *J. Vac. Sci. Technol. A* 16, 365 (1998).]

The electrical properties (dark conductivity and photoconductivity) are reported to first decrease and then increase upon increasing power [361]. The optical bandgap increases with increasing power, due to the increase of the hydrogen content [63, 82, 362, 363]. However, at very high power levels, microcrystalline silicon is formed [364], which causes the hydrogen content (and, consequently, the bandgap) to decrease.

As a result of the deterioration of film properties with increasing power levels, to obtain *device quality* material the use of low power is required, albeit with a concomitant low deposition rate. Increasing the deposition rate without altering the *device quality* material properties is a large research challenge.

1.6.1.2. Total Pressure

The deposition rate as a function of total pressure shows two different dependences. In the low-pressure, or supply-limited, regime the deposition rate is proportional to the pressure, while in the power-limited regime it is constant [365]. At low pressure the film quality is not good. The supply of SiH_3 radicals, needed for high quality films, is depleted easily, and SiH_2 and SiH radicals can reach the substrate. Moreover, ion bombardment may be too severe. Therefore, higher pressure levels are required. Usually the operating pressure (in combination with electrode distance and power lever) is set so that one works just above the Paschen curve (see Section 1.3.2.4). The pressure should not be too high, otherwise gas-phase polymerization occurs [55]. These yellow powders (which turn white upon oxygen exposure) are to be avoided, as they increase the downtime of deposition systems due to clogging of the pumps. At normal working pressures it has been found that the hydrogen content and dihydride content increase with the pressure [81].

1.6.1.3. Gas Flow

The gas flow rate is usually presented as a deposition parameter; however, it is much more instructive to report the gas residence time [6], which is determined from the flow rate and the geometry of the system. The residence time is a measure of the probability of a molecule to be incorporated into the film. The gas depletion, which is determined by the residence time, is a critical parameter for deposition. At high flow rates, and thus low residence times and low depletion [303], the deposition rate is increased [357, 365] (see Figure 39) and better film quality is obtained, as is deduced from low microstructure parameter values [366].

1.6.1.4. Hydrogen Dilution

The addition of a hydrogen flow to the silane flow often is used to improve the material quality of the deposited film. Moderate hydrogen dilution of silane

($0.15 \leq [SiH_4]/([SiH_4] + [H_2]) \leq 1$) has been found to yield a lower optical bandgap, a lower activation energy, a lower total hydrogen content, a lower microstructure factor, and a higher photoresponse [15, 82, 367–369]. Unfortunately, also the deposition rate is reduced. Device quality material is obtained for values of $[SiH_4]/([SiH_4] + [H_2])$ between about 0.3 and 0.7. Moreover, an improvement of the uniformity of deposition over a $10 \times 10\text{-cm}^2$ substrate area has been found as well [370]: for $[SiH_4]/([SiH_4] + [H_2]) = 1/3$ the variation in thickness amounted to only 1.5%.

In the intermediate regime of hydrogen dilution ($0.05 \leq [SiH_4]/([SiH_4] + [H_2]) \leq 0.15$) the hydrogen content increases again, as well as the bandgap, while the microstructure factor remains low [369]. In this regime wide-bandgap *a*-Si:H can be obtained with better optoelectronic properties than *a*-SiC:H.

Highly diluted silane ($[SiH_4]/([SiH_4] + [H_2]) \leq 0.05$) causes the deposited films to be crystalline rather than amorphous, due to selective etching of strained and weak bonds by atomic hydrogen [371, 372]. Interestingly, using hydrogen dilution to deposit *a*-Si:H films "on the edge of crystallinity" [154, 349] resulted in improved material stability. These materials are also referred to as polymorphous silicon (pm-Si) [373, 374].

The admixture of a small amount of silane to a pure hydrogen plasma drastically changes the plasma potential, the dc self-bias voltage, and the charge carrier density. Hamers [163] has found a doubling of the carrier density by changing $[SiH_4]/([SiH_4] + [H_2])$ from 0 to 0.1. The dc self-bias becomes more negative by 25%, and the plasma potential is lowered by about 10–15%. For fractions from 0.1 to 0.5 these parameters remain about constant. The parameters change again when $[SiH_4]/([SiH_4] + [H_2])$ is larger than 0.5.

1.6.1.5. Substrate Temperature

The substrate temperature is a very important deposition parameter, as it directly affects the kinetics of ad- and desorption of growth precursors, surface diffusion, and incorporation. Actual substrate temperatures may differ from substrate heater setpoints. Calibration of temperature readings is needed, so as to report the correct substrate temperature.

The deposition rate is nearly independent of temperature, while the total hydrogen content, the microstructure parameter, and the disorder decrease with increasing temperature [375]; see also Figure 40 [84, 85]. As a consequence, the optical bandgap decreases as well. An optimum deposition temperature around 250°C exists, such that the material contains only SiH bonds. At higher temperatures the hydrogen evolution from the film causes lowering of the hydrogen content, and leads to higher defect densities. The photoconductivity and photoresponse also have an optimum value around 250°C.

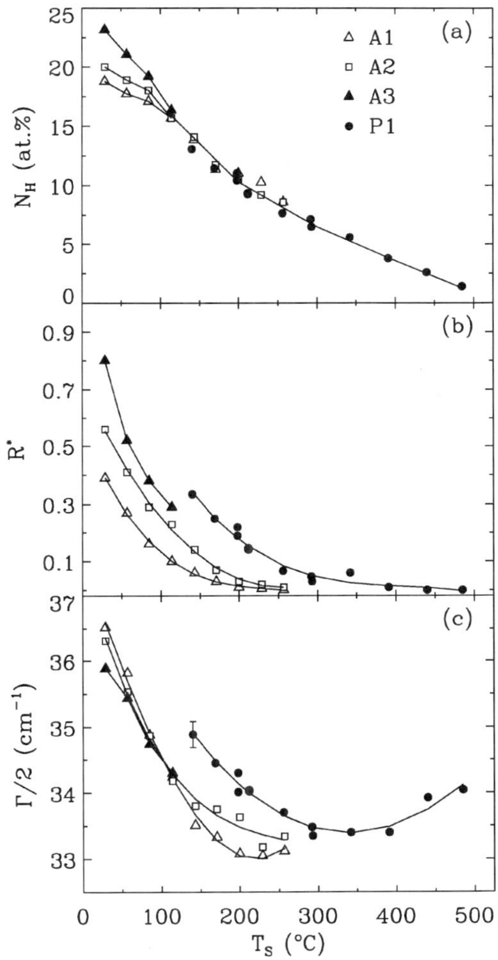

FIG. 40. The influence of deposition temperature on (a) the hydrogen concentration, (b) the microstructure parameter, and (c) the Raman half width $\Gamma/2$. The labels A and P refer to the ASTER and the PASTA deposition system. Series A1 was prepared from a SiH_4–H_2 mixture at 0.12 mbar. Series A2 and A3 were deposited from undiluted SiH_4 at 0.08 and 0.12 mbar, respectively. Series P1 was deposited from undiluted SiH_4. (From A. J. M. Berntsen, Ph.D. Thesis, Universiteit Utrecht, Utrecht, the Netherlands, 1998, with permission.)

1.6.1.6. α–γ' Transition

As was shown above, the properties of the discharge will change with process parameters such as the process pressure, the RF power, and the excitation frequency. Two different plasma regimes can be distinguished: the α- and the γ'-regime [184, 244–246, 248]. In the transition from the α-regime to the γ'-regime a change oc-

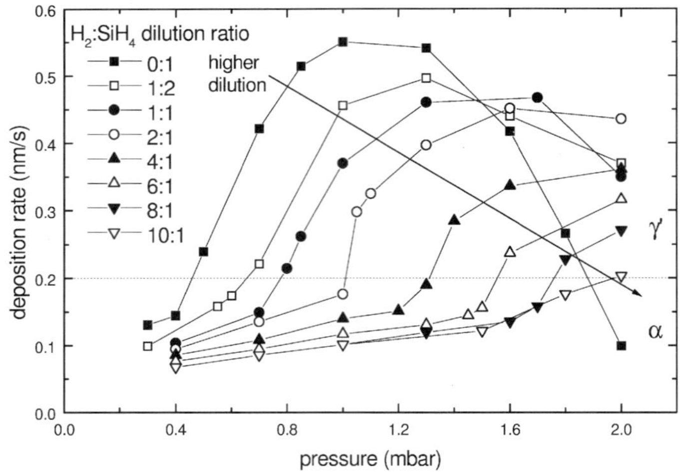

FIG. 41. The deposition rate as a function of deposition pressure for different ratios of hydrogen dilution ($0:1 < H_2:SiH_4 < 10:1$). The deposition temperature was 240°C, the SiH_4 flow was 600 sccm, and the excitation frequency was 13.56 MHz. The arrow indicates increasing hydrogen dilution. The α-regime is below the dotted line, the γ' above it. [After R. B. Wehrspohn, S. C. Deane, I. D. French, I. Gale, J. Hewett, M. J. Powell, and J. Robertson, *J. Appl. Phys.* 87, 144 (2000).]

curs in plasma properties. This includes the plasma impedance, the optical emission from the plasma [184], and the dc self-bias voltage at the powered electrode [245] in a reactor with electrodes of unequal size.

An increase in the deposition rate of the film and a change of the film properties is observed near the transition from the α- to the γ'-regime [245]. As an example, Figure 41 shows the deposition rate as a function of pressure, for various hydrogen dilution ratios [376]. It can be seen that higher hydrogen dilution pushes the α–γ' transition away, as it were. Other examples are shown in Figure 46 and Figure 64. The transition is assumed to be initiated by the formation of larger particles in the gas phase [246, 322]. Negative ions [377] are trapped in the plasma bulk, and recombine with silane radicals to become large negative ions (or small clusters). The charge of these particles can fluctuate, and can become positive. The clusters coalesce, and when the size of the coagulates is sufficiently large the charge builds up and remains negative. They grow further by deposition of *a*-Si:H on their surface. These particles are often referred to as powder or dust [378]. Due to the large amount of negative charge (up to hundreds of electrons) on them, the free electron density drops. Therefore, the electron temperature has to increase in order to sustain the discharge. The discharge becomes more resistive, causing more efficient power coupling into the plasma [322]. Most of these phenomena are observed in discharges with the conventional 13.56-MHz excitation frequency, but they can also be observed at higher excitation frequencies [301].

Interestingly, it has been argued that nanoparticulate formation might be considered as a possibility for obtaining new silicon films [379]. The nanoparticles can be crystalline, and this fact prompted a new line of research [380–383]. If the particles that are suspended in the plasma are irradiated with, e.g., an Ar laser (488 nm), photoluminescence is observed when they are crystalline [384]. The broad spectrum shifts to the red, due to quantum confinement. Quantum confinement enhances the bandgap of material when the size of the material becomes smaller than the radius of the Bohr exciton [385, 386]. The broad PL spectrum shows that a size distribution of nanocrystals exists, with sizes lower than 10 nm.

Moreover, it was found that incorporation of nanoparticles about 8 nm in diameter in *a*-Si:H led to improved properties, the most important one being enhanced stability against light soaking and thermal annealing [387]. A later study revealed a typical crystallite size of 2–3 nm, with a hexagonal close-packed structure [388]. Diamond structures can also be observed [389]. Hence the name polymorphous silicon is justified.

In view of the above, instead of avoiding powder formation regimes in discharges, careful powder management involving optimization of the ratio of radicals and silicon nanoparticles arriving on the substrate has been proposed [379].

1.6.1.7. Other Deposition Parameters

Depending on the flow pattern in the reactor, the depletion of gases can cause nonuniform deposition across the surface of the substrate. As is depicted in Figure 4, the gas usually is introduced at the top of the cylindrically symmetric reactor, and it flows into (and out of) the discharge region from the sides. Many other configurations exist, e.g., a radial flow reactor, where the gas flow is introduced underneath the powered electrode, and pumped away through the center of this electrode. A reverse radial flow reactor, where the gas flow is introduced from the center of the powered electrode and pumped away at the sides, has also been proposed [117, 370]. The best solution to overcome depletion is to inject the gas directly via a showerhead at the grounded electrode [173].

Meiling et al. [82] have investigated the effect of electrode shape. They found that in the case of a ring electrode, less dense material with a larger bandgap was obtained than in the case of a flat-plate electrode. In addition, the ring electrode induces an extra nonuniformity in thickness, as a result of the nonuniform electric field between the powered and the grounded electrode. Further, the area ratio of grounded to powered electrode in the case of the ring electrode is much higher, which yields higher dc bias voltages and higher ion bombardment energies. Hence, a ring electrode should not be used.

Many different substrates are used for *a*-Si:H deposition. Usually Corning 7059 glass [390] and crystalline silicon are used for materials research, as both have similar thermal expansion coefficients to *a*-Si:H. Devices are mostly made on glass coated with transparent conductive oxide (TCO). As TCO coatings one uses indium tin oxide (ITO), fluorine-doped tin oxide (SnO_2:F), and zinc oxide (ZnO). Ion bombardment may lead to the reduction of the ITO coating [391], while ZnO-coated glass is much more resistant to this. Polymer films have been used as substrates for flexible solar cell structures. They require lower deposition temperatures than glass or stainless steel.

It has been found that various material properties are thickness-dependent. Raman experiments show a dependence on the type of substrate (glass, *c*-Si, stainless steel, ITO on glass) and on the thickness (up to 1 μm) of the films [392, 393]. Recent transmission electron microscopy (TEM) results also show this [394]. This is in contrast to other results, where these effects are negligible for thicknesses larger than 10 nm [395, 396], as is also confirmed by ellipsometry [397] and IR absorption [398] studies.

1.6.2. INTERNAL PARAMETERS

1.6.2.1. The Role of Ions

There are many effects possible due to the interaction of ions with a surface during deposition, e.g., the enhancement of adatom diffusion, ion-induced desorption, displacement of surface or subsurface atoms, sticking or subsurface trapping of the ions, sputtering, and implantation in subsurface layers (subplantation). Light and slow ions are able to excite SiH surface bonds vibrationally, which may enhance the surface migration or desorption of weakly physisorbed species, such as SiH_3. For the nonmobile chemisorbed species, such as SiH_2, to be desorbed, a higher ion mass and/or energy is needed. At the same time, however, this may cause subplantation and resulting collision cascades or thermal spikes, which will lead to local defects and poor electronic properties [173].

Several views exist on the exact processes that occur during the deposition of *a*-Si:H. Ganguly and Matsuda [399] explain the growth of *a*-Si:H using a surface diffusion model, in which ions are ignored. Perrin [211] has formulated a model of the RF discharge in which, also, ions are not thought to be of importance to the growth. On the other hand, Vepřek et al. [400] introduce the term *ion-induced dehydrogenation*, and Heintze and Zedlitz [249, 280] show the importance of ion fluxes in plasmas, which are excited with very high frequency (VHF) electrode voltages. In addition, molecular dynamics studies of low-energy (10 eV) particle bombardment on crystalline silicon surfaces show surface dimer bond breaking [401] and the formation of interstitials [402], among other effects. Few measurements are available on the ion energies and fluxes in the sheath of a discharge under typical deposition conditions [280, 403–405]. The influence of ion bombardment on defect density [405], mobility [404], and electronic properties [403, 406] has been reported. The dependence of stress on the ion bombardment has been suggested by many authors [284, 376, 407, 408]. In the field

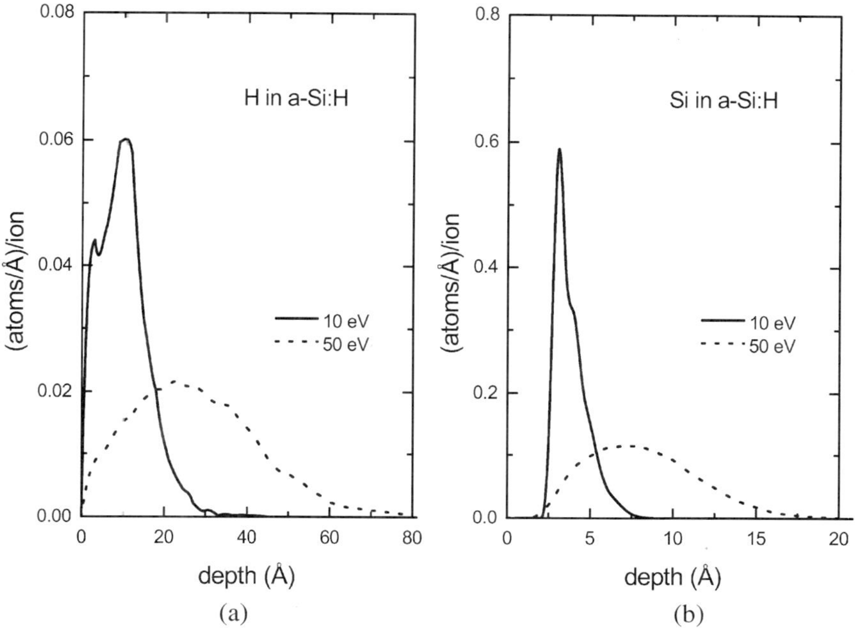

FIG. 42. Simulated depth distributions of (a) hydrogen and (b) silicon ions incident on *a*-Si:H with energies of 10 and 50 eV. Note the difference in depth scale. Simulations were performed with TRIM92.

of PECVD of amorphous or diamondlike carbon, it is assumed that ions are of prime importance for the formation of dense material [409]. Others also stress the importance of ion bombardment on film properties [410–412]. In addition, in electron beam deposition of amorphous silicon it was shown that the formation of microvoids is inhibited when Ar^+ assistance is used during deposition [413]. It was found that the energy per atom is the decisive parameter instead of the ion/atom ratio. Supplying 12 eV per Ar^+ ion is sufficient for annihilation of all microvoids. As a further example, calculation of projected ranges of low-energy particles in a solid can be performed with the Monte Carlo simulation program TRIM (transport of ions in matter) developed and supplied by Ziegler et al. [414]. The results of simulations with TRIM92 (TRIM Version 1992) of hydrogen and silicon ions incident on *a*-Si:H ($C_H = 0.1$) with energies of 10 and 50 eV, respectively, are presented in Figure 42. The projected ranges of hydrogen (1 nm at 10 eV, 2.5 nm at 50 eV) are much larger than those of silicon (0.3 nm at 10 eV, 0.75 nm at 50 eV). Clearly, these low-energy ions influence the surface and subsurface layers of the deposited film.

It is difficult to isolate the effect of ion bombardment on material properties. An external bias can be applied to the discharge in an attempt to study the effect of varying acceleration voltages over the sheath. However, this modifies the whole potential profile, and consequently the discharge chemistry. Conclusions drawn from such experiments obviously are misleading. It has been demonstrated in a multipole discharge at low pressure (<5 mTorr) that the ion flux can be controlled without disturbing the plasma properties and radical flux [358, 415–418]. At such low pressure the ion flux is monoenergetic and the relative reactive ion flux (Γ_+/Γ_0) can be as large as 0.8. Perrin [173] has summarized the effects of ion bombardment for $\Gamma_+/\Gamma_0 \approx 0.2$ and ion energy up to 150 eV: the deposited films show a densification of the microstructure, a disappearance of columnar growth, and a reduction of surface roughness. Further, the monohydride bonds in the network are favored over dihydride bonds: R^* is lowered to about zero, and the compressive stress present in the material is increased. Also the optoelectronic properties are improved.

In an RF discharge the relative reactive ion flux is much lower (0.03 [358]) and the IED is considerably broadened. Kae-Nune et al. [311] have estimated, from measurements on radicals, that ions could account for about 10% of the deposition rate. A straightforward method to directly determine the contribution of ions to the deposition process has been presented by Hamers et al. [163, 419]. The ions and radicals are separated by exploiting the difference in angular distribution of their velocities. As ions are accelerated towards the electrode, their angular distribution of ion velocities will be narrowed. Radicals diffuse through the sheath, and their angular velocity distribution will be isotropic. If one places an aperture in front of a substrate, the ions will mainly reach the substrate directly behind the aperture, while radicals will spread over a much larger area on the substrate. A similar configuration has been used to determine the reaction probability of radicals [420]. The effect is clearly demonstrated in Figure 43: the thickness (deposition rate) is enhanced by about a factor of two just behind the aperture. The deposition conditions were chosen to yield good quality material. From geometrical modeling it was found that under these discharge conditions the contribution of ions to the deposition was 10% [419], which confirms the estimate from Kae-Nune et al. [311].

1.6.2.2. The Ion Flux

In order to study the influence of ions on the deposition process, a reliable quantification of the ion flux and energy is imperative. This flux cannot be determined directly from the detected number of ions in an IED as measured by means of QMS, for three reasons [332]. First, the orifice size decreases during subsequent measurements due to deposition of a-Si:H on the edges of the orifice. Second, due to the limited acceptance angle of the mass spectrometer system, only a fraction of the ions that arrive at the substrate is actually detected. This fraction depends on the type and number of interactions that an ion experiences while traversing

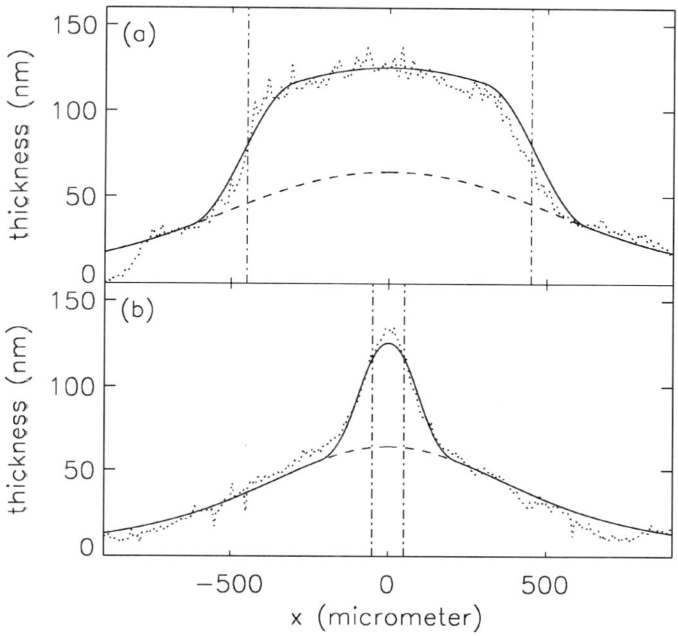

FIG. 43. Measured and simulated thickness profiles on the substrate behind the slit for a film deposited under discharge conditions that typically yield good material properties: (a) along the length of the slit, (b) across the slit. The vertical dashed–dotted lines indicate the boundaries of the apertures. The dotted lines represent the measured profiles, the solid lines the simulated profiles. The dashed lines are the simulated deposition profiles of the radicals. (From E. A. G. Hamers, Ph.D. Thesis, Universiteit Utrecht, Utrecht, the Netherlands, 1998, with permission.)

the sheath, and also on the ion species itself. In addition, this implies that the mean kinetic energy of the impinging ions cannot be determined. Third, the ion flux from a typical silane plasma consists of many different ions. All ions must be taken into account in the quantification of the ion flux.

Another approach is needed [163, 301, 332], and one follows from the definition of the ion flux Γ, i.e. the product of the density of the ions n_i and their mean velocity $\overline{v}_i$: $\Gamma = n_i \overline{v}_i$. The contribution of electrons to the time-averaged charge carrier concentration at the electrode n can be neglected, as the electron current towards the electrode is peaked in a very short time during the RF period. We thus have $n_i = n$. A further assumption is that there are no collisions in the sheath. As a result of this, the velocity $\overline{v}_i$ is equal to the maximum velocity, $v_{\max}$, that the ions gain. The value of $v_{\max}$ follows from the time-averaged plasma potential eV_{pl}: $v_{\max} = \sqrt{2eV_{\text{pl}}/\overline{m}}$, with $\overline{m}$ the average mass of the ions. If the ions reach the electrode without collisions in the sheath, the IED will show a distinct peak at eV_{pl}; see e.g. the IED of $Si_2H_4^+$ in an α-silane discharge (Figure 37).

The maximum ion flux Γ_{max} at the electrode is thus estimated to be

$$\Gamma_{max} = n v_{max} = n\sqrt{\frac{2eV_{pl}}{\overline{m}}} \tag{57}$$

The maximum *kinetic energy* flux of ions, $(\epsilon\Gamma)_{max}$ (W m^{-2}), can be defined as

$$(\epsilon\Gamma)_{max} = eV_{pl}\Gamma_{max} \tag{58}$$

where ϵ denotes the ion energy.

The collisionless-sheath assumption leads to an overestimation of the velocity. The use of an average mass $\overline{m}$, which is taken to be 31 amu/e (i.e., the mass of SiH_3^+), in addition may lead to an overestimation of the velocity, as ions with more than one silicon atom are present, albeit in lower concentrations. Hamers et al. [163, 301, 332] have argued that the velocity of the ions (and the flux) is overestimated by at most a factor of two. Also, the ion kinetic energy flux overestimates the real energy flux if collisions take place in the sheath. However, in many collisions the kinetic energy is distributed over the different particles. For instance, in a charge exchange process, the original ion is neutralized, but retains its kinetic energy. The newly formed ion will gain the energy that the neutralized ion would have gained in the rest of the trajectory. The two particles will transport the total amount of kinetic energy corresponding to the plasma potential to the surface [163].

In order to express the importance of the ions to the growth process quantitatively, two related quantities can be defined: the fraction of arriving *ions* per deposited *atom*, R_i, and the kinetic energy transferred by *ions* per deposited *atom*, E_{max}. These quantities are used in ion-beam-assisted deposition in order to relate material properties to ion flux and energy [421]. Their definition is

$$R_i = \frac{\Gamma_{max}}{\rho r_d} \tag{59}$$

$$E_{max} = \frac{(\epsilon\Gamma)_{max}}{\rho r_d} \tag{60}$$

where the number of deposited atoms per unit time and per unit area is ρr_d, with ρ the atomic density of the amorphous network (5×10^{22} cm^{-3}) and r_d the growth rate.

1.6.2.3. Relation between Ion Flux and Material Quality

A systematic study of the role of the ions in the deposition process and their influence on the quality of the layers has been performed by Hamers et al. [163, 301, 332] in the ASTER deposition system. More specifically, a study has been made on the relation between the plasma parameters and the material properties in both the α- and the γ'-regime at typical deposition conditions. Here, the results for power and pressure variation are summarized. Details can be found elsewhere [163, 301, 332].

First, results on power variation are described, for two different discharges: (1) a silane–hydrogen discharge in the powder-free α-regime, and (2) a silane–argon discharge in both the α- and the powder-producing γ'-regime.

The process conditions of the series in silane–hydrogen are an excitation frequency of 13.56 MHz, a pressure of 0.20 mbar, gas flows of 30 sccm SiH_4 and 30 sccm H_2, and a substrate temperature of 250°C. The power was varied between 5 and 20 W. It was found that the absolute magnitude of the dc self-bias voltage increases linearly with increasing power, and the trend of the ion flux is similar (0.8×10^{18} m^{-2} s^{-1} at 5 W, and 2.1×10^{18} m^{-2} s^{-1} at 20 W). This linear relationship between V_{dc} and growth rate had been observed earlier [15, 151, 422], and in practice can even be used as a calibration curve. The plasma potential was 22.7 V at 5 W and was only 3 V higher at 20 W. In this series only two films were deposited, one at 10 W and one at 15 W. The refractive index in both cases was 4.33, representative of good material. The fractions of arriving ions per deposited atom, R_i, as defined in Section 1.6.2.2, are 0.26 and 0.20, respectively, while the kinetic energy transferred by ions per deposited atom, E_{max} is about 5 eV in both cases.

The process conditions of the series in silane–argon are an excitation frequency of 50 MHz, a pressure of 0.40 mbar, gas flows of 30 sccm SiH_4 and 60 sccm Ar, and a substrate temperature of 250°C. The power is varied between 3 and 50 W. Plasma parameters deduced from measurements of the IED and material properties are shown in Figure 44 and Figure 45. The discharge changes from the α-regime to the γ'-regime around 10 W, which can best be observed by the decrease of the dc self-bias and an increase in sheath thickness (see Figure 44a and b). The bias is very low in the γ'-regime, typically between -2 and -1 V, indicative of a more symmetrical discharge. The plasma potential increases with increasing power and tends to saturate at the highest powers. Visual inspection of the discharge shows that the optical emission from the plasma becomes brighter and more homogeneously distributed on going from the α- to the γ'-regime. In the α-regime the emission intensity is highest at the boundary between plasma bulk and sheaths.

The growth rate, plotted in Figure 44d, increases with increasing power, which is ascribed to the higher degree of dissociation in the discharge. The growth rate reaches a maximum around 30 W, and decreases somewhat at higher powers: the growth rate then is limited by the flow rate. It has been estimated that about 60% of the silane is used for deposition [163]; the rest may well (partially) partake in the processes that lead to the creation of powder. This yellow powder is always observed in the reactor after experiments performed in the γ'-regime. It is always found outside the plasma confinement, and never on the substrate.

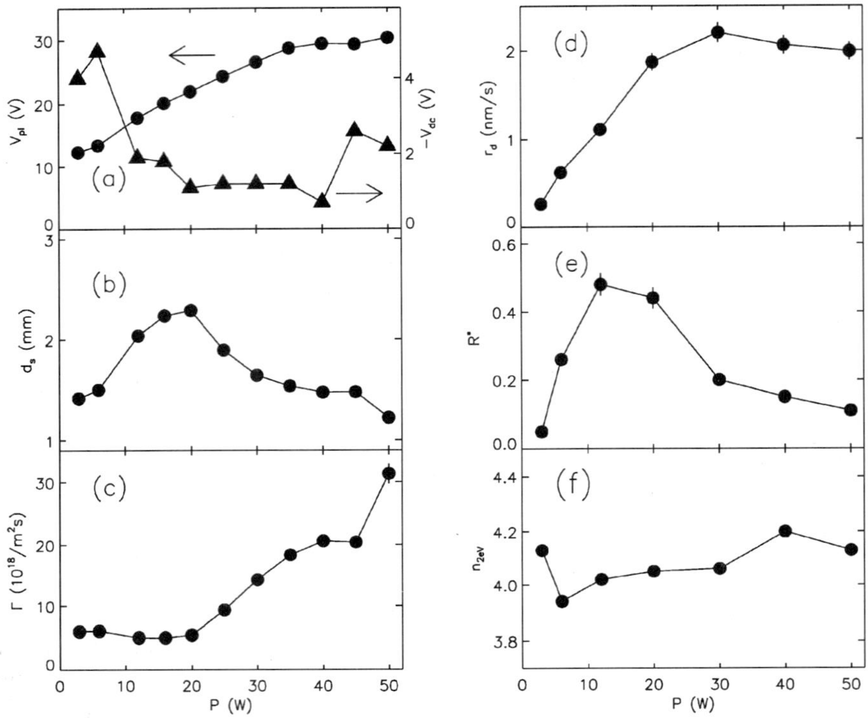

FIG. 44. Plasma parameters as deduced from the IEDs and material properties as a function of power delivered to the SiH_4–Ar discharge at an excitation frequency of 50 MHz and a pressure of 0.4 mbar: (a) the plasma potential V_{pl} (circles) and dc self bias V_{dc} (triangles), (b) the sheath thickness d_s, (c) the maximum ion flux Γ_{max}, (d) the growth rate r_d, (e) the microstructure parameter R^*, and (f) the refractive index $n_{2\,eV}$. (Compiled from E. A. G. Hamers, Ph.D. Thesis, Universiteit Utrecht, Utrecht, the Netherlands, 1998.)

At the transition from the α- to the γ'-regime the enhancement of the growth rate (Fig. 44d) is larger than the change in ion flux (Fig. 44c), as shown in Figure 45a, where R_i is plotted. An R_i-value of 0.25 or larger is commonly found in the α-regime, whereas R_i is typically 0.10 or lower in the γ'-regime. In addition, the amount of kinetic ion energy per deposited atom, E_{max}, shows a minimum of about 2 eV between 10 and 20 W; see Figure 45b.

The microstructure parameter is low in the material deposited at the lowest power (Fig. 44e); it increases rapidly with increasing power up to 20 W, and then decreases again with further increasing power. The opposite holds for the refractive index (Fig. 44f), although that is less clear. A high value of the microstructure indicates a large fraction of $Si-H_2$ bonds in the material, corresponding to an open material structure and a low refractive index.

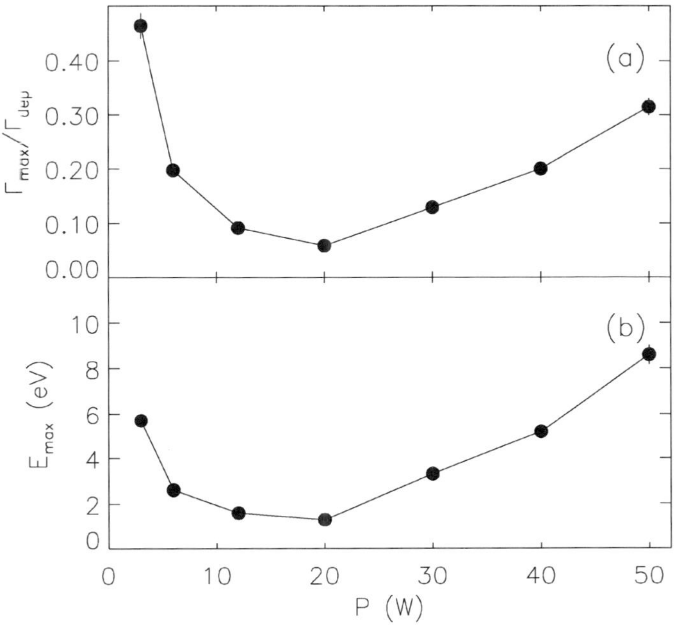

FIG. 45. (a) The ratio of the ion flux to the deposition flux, R_i, and (b) the ratio of the energy flux to the deposition flux, E_{max}, as a function of power delivered to the SiH_4–Ar discharge at an excitation frequency of 50 MHz and a pressure of 0.4 mbar. (Compiled from E. A. G. Hamers, Ph.D. Thesis, Universiteit Utrecht, Utrecht, the Netherlands, 1998.)

From a comparison between the behavior of the microstructure parameter R^* (Fig. 44e) and the ion kinetic energy per deposited atom, E_{max} (Fig. 45b), it can be concluded that a one-to-one relation appears to exist between the relative strength of the ion bombardment, expressed in terms of E_{max}, and the microstructure parameter. This has also been suggested by others [246, 422].

Both the dark conductivity and the photoconductivity of the sample deposited at 40 W have been measured. The photoconductivity has been measured under AM1.5 conditions. The dark conductivity and photoconductivity of the material deposited at 40 W are 1×10^{-10} and $1.2 \times 10^{-4}\ \Omega^{-1}\ cm^{-1}$, respectively. These values are indicative of good electronic properties, such as a low defect density and a high carrier mobility. The improvement of electrical properties of the deposited films at high RF power densities has also been reported by Nishikawa et al. [361].

The influence of power variation on the material properties is in agreement with the trends observed by Andújar et al. [246], who studied the α–γ' transition in pure silane discharges at 13.56 MHz. Further, this has also been observed for pure silane discharges at 50 MHz by Meiling et al. [423].

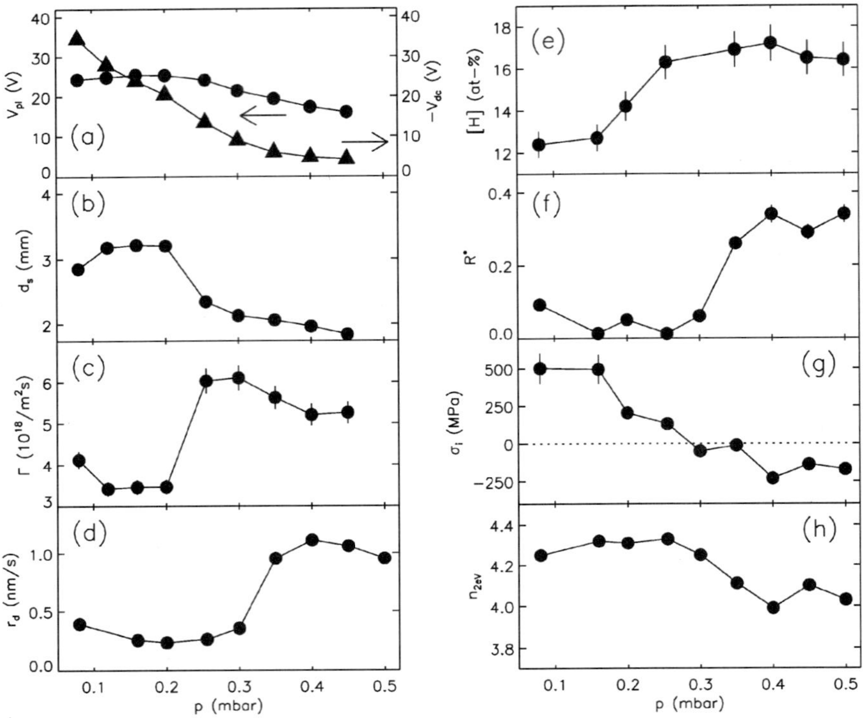

FIG. 46. Plasma parameters as deduced from the IEDs and material properties as a function of process pressure of a SiH_4–H_2 discharge at an excitation frequency of 50 MHz and a power of 10 W: (a) the plasma potential V_{pl} (circles) and dc self-bias V_{dc} (triangles), (b) the sheath thickness d_s, (c) the maximum ion flux Γ_{max}, (d) the growth rate r_d, (e) the hydrogen content, (f) the microstructure parameter R^*, (g) the internal stress σ, and (h) the refractive index $n_{2\,eV}$. (Compiled from E. A. G. Hamers, Ph.D. Thesis, Universiteit Utrecht, Utrecht, the Netherlands, 1998.)

The α–γ' transition can also be induced by changing the process pressure. At low pressures the discharge is in the α-regime. At a certain pressure it changes to the γ'-regime on increasing the pressure by only 0.02–0.04 mbar. The experiments presented here span a pressure range from 0.08 to 0.5 mbar, and, as above, plasma properties that are deduced from IED measurements are compared with material properties. The other process conditions are an excitation frequency of 50 MHz, a power level of 10 W, gas flows of 30 sccm SiH_4 and 30 sccm H_2, and a substrate temperature of 250°C. Data are summarized in Figure 46 and Figure 47. The self-bias voltage decreases quite rapidly in magnitude with increasing pressure, as shown in Figure 46a. The plasma potential slowly increases with pressure up to a pressure of 0.20 mbar and decreases towards the highest pressures (Fig. 46a). As both dc self-bias and plasma potential decrease, V_{rf} will also de-

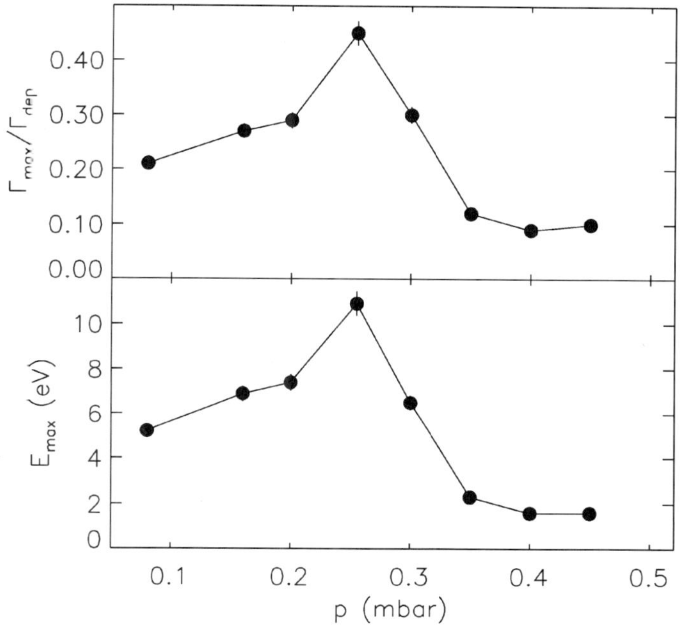

FIG. 47. (a) The ratio of the ion flux to the deposition flux, R_i, and (b) the ratio of the energy flux to the deposition flux, E_{max}, as a function of process pressure of a SiH_4–H_2 discharge at an excitation frequency of 50 MHz and a power of 10 W. (Compiled from E. A. G. Hamers, Ph.D. Thesis, Universiteit Utrecht, Utrecht, the Netherlands, 1998.)

crease [134]. Such a decrease has been measured by others [184, 245]. The sheath thickness falls from around 3 to 2 mm at around 0.2 mbar (Fig. 46b). The ion flux (Fig. 46c) is rather constant in the region up to 0.2 mbar, whereas it is nearly doubled on the high-pressure side. A slight decrease of the ion flux in the α-regime has been observed in 13.56-MHz discharges by Roca i Cabarrocas [403]. The growth rate increases by about a factor of 4 upon a pressure change from 0.2 to 0.3 mbar (Fig. 46d). The fraction R_i is shown in Figure 47a to be around 0.25 in the α-regime, whereas it drops to about 0.10 at pressures higher than 0.3 mbar, because the deposition rate increases a factor of 2.5 more than the ion flux. Since the plasma potential decreases with increasing pressure, the ions arriving at the film surface are expected to have a lower energy per ion in the γ'-regime than in the α-regime. This is clearly seen in Figure 47b, where E_{max} changes from about 5 eV at pressures below 0.3 mbar to about 2 eV at the higher pressures.

As can most clearly be seen in Figure 46d, the α–γ' transition occurs at a pressure of about 0.3 mbar for these experimental conditions. The impedance of the plasma, as well as the optical emission from the plasma, changes on going through the transition. The depletion of SiH_4 during deposition was already shown and compared with the deposition rate in Figure 31. The effect of the α–γ' transition on the partial pressures of disilane and trisilane was already presented in Figure 32 (see Section 1.5.3.1). The amount of silicon in these higher-mass neutral species is only about 5% of the total amount of silicon in SiH_4, as can be concluded from the measurements of the partial pressures of these gases. The transition however effects the production of these higher silanes. The increase of the partial pressures is larger than the increase in depletion. This means that the amount of these higher silanes produced per consumed quantity of silane increases at the higher pressures.

The material properties are also affected on going through the α–γ' transition. The hydrogen content is about 12% up to 0.2 mbar, and increases to over 16% at higher pressures (Fig. 46e). The increase in the hydrogen content occurs at a lower pressure than the increase in deposition rate, much like the behavior of the ion flux. The microstructure parameter is lower than 0.1 up to a pressure of 0.3 mbar, and is around 0.3 above this pressure (Fig. 46f). The internal stress in the layers is about 500 MPa compressive at low pressures, changes to a tensile stress around 0.3 mbar, and remains tensile at 200 MPa at higher pressures (Fig. 46g). The refractive index is around 4.30 up to a pressure of 0.3 mbar, but decreases towards the highest pressures (Fig. 46h). In summary, the material becomes of poorer quality at higher pressures. The γ'-regime is associated with the formation of particulates in the plasma [322]. However, these particles are assumed not to be incorporated in the film and thus not to be the direct cause of the different material properties.

In the presented range of pressure variation, Hamers [163] also has studied the influence of the substrate temperature on the plasma and the material. It was found that in the temperature range of 200 to 300°C the trends of the bias voltage, the plasma potential, and the growth rate as functions of pressure all are the same, while the absolute magnitude depends on the temperature. The trends in material properties are similar to the ones reported above: at a temperature of 200°C the material quality is worse than at higher temperatures. The α–γ' transition occurs at a lower pressure than at a temperature of 250°C. This has been observed before [248].

In all the experiments reported here it is observed that the value of R_i is rather constant at a value of 0.25 in the α-regime, whereas in the γ'-regime it typically amounts to 0.10. In other words, if all the Si atoms that originate from ions contribute to the deposition, the contribution of ions to the deposition in the α-regime is 25%, and in the γ'-regime 10%. The observation that these values are rather constant in each regime strongly indicates that the deposition rate is limited by the ion flux.

In an attempt to relate ion bombardment to material structure it is very illustrative to correlate the refractive index $n_{2\ \mathrm{eV}}$ and the microstructure parameter R^* with the kinetic ion energy per deposited atom, $E_{\max}$. The data presented above

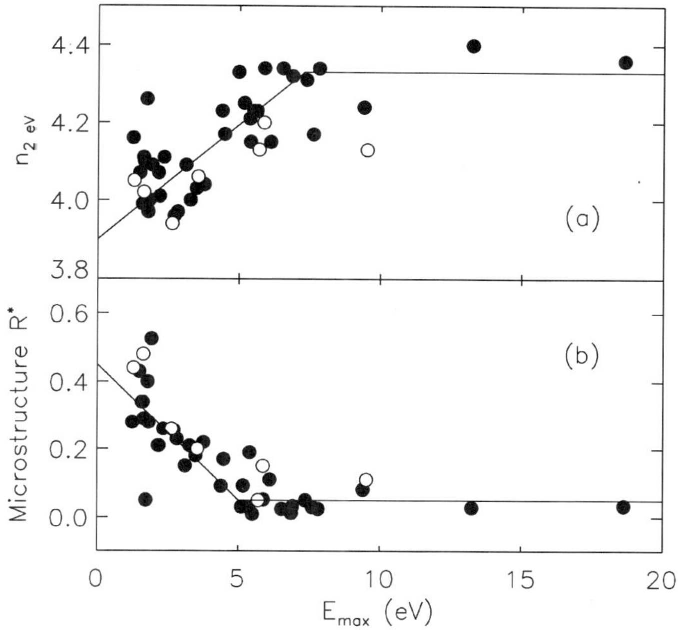

FIG. 48. Material properties as functions of E_{max} at 250°C: (a) refractive index $n_{2\ eV}$, and (b) microstructure parameter R^*. The closed circles represent pure silane and silane–hydrogen plasmas. The open circles refer to silane–argon plasmas. Lines are to guide the eye. (Adapted from E. A. G. Hamers, Ph.D. Thesis, Universiteit Utrecht, Utrecht, the Netherlands, 1998.)

for various discharges (pure silane, silane–argon, and silane–hydrogen) at a temperature of 250°C are thus summarized in Figure 48. It is very clear that the structural properties of the layers are poor (small $n_{2\ eV}$ and large R^*) for small E_{max}. The structural properties improve rapidly with increasing E_{max} up to a value of about 5 eV. All samples with E_{max} below 5 eV are deposited in the γ'-regime. Above 5 eV a dense network with only a small fraction of $Si{-}H_2$ bonds is produced. The structural properties do not change further with increasing E_{max}.

The role of the substrate temperature can be inferred from a plot of $n_{2\ eV}$ and R^* versus E_{max} at the three temperatures mentioned: 200, 250, and 300°C (see Figure 49). At a substrate temperature of 200°C the refractive index is lower at every E_{max} than at a substrate temperature of 250°C. Further, the threshold at which dense material is obtained is observed to be a few electron volts higher than at 250°C. The refractive index at 300°C is high and independent of E_{max}. The microstructure parameter R^* as a function of E_{max} behaves similarly for material deposited at 200 and at 250°C. At 300°C the value of R^* is less than 0.1 and independent of E_{max}. It is noteworthy to show the relation between the internal stress and E_{max} as a function of temperature (Fig. 50). The stress is linearly

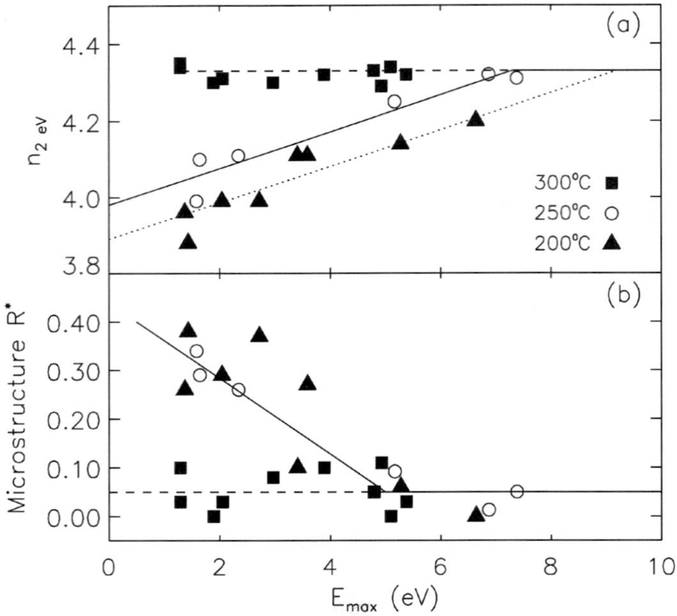

FIG. 49. Material properties as function of $E_{\max}$ at three temperatures: (a) refractive index $n_{2\text{ eV}}$, and (b) microstructure parameter R^*. Lines are to guide the eyes. (Adapted from E. A. G. Hamers, Ph.D. Thesis, Universiteit Utrecht, Utrecht, the Netherlands, 1998.)

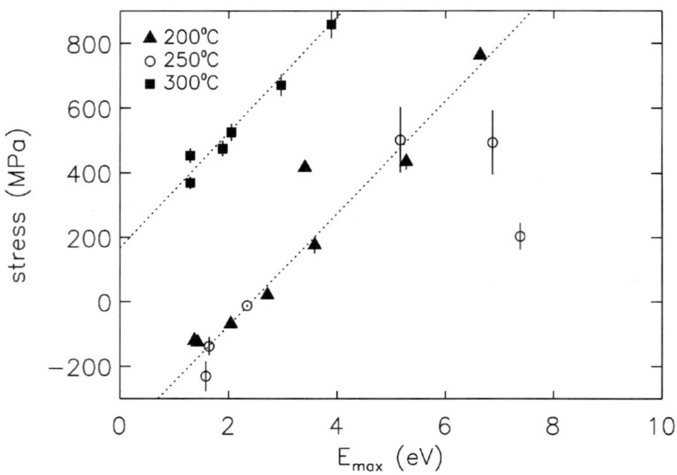

FIG. 50. Internal stress as a function of $E_{\max}$ at three temperatures, for the pressure series in Figure 46 in the SiH_4–H_2 discharge running at an excitation frequency of 50 MHz and a power of 10 W. (From E. A. G. Hamers, Ph.D. Thesis, Universiteit Utrecht, Utrecht, the Netherlands, 1998, with permission.)

dependent on E_{max} (between 1 and 7 eV) with a slope of about 175 MPa/eV for material deposited at 200 and at 250°C. For the 300°C data series this relation is shifted upwards by about 600 MPa, and at values of the stress larger than about 1000 MPa (E_{max} was about 4.5 eV) the deposited layers started to peel off from the silicon substrate directly after exposure to air. In summary, the ion bombardment clearly is needed to create or promote a dense amorphous network at temperatures up to 250°C. At higher temperatures energetic ion bombardment results in too high intrinsic stress.

1.6.2.4. Ion–Surface Interactions

The different processes in which the ions might well be involved at the surface can be indentified on the basis of their chemical reactivity at the surface and the amount of energy they transport towards the surface.

It is assumed that SiH_3 radicals are responsible for the production of deposition sites at the hydrogen-covered surface when no ions are present [138]. If ions contribute to the deposition process, a higher deposition rate has been reported; see e.g. [419]. The fraction of arriving ions per deposited silicon atom varies between 0.1 and 0.3, depending on the discharge regime (see Section 1.6.2.3). Assuming equal sticking probabilities of ions and radicals (0.3 for the SiH_3 radical [192]), up to about 10% of the deposition is attributed directly to the ions. This value has also been found by Kae-Nune et al. [311]. On the other hand, ions can penetrate into the subsurface layers, and hence their sticking probability will be closer to one. In addition, direct incorporation of ions in the amorphous network can only partly explain the increase in deposition rate. The creation of deposition sites by the ions possibly is of greater importance. It has been suggested that a direct relation between the ion flux and the deposition rate exists, if the availability of deposition sites is the limiting factor in the deposition process [236, 400, 424]. Also, partial dissociation of a polyatomic ion upon impact may occur [412, 425–427]. The dissociation of a hydrogen-containing ion may lead to the production of more than one hydrogen atom per ion. These atoms create molecular hydrogen and dangling bonds upon reaction with hydrogen bonded to the surface. The probability γ of the recombination process of atomic hydrogen has been estimated to be between 0.4 and 1 [317].

In discharges operating in the γ'-regime large particles consisting of several ions are present, in contrast to discharges in the α-regime. In both regimes the same relative amount of ions contributes to the deposition, as is deduced from the measured amount of radicals that contribute to the deposition, which is regime-independent [317]. These larger ions in the γ'-regime are thought to create more deposition sites per arriving ion than the smaller ions in the α-regime [163]. This results in a smaller number of ions per deposited atom in the γ'-regime.

The kinetic ion energy flux, $(\epsilon\Gamma)_{max}$, which is typically 20 W m^{-2} [163, 301], will raise the substrate temperature by only a few degrees. Therefore, the influence of ions will be limited to the vicinity of impact. Furthermore, typical ion energies are below the sputtering threshold of silicon [134].

Enhancement of surface diffusion of the growth precursors is considered as one of the beneficial effects of ion bombardment [246, 428]. The potential energy of ions, which is released when the ion is neutralized, is typically 10 eV. This energy can be a substantial fraction of the total energy transferred. The release of this ionization energy is sufficient to excite atoms into excited electronic states, thereby weakening their bonds and enhancing their mobilities [429].

The influence of ion bombardment on the structure of deposited materials has been studied by Müller [428, 430] by performing molecular dynamics calculations on ion-assisted vapor phase growth of nickel. Nickel, of course, is very different from hydrogenated amorphous silicon, but the resemblance of the two deposition processes and some material properties is remarkable. In both cases, the growth precursors stick to the surface or diffuse along the surface, and the material can have a columnar-like structure or a structure with voids. Ion bombardment during the deposition of Ni promotes the formation of a dense film. The ion energies and fluxes in the molecular dynamics studies were similar to those in typical *a*-Si:H deposition conditions. Two important processes modify the structure of the nickel film, viz., forward sputtering and diffusion enhancement of surface species. Similar processes are likely to occur in *a*-Si:H deposition. In the deposition of amorphous germanium by ion-assisted electron-beam evaporation, it was also found that a kinetic energy of about 5 eV per deposited atom is needed to produce a dense amorphous germanium network, irrespective of the amount of kinetic energy of the ions: the effect of five ions with 20 eV was the same as that of one ion of 100 eV [431]. These two additional examples clearly indicate the beneficial role that ion bombardment apparently has on the density of the deposited film.

In the formation of a hydrogenated amorphous silicon network, the presence of the bonded hydrogen must be taken into account, as it was shown in the preceding that both the hydrogen content and the microstructure parameter are influenced by the ion bombardment. The high value of the microstructure parameter and the concomitant low density of the material for low values of E_{max} at 250°C (Fig. 48) indicate that the cross-linking is not sufficiently fast to create a dense network. This cross-linking process is activated by the locally released energy of chemisorption [432, 433], and involves the temperature-activated evolution of H_2 from the surface. Hydrogen desorption due to cross-linking, which is known to occur from thermal desorption measurements [434], starts to become important at temperatures higher than 300°C. At this temperature a drastic change in the importance of ion bombardment for achieving a dense network was observed (see Fig. 49). At lower deposition temperatures the cross-linking during the deposition is temperature-limited and the kinetic energy of the ions is needed to achieve cross-linking. If the amount of ion kinetic energy is insufficient, cross-linking will be incomplete and will result in an open network structure with a large amount of Si$-H_2$ bonds. At larger values of the ion kinetic energy the cross-linking is stimulated and a dense network is formed. Ion bombardment rather than the temperature induces cross-linking, as the microstructure parameter, the stress, and (to a lesser extent) the refractive index have the same dependence on E_{max} at both 200 and 250°C. At 300°C the thermally activated cross-linking process is becoming important.

Tensile stress can be related to the presence of voids in the material. The large microstructure parameter at low E_{max} is indicative of voids. A kinetic energy of the ions of 20 eV is typically enough to implant some of these ions a few monolayers below the surface of the growing film [400]. The extra Si atoms deposited in the layer at this depth will cause a compressive stress. Light hydrogen ions, which are less abundant than e.g. SiH_3^+ ions, penetrate deeper into the material (see also the TRIM simulation results in Fig. 42). The stress in the films deposited at 200 and 250°C depends similarly on E_{max}. At these temperatures the ion energy is needed to densify and to cross-link the network. At a temperature of 300°C the stress is much higher, which is explained by the fact that the ions reach an already dense network and some will be implanted, resulting in excess-stress in the material.

1.7. Deposition Models

A conceptual view of the processes that occur in PECVD of *a*-Si:H has been given by Perrin [173] and is shown in a slightly adapted form in Figure 51. Primary gas phase reactions are electron-impact excitation, dissociation, and ionization of the source gas molecules (SiH_4), thereby producing radicals and positive and negative ions. Secondary reactions in the gas phase between (charged) molecules and radicals produce other species. Diffusion of reactive neutral species leads to material deposition. Positive ions are accelerated to the substrate and bombard the growing film. Negative ions are confined in the bulk of the discharge, which can lead to particulate formation. On the surface, species diffuse to growth sites and contribute to the film after sticking to the underlying material. Also, surface species recombine with other species and desorb from the surface (see also Fig. 14). Subsurface reactions are the release of hydrogen from this hydrogen-rich layer and the relaxation of the silicon network.

Nowadays, it is generally accepted that *device quality a*-Si:H is obtained under PECVD conditions where the SiH_3 radical is the predominant growth precursor [123, 137, 192]. At typical deposition temperatures and pressures the *a*-Si:H surface is almost completely terminated by hydrogen atoms. The SiH_3 radical

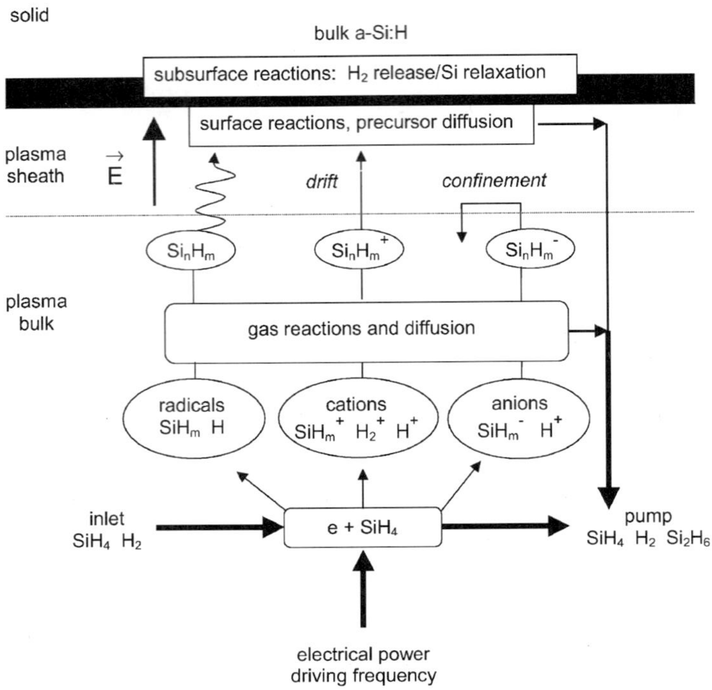

FIG. 51. Schematic representation of the processes occurring in a SiH_4–H_2 discharge and the various particles present in the energy and material balances. [After J. Perrin, in "Plasma Deposition of Amorphous Silicon-Based Materials," (G. Bruno, P. Capezzuto, and A. Madan, Eds.), Chap. 4, p. 177. Academic Press, Boston (1995).]

is thought to physisorb on this hydrogen-terminated surface before it is incorporated into the film [136, 137, 317, 435–438]. The excess hydrogen is eliminated from the hydrogen-rich (40–50%) subsurface, so as to form "bulk" *a*-Si:H with a hydrogen content of about 10%.

Robertson has summarized the three recent classes of models of *a*-Si:H deposition [439]. In the first one, proposed by Ganguly and Matsuda [399, 440], the adsorbed SiH_3 radical reacts with the hydrogen-terminated silicon surface by abstraction or addition, which creates and removes dangling bonds. They further argue that these reactions determine the bulk dangling bond density, as the surface dangling bonds are buried by deposition of subsequent layers to become bulk defects.

The second class of models was formulated by Winer [441] and Street [442, 443]. Here the notion that hydrogen atoms are more mobile than silicon atoms forms the basis of the model. The silicon network is fixed up to a temperature near the crystallization temperature (650°C). Hydrogen is mobile above a temperature of about 200°C, and its motion allows for the interconversion of defects, weak bonds, and strong bonds. These processes occur in the subsurface layer, and surface processes are not explicitly included. In this model, the optimum deposition temperature is the one at which hydrogen atoms can diffuse one atomic spacing as to remove the weak bonds. After deposition, the weak bond distribution can only be changed irreversibly while the defect distribution can be changed reversibly, as described by the defect pool models [444–446].

In the third class of models, computer simulations try to fully incorporate all processes in the discharge, the interaction of species created in the discharge with the wall (i.e., the substrate), and the network formation [190, 191, 232, 447–449]. These models to date do not treat the formation of disorder or defects, but aim at the understanding of the deposition rate, hydrogen content, and other macroscopic properties in relation to the discharge conditions (see also Section 1.4).

Robertson has combined and extended the first and second classes of models [439] by focusing on the origin of weak bonds. A surface adsorption model is used to describe surface coverage of the silyl radical SiH_3. Then, the processes that cause hydrogen to be expelled from the subsurface layer lead to the formation of weak bonds, which are frozen in. The dangling bond defects arise from these weak bonds by the defect pool process. In the following, a description is given of the Robertson deposition model (details can be found in [439, 450]).

1.7.1. SURFACE ADSORPTION

On the hydrogen-terminated surface, the surface species are mainly $\equiv$SiH, but at lower temperature (<200°C) $=SiH_2$ and $-SiH_3$ become increasingly important. The SiH_2 and SiH_3 radicals react differently with this surface. The SiH_2 radical can insert directly into Si$-$H or Si$-$Si bonds. The sticking coefficient is close to 1, which leads to columnar growth (physical vapor deposition regime [451]). Also, the SiH_2 can react with SiH_4 to form Si_2H_6. The SiH_3 radical cannot react directly with the surface. It has a small sticking coefficient, and it diffuses over the surface to find a reactive site. This reactive site has been suggested to be a dangling bond at kink and step edges [452]. As a result, this leads to conformal coverage (chemical vapor deposition regime [451]). The SiH_3 radical is physisorbed onto the surface [436]. A contrasting view has been proposed by Von Keudell and Abelson [453], who conclude from their experiments that SiH_3 can insert into strained Si$-$Si bonds on the surface to form a fivefold-coordinated bond. This was also observed by Ramalingam et al. [454] in molecular dynamics simulations of SiH_3 impinging on a 2×1 Si(100) reconstructed surface. Nevertheless, as these observations do not as yet explain film deposition, they are considered at least questionable.

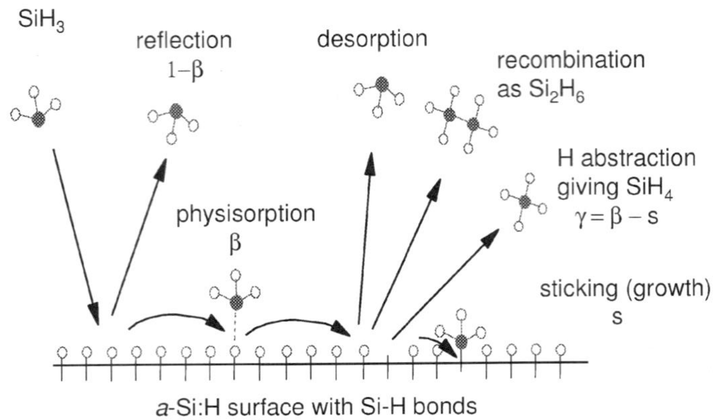

FIG. 52. Schematic respresentation of the processes that a SiH_3 may undergo at a hydrogen-terminated *a*-Si:H surface. [After J. Robertson, *J. Appl. Phys.* 87, 2608 (2000).]

TABLE VIII
REACTIONS OF SURFACE SITES $\equiv$SIH, $\equiv$SI·, AND $\equiv$SIHSIH$_3$ WITH THE RADICAL ·SIH$_3$, WITH CORRESPONDING RATE COEFFICIENTS

Symbol	Reaction		Rate coefficient
R1	$\equiv SiH + \cdot SiH_3$	$\rightarrow \equiv SiHSiH_3$	s_1
R2	$\equiv SiHSiH_3$	$\rightarrow \equiv SiH + \cdot SiH_3$	ν_d
R3	$\equiv SiHSiH_3 + \equiv SiH$	$\rightarrow \equiv SiH + \equiv SiHSiH_3$	ν_h
R4	$\equiv SiH + \cdot SiH_3$	$\rightarrow \equiv Si\cdot + SiH_4$	ν_a
R5	$\equiv Si\cdot + \cdot SiH_3$	$\rightarrow \equiv Si\text{–}SiH_3$	s_0
R6	$2 \equiv SiHSiH_3$	$\rightarrow 2 \equiv SiH + Si_2H_6$	ν_r
R7	$2 \equiv SiH$	$\rightarrow 2 \equiv Si\cdot + H_2$	ν_b

The possible reactions with a $\equiv$SiH surface site are schematically shown in Figure 52, and summarized together with their reaction rates in Table VIII. As a simplification, the surface site is $\equiv$SiH, the dangling bond is $\equiv$Si·, and the adsorbed SiH_3 radical is $\equiv SiHSiH_3$. The possible reactions include adsorption (R1 in Table VIII), desorption (R2), hopping to an adjacent $\equiv$SiH site (R3), abstraction of an hydrogen atom from $\equiv$SiH to create $\equiv$Si· (R4), addition to $\equiv$Si· (R5), surface recombination with another SiH_3 radical to form a desorbing Si_2H_6 molecule (R6), and elimination of hydrogen molecules from two adjacent $\equiv$SiH sites to yield surface dangling bonds (R7).

Several simplifications are further made [439]: (1) the SiH_3 adsorption (R1) is temperature-independent, and s_1 is constant; (2) the SiH_3 addition to a dangling bond is temperature-independent, and $s_0 = 1$, but the SiH_3 must hop to this dangling bond site from a neighboring site, so the net rate is $v_h s_0$; (3) the abstraction reaction R4 is faster than the desorption reaction R2, so the latter is ignored; and (4) the recombination reaction R6 is ignored.

Perrin et al. have denoted the fractional coverages of the surface sites as θ_1, θ_0, and θ_4 for $\equiv$SiH, $\equiv$Si, and $\equiv SiHSiH_3$, respectively [317], with $\theta_1 + \theta_0 + \theta_4 = 1$. These three fractional coverages can be found by expressing them as rates in terms of creation and destruction. This gives, with ϕ the incident SiH_3 flux density per surface site [439],

$$\frac{d\theta_0}{dt} = \nu_0\theta_4(1-\theta_0) - \nu_4\theta_4\theta_0 + \nu_b(1-\theta_0) \tag{61}$$

$$\frac{d\theta_4}{dt} = \phi - \nu_a\theta_4(1-\theta_0) - \nu_h\theta_4\theta_0 \tag{62}$$

In the steady state, and ignoring multiple hopping, the coverages are approximated as

$$\theta_0 \approx \frac{\nu_a}{\nu_h} + \frac{1}{1+\phi/\nu_b} \tag{63}$$

$$\theta_4 \approx \frac{\phi}{2\nu_a\,(1+\nu_b/\phi)} \tag{64}$$

At temperatures lower than 350–400°C the expressions for θ_0 and θ_4 reduce to $\theta_0 \approx \nu_a/\nu_h$ and $\theta_4 \approx \phi/(2\nu_a)$. The residence time of SiH_3 radicals at the surface, τ_4, is given by the coverage θ_4 divided by the arrival flux ϕ: $\tau_4 \approx 1/(2\nu_a)$. The SiH_3 surface diffusion length is $L = \sqrt{D\tau_4}$, where $D = \nu_h a^2$ is the SiH_3 diffusion coefficient, with a the Si–Si bond length. We thus have $L = a\sqrt{\nu_h/2\nu_a} = a/\sqrt{\theta_0}$.

The reaction rates ν_i ($i = h, a, b$) for each reaction are thermally activated as $\nu_i = \nu_{i0}\exp(-E_i/k_BT)$. Robertson has critically assessed the values ν_{i0} and E_i for each reaction [439], and used for the model $\nu_{h0} = 10^{13}$ s^{-1}, $E_h = 0.2$ eV, $\nu_{a0} = 3\times10^{11}$ s^{-1}, $E_a = 0.4$ eV, $\nu_{b0} = 10^{13}$ s^{-1}, and $E_b = 2.2$ eV. These values give a surface dangling bond density that increases monotonically with temperature from about 10^{-5} at room temperature to about 10^{-3} at 400°C. Thermal desorption of hydrogen then sets in, and the surface dangling bond density is suddenly increased and approaches unity around 600°C. This increase in θ_0 is due to the fact that $E_a > E_h$. The critical parameter is the difference $E_a - E_h$, not their absolute values. At typical deposition temperatures of 200–250°C the surface dangling bond density is about 3×10^{-4}, which is two orders of magnitude larger than the bulk dangling bond density. The SiH_3 surface diffusion length decreases from about 30 nm at room temperature to about 8 nm at 250°C.

These findings from Robertson [439, 450] are in contrast to the ones from the model proposed by Ganguly and Matsuda [399, 440]. They found a minimum in the surface dangling bond density, and a maximum in the surface diffusion coefficient, simultaneously occurring at about 250°C ($\theta_0 \approx 10^{-5}$, $L \approx 400$ nm). This would explain the optimum in material quality around 250°C. Surface defect densities have been measured during deposition by *in situ* electron spin resonance by Yamasaki et al. [455]. They found that the surface defect density was much higher than the bulk defect density. Therefore, the hypothesis of surface dangling bonds that are buried by deposition of subsequent layers is questionable. Robertson argues that as most of the hydrogen is eliminated from the hydrogen-rich *a*-Si:H surface layer, about half of the bonds in this layer are changed, from Si$-$H to Si$-$Si. It is very unlikely that this enormous change in bonding configurations would leave a surface dangling bond density of 10^{-6} unchanged [450].

1.7.2. SOLUBILITY OF HYDROGEN IN SILICON

The elimination of excess hydrogen from the surface layer seems counterintuitive, the hydrogen leaves the *a*-Si:H network towards the surface, where the hydrogen content is larger. This can be understood by considering the thermodynamics of hydrogen in silicon [439]. In general, the Gibbs free energy of a mixture of two components A and B, with mole fractions x_A and x_B (with $x_A + x_B = 1$), is given as [456]

$$G = x_A G_A + x_B G_B + x_A x_B \Omega_{AB} + RT(x_A \ln x_A + x_B \ln x_B) \qquad (65)$$

with G_A and G_B the Gibbs free energies of the pure species, Ω_{AB} the heat of mixing, and $RT(x_A \ln x_A + x_B \ln x_B)$ the entropy of mixing. The chemical potential μ_A is given by

$$\mu_A = \left(\frac{\partial G}{\partial x_A}\right)_{T,B} = G_A + \Omega_{AB}(1 - x_A)^2 + RT \ln x_A \qquad (66)$$

and similarly for μ_B. If Ω_{AB} is negative, mixing is favored. If Ω_{AB} is positive, two stable mixtures (A-rich and B-rich) exist with a miscibility gap in between. Any mixture with a composition between the two stable phases will decrease its free energy by decomposing into an A-rich and a B-rich phase. Such a process is known as spinoidal decomposition, and in it species diffuse up a concentration gradient [456].

In the case of *a*-Si:H, it turns out that Ω_{AB} is positive [439]. This can be inferred from the bond energies of Si$-$Si (2.35 eV), Si$-$H (3.3 eV), and H$-$H (4.5 eV). A Si$-$H network is thus unstable compared to pure Si and H_2. A mixture of Si$-$Si and Si$-$H bonds will be driven towards their two stable mixture phases, which leads to the solubility limit of H in Si. Acco et al. [69, 70] have determined this limit to be about 4%. Excess hydrogen in *a*-Si:H forms SiH_2 or SiH around microvoids.

1.7.3. ELIMINATION OF HYDROGEN FROM *a*-SI:H

Successive burial of hydrogen-rich surface layers leads to the formation of the *a*-Si:H material. The large amount of hydrogen at the surface is to saturate the surface dangling bonds. The much lower hydrogen content in the bulk is due to the solubility limit of hydrogen in silicon.

The hydrogen content in the bulk *a*-Si:H depends on deposition temperature and plasma conditions (e.g. [375, 457]; see also Section 1.6.1). In the γ'-regime the hydrogen content increases considerably with decreasing deposition temperature, while in the α-regime the hydrogen content also increases with decreasing temperature, but this increase sets in at lower temperatures. Robertson has defined the hydrogen elimation rate R_{H}, expressed as the number of hydrogen atoms per unit cell, as the sum of a thermal (R_T) and an athermal, plasma-driven (R_P) elimination process [450]: $R_{\mathrm{H}} = R_T + R_P$. Also, R_{H} equals the growth rate r_d multiplied by the fractional loss of hydrogen from the surface layer to the bulk: $R_{\mathrm{H}} = r_d(C_{\mathrm{H},s} - C_{\mathrm{H},b})$, where $C_{\mathrm{H},s}$ and $C_{\mathrm{H,b}}$ are the hydrogen concentration at the surface and in the bulk, respectively.

1.7.3.1. Thermal Reactions

The thermal processes can be derived starting with the behavior of hydrogen content as a function of reciprocal temperature [439, 450]. There are two regimes: a high-temperature regime ($T_s > 400°\mathrm{C}$), characterized by an apparent activation energy of about 1.6 eV, and a low-temperature regime ($T_s < 250°\mathrm{C}$) with an apparent activation energy of about 0.15 eV. Various hydrogen elimination processes are summarized in Figure 53 [439].

In the high-temperature regime it is most likely that hydrogen is eliminated to form H_2 at the surface, with an activation energy of 2.1 eV (Fig. 53a). This is similar to H evolution from crystalline silicon [439, 458]. At intermediate temperatures ($250°\mathrm{C} < T_s < 400°\mathrm{C}$) the elimination of hydrogen occurs via diffusion of atomic hydrogen towards the surface with an activation energy of 1.5 eV to recombine as H_2, i.e., the standard effusion process [459–461]. This is described with the hydrogen density of states (HDOS) diagram [442, 443, 446, 459, 462]; see Figure 53b. The HDOS diagram displays the energy of a hydrogen atom in *a*-Si:H compared to that of a free hydrogen atom in vacuum. A hydrogen bound as Si—H has an energy of -3.3 eV with respect to the vacuum level, while a hydrogen at a Si—Si bond center (BC in Fig. 53b) has an energy of about -1 eV. In *a*-Si:H hydrogen is more stable at the bond center of a weak bond, which is represented as a tail in the HDOS below -1 eV. In the HDOS there exists a mobility edge (H_M), above which the hydrogen states form a percolation path. Hydrogen diffusion occurs by excitation from μ_{H} to H_M. The difference $\mu_{\mathrm{H}} - H_M$ amounts to 1.5 eV which is less than 2.2 eV, but much larger than the apparent activation energy of 0.15 eV.

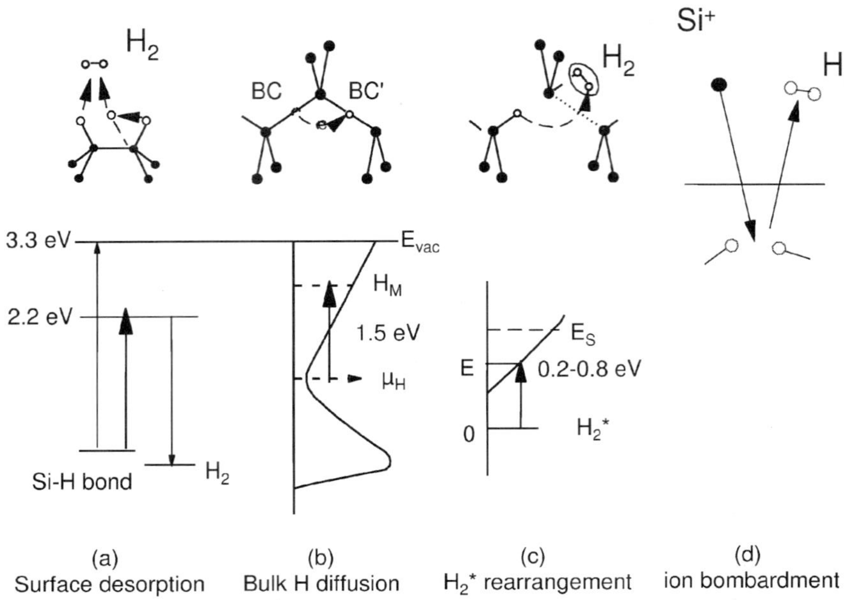

FIG. 53. Possible hydrogen elimination processes: (a) surface loss from adjacent Si$-$H groups by prepairing on a single Si site, (b) bulk diffusion of atomic H by activation to a H mobility level μ_M, as described by the hydrogen density of states (HDOS), (c) rearrangement of H_2^*, and the lowering of the barrier energy below the normal value E_s by excitation to a weak bond, expressed in the HDOS, (d) subsurface ion-induced displacement and desorption. [After J. Robertson, *J. Appl. Phys.* 87, 2608 (2000).]

Robertson proposed that the local bond rearrangement close to the surface can account for this low activation energy [439]. It is assumed that hydrogen exists largely in pairs, analogous to the H_2^* in crystalline silicon [463]. This H_2^* consists of two Si$-$H bonds in the same direction [464, 465]. It can undergo a local rearrangement with one hydrogen atom passing through two bond center positions, at a cost of 1 eV [245]. Another local rearrangement is possible in which the H passes through the bond center to combine with the other hydrogen atom to form the interstitial molecule H_2 (Figure 53c). The Si$-$Si bond center site is the transition state of the reaction. As hydrogen is more strongly bound at a weak Si$-$Si bond than at a normal Si$-$Si bond [466, 467], the energy barrier is lowered. As a consequence, the elimination of hydrogen to form H_2 is eased in the presence of weak bonds.

An alternative hydrogen elimination process has been proposed by Severens et al. [432, 468]. They argue that hydrogen can be eliminated by a cross-linking step [469] immediately after a physisorbed SiH_3 radical has chemisorbed on a surface dangling bond. This cross-linking probability is thermally activated, but due to the energy released at the chemisorption of SiH_3 (about 2 eV), the activation is quite low [432]. If this cross-linking process does not occur immediately after chemisorption, it is assumed that hydrogen is incorporated into the film. An activation energy of 0.15 eV is deduced from fitting hydrogen incorporation data [468].

A detailed description of the local bond rearrangement has been derived [439], using the concept of the HDOS with a low-energy tail that corresponds to the H present at weak Si—Si bonds. The width of this tail is $2E_{v0}$, i.e., twice the width of the valence band tail in the electronic density of states, which in turn is about equal to the Urbach energy E_0 [442, 443]. The HDOS then is [439]

$$N(E) = \frac{N_0}{2E_{v0}} \exp\left(\frac{E - E_s}{2E_{v0}}\right) \tag{67}$$

with E_s at the top of the HDOS, i.e., the energy barrier for rearrangement of the ideal H_2^* (1 eV). The rearrangement occurs by activation of H to a state E in the tail. The reaction rate in atoms per unit surface area is the product of this HDOS, the Maxwell–Boltzmann factor, and an attempt frequency ν_1 ($\sim 10^3\ s^{-1}$), and is expressed in the low-temperature regime as [439]

$$R_T = \frac{\nu_1 k_B T}{2E_{v0}} \exp\left(-\frac{E_s}{2E_{v0}}\right) \tag{68}$$

The values obtained from this equation compare well with the experimental values of E_{v0} as a function of temperature.

1.7.3.2. *Athermal Reactions*

Next, athermal, plasma-driven hydrogen elimination reactions are considered. Ions play an important role in the deposition. They can account for 10–25% of the total deposition rate [301, 311], depending on the regime of the discharge. Moreover, the energy that they carry to the (sub)surface [163] can be used to overcome energy barriers (see also Section 1.6.2.1). Ions penetrate the film and eliminate hydrogen by displacing hydrogen atoms from subsurface Si—H bonds. They subsequently recombine as hydrogen molecules and effuse back to the surface (see Fig. 53d). The displacement yield Y_D is given by the modified Kirchin–Pease equation $Y_D = E_i/2E_D$, where E_i is the ion energy and E_D the displacement energy [412]. Here, E_D is equal to the Si—H bond energy, i.e., 3.3 eV. In the α-regime the contribution of ions to the deposition rate is 25% [301], and a typical average ion energy is 10 eV (see Section 1.6.2.1). This gives a displacement yield Y_D of about 40% of the deposition rate. This illustrates the beneficial role of ions in improving the material quality [284, 301, 358, 417, 470]. For the γ'-regime about 10% of the ions contribute to the deposition rate, and typical average ion energies are 20–30 eV [301]. This gives a displacement energy Y_D of about 30–45% of the deposition rate. In this regime average ion energies are larger than the bulk displacement energy of silicon of 22 eV [285, 376], which degrades the material quality.

It is this athermal dehydrogenation process that reduces the excess weak-bond density associated with a high polymeric content, rather than ions' densifying the material by removing polymeric SiH_x groups. Low energy ions are especially effective at dehydrogenation because the displacement energy E_d is low. This favors the use of VHF PECVD reactors, where the ion energies are much lower and the ion densities much higher than at 13.56 MHz.

Further, atomic hydrogen in hydrogen-diluted discharges can contribute to abstraction of surface and subsurface hydrogen [471]. The atomic hydrogen reduces the fraction of higher SiH_x groups [466]. Only a small amount is incorporated directly [472]. Atomic hydrogen does not reduce the surface hydrogen content below 50%, because this would result in surface dangling bonds that would be passivated by incident hydrogen [472, 473]. Atomic hydrogen can also diffuse into the *a*-Si:H with an activation energy of 0.4–0.7 eV [461], which allows subsurface hydrogen abstraction. However, this would be less important at low temperatures, due to the considerable activation energy.

1.7.4. Dangling-Bond and Weak-Bond Density

The hydrogen elimination processes described above create the distribution of weak bonds. The dangling-bond distribution is formed from this by interconversion from weak bonds to dangling bonds. For substrate temperatures above the hydrogen equilibration temperature of 200°C, hydrogen is mobile, and the defect pool operates. This allows the weak-bond and dangling-bond distributions to equilibrate [439]. At temperatures below 200°C hydrogen motion is inhibited, and equilibration should not be possible. Nevertheless, Stutzmann [474] reported that the dangling-bond density as a function of temperature shows the the same behavior as the weak-bond density, for all temperatures. This suggests that a defect pool of sorts operates below 200°C in the surface layer [439].

It is assumed that a large athermal, plasma-driven hydrogen flux is present in the hydrogen elimination zone at temperatures below 200°C, which mediates the equilibration of the dangling-bond and weak-bond distributions. The dangling-bond density and weak-bond density thus remain linked in the elimination zone by a largely temperature-independent relationship, given by the defect pool [445, 446]. The dangling-bond density in the elimination layer becomes the bulk defect density as the layer is buried deeper and becomes frozen in. This proposed process allows the weak-bond density to determine the dangling-bond density at all temperatures [439].

1.8. Modifications of PECVD

Of the great number of possible and reported modifications, in this section only the most important ones are described, viz., the use of higher excitation frequency (VHF), the use of gas flow modulation (known as chemical annealing or layer-by-layer deposition), and the use of RF modulation.

1.8.1. VHF

1.8.1.1. General

Deposition of hydrogenated amorphous silicon employing the VHF PECVD technique (typical frequency range 20–110 MHz) has been reported to yield an increase in deposition rate by one order of magnitude over the conventionally used frequency of 13.56 MHz [16, 146, 250, 280], without adversely affecting material quality [183, 280, 475]. This is of great importance for lowering the production cost of *a*-Si:H solar cells.

The explanation of the VHF influence on the deposition rate is still a topic of discussion. It has been proposed theoretically that the high-energy tail in the EEDF is increased with an increasing ratio ω/ν of the excitation frequency to the energy collision frequency [276, 476]. This leads to an increased ionization rate [146, 250, 280]. However, mass spectrometry results on the decomposition of silane as a function of frequency show that the increase of the deposition rate cannot solely be attributed to the enhancement of radical production [119, 120]. It has also been found or deduced that the flux of ions towards the surface is increased with increasing frequency [146, 183, 279, 284] while at the same time the ion energy is decreased. This low-energy, high-flux ion bombardment enhances surface mobilities of adsorbed species. It was shown in Section 1.6.2.3 that a kinetic energy of 5 eV per deposited atom is needed to produce good quality *a*-Si:H [301].

Because the wavelength of the RF signal is of the order of the substrate dimensions (3 m at 100 MHz), it can be expected that uniform deposition is more difficult at these high frequencies [477]. In fact, a practical optimum frequency around 60–70 MHz is used [478, 479], which provides a good compromise between high deposition rate and attainability of uniform deposition. Further, the use of a distributed RF electrode network where all nodes have identical amplitude and phase improves the homogeneity of deposition [480].

1.8.1.2. Optimization of Deposition Conditions

A change of excitation frequency prompts for optimization of other deposition parameters, such as pressure, power, and geometry. Chatham and Bhat have shown that the maximum of SiH* emission at 414 nm is shifted towards lower pressure

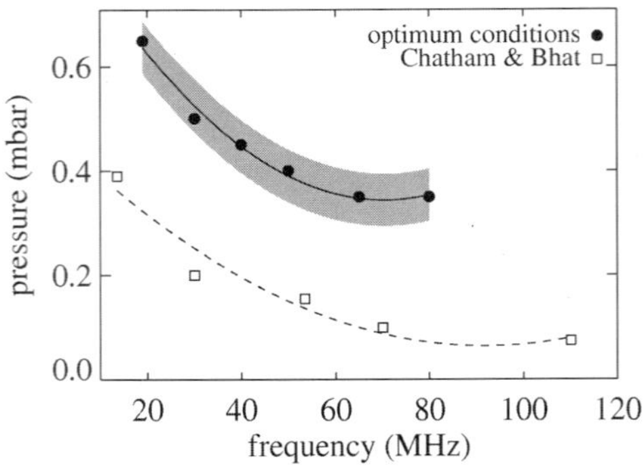

FIG. 54. Optimum values for pressure and frequency for uniform (thickness variation <5%) film deposition. The shaded area signifies the process window. Also shown are similar data by Chatham and Bhat [146].

upon increasing the frequency [146]. This means that the maximum in radical production also shifts to lower pressure. Experimental results obtained in the ASTER deposition system corroborate this [247].

A 20-fold increase in deposition rate was reported for an increase in frequency from 13.56 to 100 MHz [119, 120]. Here the electrode distance was about 4 cm, while the pressure was 0.12 mbar. A reasonable thickness uniformity was observed. For a smaller electrode distance (2.7 cm) higher pressures are needed in order to maintain good homogeneity in thickness. This is demonstrated in Figure 54, where the conditions to obtain a homogeneity deviating less than 5% over a $10 \times 10\text{-cm}^2$ area are presented. The power density in this case was 57 mW/cm^2. The shaded area represents the process window. It is clear that for this electrode distance the homogeneity requirement can be met, if one decreases the pressure according to the increase in frequency. This observation also explains the improvement in homogeneity as reported by Howling et al. [250]: at constant pressure the deposited layers become more homogeneous upon increasing the frequency. A similar effect was observed by Chatham and Bhat [146]. For comparison, their data are also shown in Figure 54. The electrode distance in their work amounted to about 1 cm, while the reported power density was 31 mW/cm^3.

The fact that pressure and frequency are related by the need to obtain homogeneous layers leads to the suggestion that the sheath thickness d_s between the grounded electrode and plasma bulk plays an important role. It is known that an increase of frequency yields a decrease in sheath thickness, while a decrease in pressure enlarges it. The combination of these effects suggests that an optimum sheath thickness is required.

The dependence of sheath thickness on frequency or pressure has been reported by several authors [179, 328, 481–485]. Their results are based on both theoretical studies of argon and experimental studies of argon, helium, and silane, using parallel plate configurations. It was found in these studies that $d_s \propto p^a$, where a ranges from -0.3 to -0.55, and $d_s \propto \omega^b$, where b ranges from -0.5 to -1.0. The data represented in Figure 46 and Figure 63 (Section 1.8.1.3) give $a = -0.28 \pm 0.08$ and $b = -0.26 \pm 0.02$.

Fitting the data in Figure 54 using the relationship $p^a f^b = c$, with c a constant that could represent d_s, shows that several combinations of the powers a and b can be used, as long as the exponents a and b are related as $b \approx 0.45a$. These fit results are inconsistent with the mentioned values for a and b; hence the sheath thickness probably is not constant. This was also deduced from direct measurement of the IEDs [161, 235, 486, 487]. In contrast, a recently developed method to determine the sheath thickness directly from deposition experiments [150, 151, 488] revealed that the value of d_s is constant for the conditions where $p\omega^{1/2}$ is constant [487].

The deposition rate as a function of pressure at various frequencies and at a power density of 57 mW/cm^2 is shown in Figure 55a. The shaded area represents the conditions for which uniform deposition is observed. At all frequencies the deposition rate at low pressure is lower than at high pressure by a factor of 5 to 6. This effect is well known, and it is explained by the transition from the α- to the γ'-regime of the plasma [245]. It follows from the figure that the transition region shifts to lower pressure with higher frequencies. Fitting the deposition rate as a function of the scaling parameter $p^a f^b$ shows that a good fit is obtained for $a = 1$ and $b = 2/3$; see Figure 55b. This holds for all frequencies. The same behavior has been found for the plasma potential [163]. As is illustrated in Figure 56, the plasma potential slightly increases in the low-pressure α-regime, but clearly decreases with increasing pressure at all three frequencies. The pressure at which the plasma potential starts to decrease occurs at a lower pressure at the higher frequencies. The rapid increase of deposition rate as a function of frequency, as reported in the literature, e.g. [16, 119, 120, 280, 475], is observed in the α-regime. In or near the γ'-regime the enhancement in deposition rate as a function of frequency is not pronounced. This is illustrated in Figure 57 by results at three power densities. Chatham and Bhat's data [146] also show only a slight increase with frequency.

In Figure 58 optical and structural material properties are shown. The refractive index at 2.07 eV (600 nm), $n_{2.07\ \mathrm{eV}}$, decreases with increasing applied RF power P, and is more or less independent of the frequency (Fig. 58a). The absorption coefficient at 2.07 eV behaves similarly. Values are around 2 to 3 $\times$ 10^4 cm^{-1}. The cubic bandgap E_g increases with power, and is also not dependent on frequency (Fig. 58b). The hydrogen content C_{H} increases from 13 at.% at low power to 19 at.% at high power, again being nearly independent of frequency (Fig. 58c). The microstructure parameter R^* increases with power, with only a slight dependence on frequency (Fig. 58d). The character of the intrinsic stress is changed

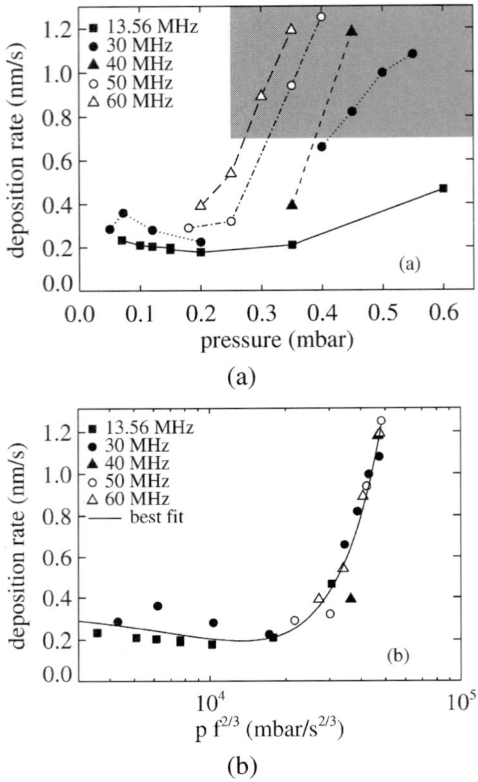

FIG. 55. (a) Deposition rate as a function of pressure at various frequencies. The shaded area shows the 5% uniformity conditions. (b) Best fit of deposition rate versus $pf^{2/3}$, clearly showing the α–γ' transition.

from compressive (about 600 MPa) at low power to tensile (100 MPa) at high power. Only a slight influence of frequency is observed, and the dependence on power is much clearer. Small–angle X-ray spectroscopy (SAXS) data show that voids are absent in the samples deposited at low power densities [66]. For higher powers voids are observed, and there appears to be a sudden change between 27 and 58 mW/cm^2.

Electrical data are shown in Figure 59 as a function of deposition rate for all frequencies, using the relation between deposition rate and power density as depicted in Figure 54. Both dark conductivity and photoconductivity decrease exponentially with increasing deposition rate. The data in this range of deposition rates can be fitted with $\sigma_\mathrm{d} = 9 \times 10^{-12} \exp(-1.5 r_d)$ and $\sigma_\mathrm{ph} = 8 \times 10^{-6} \exp(-0.5 r_d)$, with σ_d and σ_ph expressed in $\Omega^{-1}\,\mathrm{cm}^{-1}$, and r_d in nanometers per second.

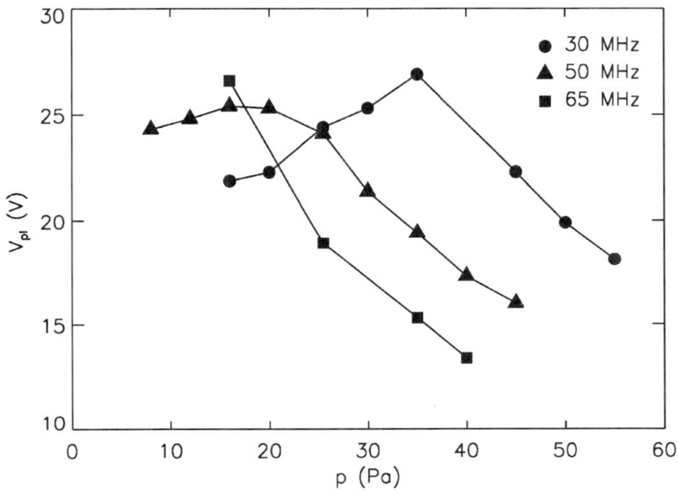

FIG. 56. The plasma potential as a function of pressure for a frequency of 30, 50, and 65 MHz. (From E. A. G. Hamers, Ph.D. Thesis, Universiteit Utrecht, Utrecht, the Netherlands, 1998, with permission.)

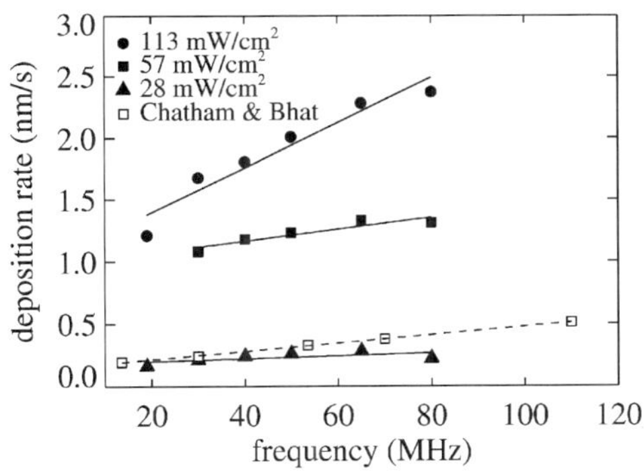

FIG. 57. Deposition rate as function of frequency at optimum pressure at each frequency (see Fig. 54) at three power densities. Also shown are data by Chatham and Bhat [146].

Consequently the photoresponse σ_{ph}/σ_d increases with deposition rate as about $10^6 \exp(r_d)$. Activation energies amounted typically to 0.7–1.0 eV. From thermally stimulated conductivity (TSC) measurements [489–492] a midgap density of states (DOS) of 1.5×10^{16} cm^{-3} eV^{-1} is determined. The product $\mu\tau$ at 300 K is 9×10^{-5} cm^2 V^{-1}. Both DOS and $\mu\tau$ are independent of frequency.

Summarizing the results, it is clear that most of the material properties are not strongly dependent on frequency in the range of 30 to 80 MHz. The effect of power density is much more important. The material deposited at the lowest power

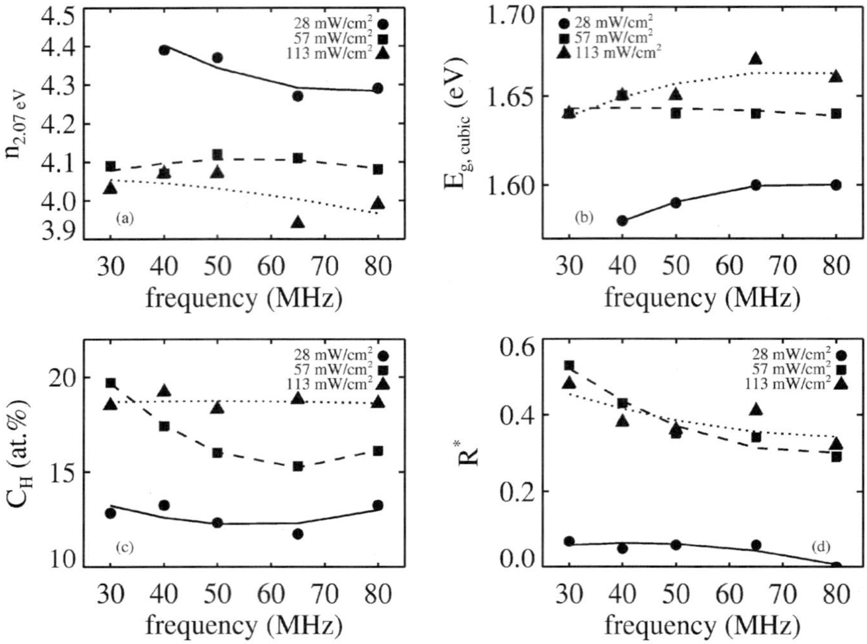

FIG. 58. (a) Refractive index $n_{2.07\ \mathrm{eV}}$, (b) cubic bandgap E_g, (c) hydrogen content C_{H}, and (d) microstructure parameter R^* as function of frequency at optimum pressures (see Fig. 54) at three power densities.

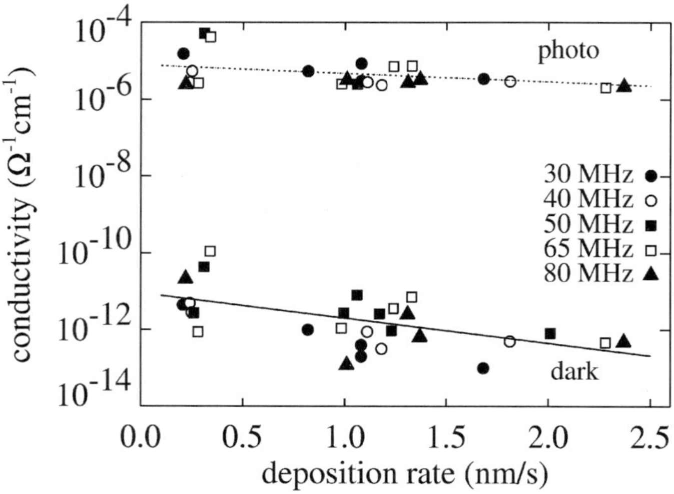

FIG. 59. Dark conductivity (σ_{d}) and photoconductivity (σ_{ph}) as functions of deposition rate for all frequencies.

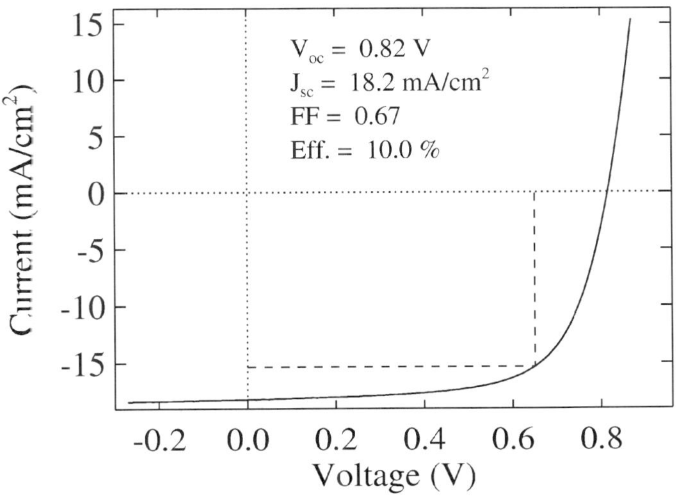

FIG. 60. Current–voltage characteristics of a solar cell made at 65 MHz and 42 mW/cm^2. The dashed line indicates the maximum-power point.

is of *device quality*. Using larger power densities results in less dense material. This can be inferred from the decrease of the refractive index and the increase of the microstructure parameter, in combination with the increase in hydrogen content.

Perrin et al. [245] have shown that the increase in deposition rate is more pronounced above a certain threshold power density. This power density is close to the lowest power density in the experiments shown here. The deposition rate is probably the main parameter causing the observed difference between the low-power and high-power material and the deterioration of layers at higher power levels. This confirms the known observation that in general, the quality of material deposited in the γ'-regime is worse than in the α-regime [245].

The presented material quality results are similar to results reported by others [119, 120, 280, 475], although those frequency series were obtained in the α-regime at constant low pressures. Here, good quality homogeneous layers were deposited *at* the α–γ' transition region. Especially the low power results compare well with results by others.

The intrinsic material fabricated at the frequencies reported above was incorporated in p^+–i–n^+ solar cells [493]. The p- and n-layer were prepared by the conventional 13.56-MHz discharge. The *device quality* films indeed yield good solar cells, of 10% efficiency, as is shown in Figure 60. This cell is manufactured with a 500-nm-thick i-layer made at 65 MHz with a power density of 42 mW/cm^2, resulting in excellent properties. The deposition rate still is 2–3 times higher than

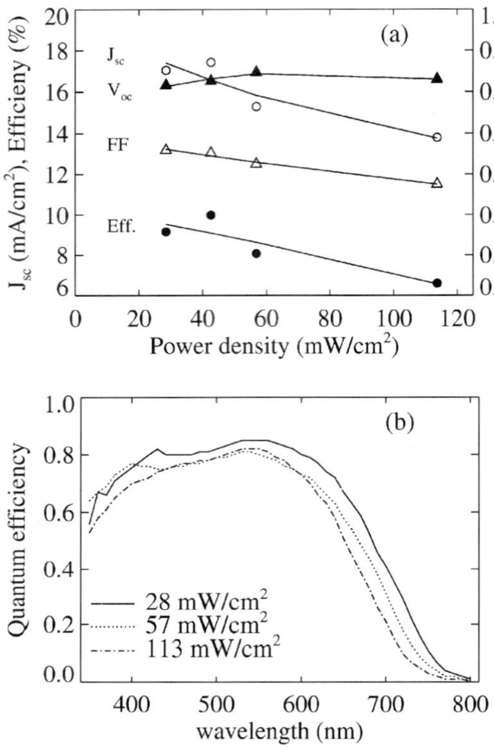

FIG. 61. Solar cell performance parameters as functions of power density for cells made at 65 MHz: (a) J_{SC}, efficiency, V_{OC}, and fill factor (FF); (b) spectral response.

at the conventional 13.56 MHz; 0.2–0.5 nm/s versus 0.1–0.15 nm/s [494]. Similar results were reported by Jones et al. [495].

The effects of power density on the material quality are also reflected in the cell quality, e.g. the efficiency. Figure 61a shows a decrease in efficiency with increasing power density or deposition rate. This is also seen in the spectral response measurements (Fig. 61b). A shift in the red from long (700 nm) to shorter (650 nm) wavelength is observed if one compares cells deposited at low and high power densities. This can be explained by the fact that the material is less dense, with a higher hydrogen content, yielding a larger bandgap. The observed decrease in the blue part of the curve as a function of power density may well be due to changes in the *p–i* interface, which are induced by a different ion bombardment condition.

The differences in material properties are also reflected in the degradation behavior of the cells. Figure 62 shows the normalized efficiencies as a function of

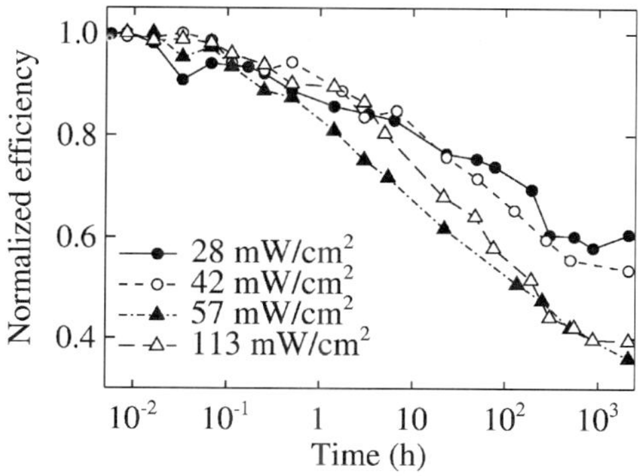

FIG. 62. Normalized solar cell efficiency as a function of illumination time for different power densities as obtained by continuous illumination of 1000-W/m^2 AM1.5 light. The initial efficiencies of the four cells were 9%, 10%, 8%, and 6% for 28-, 42-, 57-, and 113-mW/cm^2 power density, respectively.

illumination time. The efficiencies of the cells with the *i*-layer deposited at low power densities stabilize at around 60% of their initial value, while the cells with the *i*-layer deposited at high power densities stabilize at 40%. This correlates with the hydrogen content of the *i*-layers, which is around 12% and 19% for the low and the high power densities (see Fig. 58) and which is indicative for the low-density material at obtained at these high power densities.

Also a frequency dependence was observed, especially in V_{oc}, which was not expected from the material study. As only the deposition conditions of the i-layers are varied, a change at the *p–i* interface must be responsible for the change in V_{oc}. The lower value of V_{oc} at low frequency was attributed to the difference in ion bombardment at the *p–i* interface [493].

1.8.1.3. Discharge Analysis

In the ASTER deposition system, experiments have been carried out in which the excitation frequency was varied between 13.56 and 65 MHz [169]. The other process conditions were kept constant at a power of 10 W, a pressure of 0.16 mbar, gas flows of 30 sccm SiH_4 and 30 sccm H_2, and a substrate temperature of 250°C. As in Section 1.6.2.3, plasma properties that are deduced from IED measurements are compared with material properties in Figure 63. The IEDs of SiH_2^+ at four frequencies are shown in Figure 64.

The plasma potential is about 25 V (Figure 63a). This value of the plasma potential is typical for the silane plasmas in the *asymmetric* capacitively coupled RF reactors as used in the ASTER deposition system, and is also commonly found in argon or hydrogen plasmas [170, 280, 327]. From the considerable decrease of the dc self-bias with increasing frequency (Figure 63a) it is inferred that the

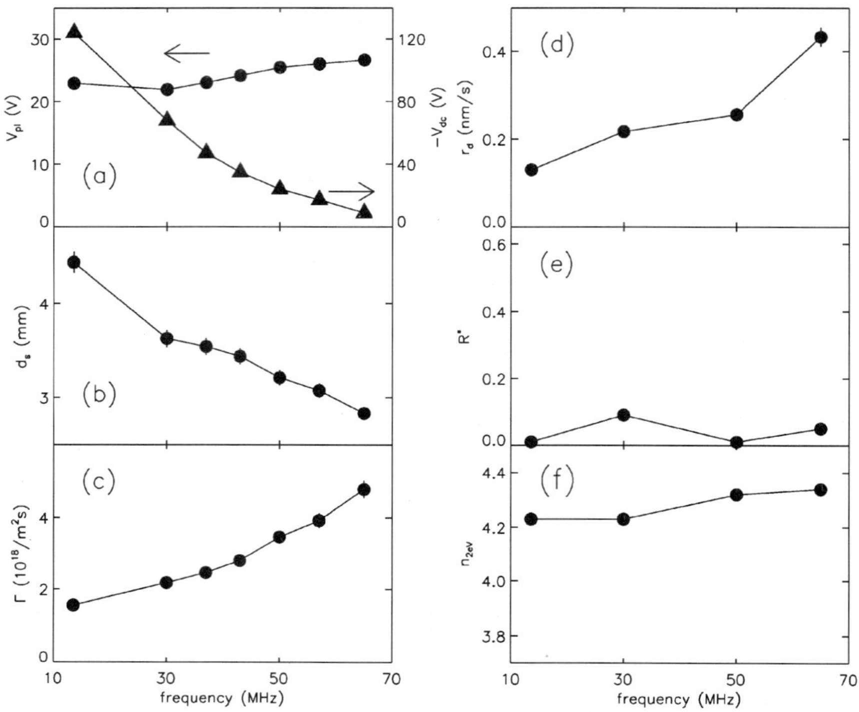

FIG. 63. Plasma parameters as deduced from the IEDs and material properties as a function of excitation frequency of the SiH_4–H_2 discharge at a power of 10 W and a pressure of 0.16 mbar: (a) the plasma potential V_{pl} (circles) and dc self-bias V_{dc} (triangles), (b) the sheath thickness d_s, (c) the maximum ion flux Γ_{max}, (d) the growth rate r_d, (e) the microstructure parameter R^*, and (f) the refractive index $n_{2\,eV}$. (Compiled from E. A. G. Hamers, Ph.D. Thesis, Universiteit Utrecht, Utrecht, the Netherlands, 1998.)

potential drop over the sheath of the grounded electrode (V_{pl}) and the one over the sheath of the powered electrode ($V_{pl} - V_{dc}$), become comparable in magnitude. Hence, the discharge is becoming more symmetric with increasing frequency.

In contrast, Heintze and Zedlitz [236] also presented data on the plasma potential as function of frequency in silane plasmas: the plasma potential varies from about 27 V at 35 MHz to about 20 V at 180 MHz. Moreover, Dutta et al. [284] used a *symmetric* capacitively coupled RF reactor and estimated the plasma potential in their system from the applied voltage at the powered electrode. A decrease of the plasma potential from 45 V at 13.56 MHz to only 15 V at 70 MHz is observed. This difference in behavior is thought to be solely due to the different reactor geometries.

The charge carrier density in the sheath increases by a factor of 3 on increasing the excitation frequency from 13.56 to 65 MHz. As a consequence, the

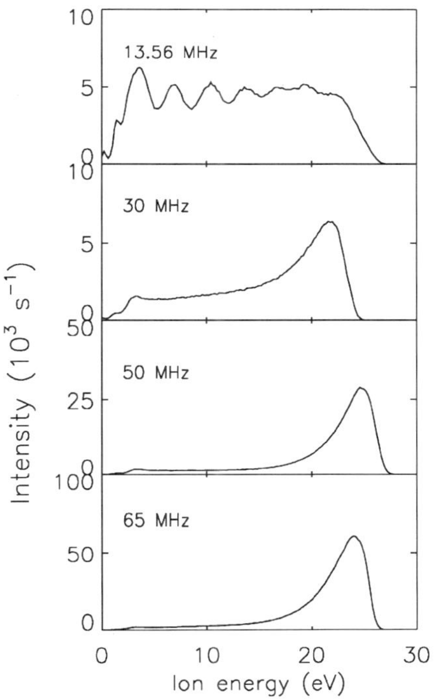

FIG. 64. The ion energy distributions of SiH_2^+ at several frequencies. (From E. A. G. Hamers, Ph.D. Thesis, Universiteit Utrecht, Utrecht, the Netherlands, 1998, with permission.)

sheath thickness d_s decreases from 4.4 mm at 13.56 MHz to 2.8 mm at 65 MHz (Fig. 63b), which results in a reduced number of collisions in the sheath. This is nicely illustrated in Figure 64, where the IEDs of SiH_2^+ are shown for several frequencies. Clearly, the contribution of low-energy ions decreases upon increasing the frequency.

The increase in the deposition rate r_d (Fig. 63d) corresponds to the increase in the ion flux (Fig. 63c): the fraction of arriving ions per deposited atom, R_i, is constant at about 0.25. Such observations have also been reported by Heintze and Zedlitz [236], who furthermore suggested that the deposition rate may well be controlled by the ion flux. The kinetic ion energy per deposited atom, E_{max}, is also constant and amounts to about 5 eV. As was shown in Section 1.6.2.3, the material quality as reflected in the refractive index $n_{2\ eV}$ (Fig. 63e) and the microstructure parameter R^* (Fig. 63f) is good: $n_{2\ eV}$ is around 4.25, and R^* is low (<0.1). The depletion of the silane stays constant at a value of 4.0 ± 0.4 sccm in this frequency range. The partial pressures of silane, hydrogen, disilane (1.3×10^{-3} mbar), and trisilane (2×10^{-5} mbar) in the plasma are also independent of frequency. Similar results have been obtained by Heintze and Zedlitz [280]. As a result of varying the frequency from 50 to 250 MHz, the dissociation rate only increased by a factor of 1.5, while the deposition rate on the electrodes increased by nearly an order of magnitude.

From the constant depletion, assuming a homogeneous deposition throughout the reactor, a frequency-independent deposition rate of 0.44 nm/s is deduced [163]. At 13.56 MHz the observed deposition rate is a factor of 3 lower, whereas at 65 MHz the measured deposition rate equals the estimated value of 0.44 nm/s. This suggests that the increased deposition rate at higher frequencies, as measured at the center of the grounded electrode, is partly due to a more homogeneous deposition profile throughout the reactor at higher frequencies [163, 280]. This suggestion is further supported by the fact that the discharge is more symmetric at higher frequencies, as deduced from the low dc self-bias at high frequencies.

1.8.2. CHEMICAL ANNEALING

It was recognized by, e.g., Tsai et al. [496] that high hydrogen dilution resulted in microcrystalline films as a result of atomic hydrogen, which preferentially etches weak bonds during deposition. This prompted the use of fluorine-containing precursors, with which enhanced etching is achieved during deposition. Etching by means of periodically interrupting the deposition process has been reported by Tsuo et al. [497], who used a constant XeF_2 flow and a periodic SiH_4 flow. The use of an RF hydrogen discharge for the etching cycles and a diluted silane discharge for the deposition cycles was tried by Xu, as reported by Luft and Tsuo [6]. With both methods good material is obtained.

The *chemical annealing* technique was proposed by Shirai et al. [17, 471, 498] as a means to tackle the metastability problem in a-Si:H solar cells. It aims at the lowering of the hydrogen content in a-Si:H by alternating deposition and exposure to atomic hydrogen. It was demonstrated that a lower hydrogen content indeed leads to more stable material. A two-discharge-zone apparatus was used by Asano et al. [499, 500]. In one zone SiH_4 and H_2 were mixed to deposit a-Si:H, while in the other zone pure H_2 was used for the etching process. The substrate was rotated between the two zones. Chemical annealing alternatively is termed alternately repeating deposition and hydrogen plasma treatment (ADHT) and layer by layer (LBL).

The application of this technique in conventional RF deposition equipment revealed the possibility of producing microcrystalline silicon at relatively low power densities and temperatures [501]. Depending on deposition conditions, (i.e., deposition cycle time, hydrogen exposure time, and power density), different microcrystalline fractions are found; see [165, 251, 502].

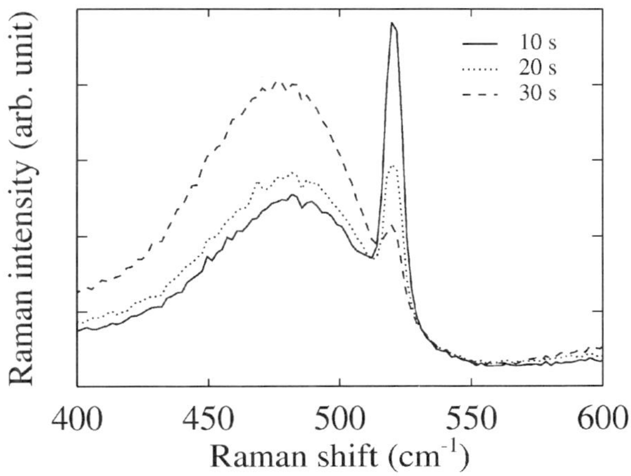

FIG. 65. Raman spectra for films deposited employing the chemical annealing technique for different hydrogen exposure times, 10, 20, and 30.5 seconds.

As an example, the Raman spectra of three films deposited in the ASTER deposition system are shown in Figure 65, for which only the hydrogen exposure time was varied, 10–30 s. In the deposition cycle a 1 : 1 mixture of silane and hydrogen was used. Conditions were chosen so that *device quality* material was obtained [370]. The duration of the deposition cycle was kept constant at 10 s. At the start of the H_2-plasma cycle the SiH_4 flow was stopped and extra H_2 was introduced to keep the pressure constant. A practical problem to be addressed is the matching of the discharge when changing from H_2 to SiH_4 and vice versa. A sudden change in dc self-bias is observed, as well as a change in reflected power. Fast automated matching networks are to be used to minimize these effects.

The broad TO peak, characteristic for *a*-Si:H, is observed at 480 cm^{-1}, while the crystalline peak is at 520 cm^{-1}. The ratio between integrated peak intensities clearly depends on the hydrogen exposure time. The hydrogen content increases linearly with exposure time; the extra hydrogen forms SiH_2 bonds. These effects were thought to originate from the hydrogen bombardment during the hydrogen exposure cycle [165].

Layadi et al. have shown, using *in situ* spectroscopic ellipsometry, that both surface and subsurface processes are involved in the formation of μc-Si [502, 503]. In addition, it was shown that the crystallites nucleate in the highly porous layer below the film surface [502, 504], as a result of energy released by chemical reactions [505, 506] (chemical annealing). In this process four phases can be distinguished: incubation, nucleation, growth, and steady state [507]. In the incubation phase, the void fraction increases gradually while the amorphous fraction decreases. Crystallites start to appear when the void fraction reaches a maximum (at a thickness of about 10 nm). Subsequently, the crystalline fraction increases exponentially in the nucleation phase. Then a linear increase in the crystalline fraction sets in, corresponding to the growth of crystallites in the whole film. At a certain thickness, the crystalline fraction remains constant, and a steady state is reached. *In situ* conductivity measurements revealed that the dark conductivity increases by more than an order of magnitude at the onset of the growth phase [507]. These films have been employed in thin-film transistors, and excellent stability was demonstrated [508].

It was argued by Saito et al. [509] that a hydrogen plasma treatment as used in the LBL technique causes *chemical transport*: the *a*-Si:H film deposited on the cathode prior to the hydrogen treatment is etched and transferred to the anode. Hence, in the hydrogen plasma silane-related molecules and radicals are present. In fact, material is deposited on the anode under the same conditions as in the case of high dilution of silane with hydrogen.

1.8.3: RF MODULATION

Modulation of the RF excitation has been used in an attempt to increase the deposition rate. Increasing the gas pressure or raising the power generally leads to dust formation and deterioration of material properties. To overcome this problem, one can pulse the plasma by modulating the RF signal in amplitude with a square wave (SQWM). Depending on the regime (α or γ'), different effects are observed.

A considerable decrease both in the deposition rate and in powder formation was found when the 13.56-MHz excitation was modulated with a square wave of 2 Hz [510, 511]. In the γ'-regime, Bieberichер et al. [512] have observed a decrease in deposition rate from 1.0 nm/s in a continuous-wave (cw) 50-MHz SiH_4–H_2 plasma, to 0.2 nm/s in a similar (i.e. with the same *average* power of 10 W) plasma, modulated by a frequency of 100 kHz.

When modulating the power in the α-regime, an increase in deposition rate has been observed with SQWM, with an optimum at a frequency of about 100 kHz [513–515]. It was reported that the time-averaged electron density for a modulated discharge was higher than in the case of a continuous discharge [18]. Moreover, the amount of dust generated in the discharge was much smaller, due to the low density of negative ions.

At the start of each modulation pulse, a sharp peak in optical emission is seen. Similar SiH^* emission peaks in pulsed plasmas have been found by Scarsbrook et al. [516] and Howling et al. [321]. The sharp peak was claimed to be caused by a pulse of high-energy electrons. Overzet and Verdeyen [517] measured electron densities at a 2.9-MHz excitation frequency and modulation frequencies up to 20 kHz. The optical emission of a SQWM argon plasma was measured by Booth et al. [518], who also performed particle-in-cell modeling.

Here, results are shown from experiments performed in ASTER, reported by Biebericher et al. [512, 519]. A SiH_4–H_2 (50 : 50 flow ratio, total flow 60 sccm) plasma was generated at an RF excitation frequency of 50 MHz. The substrate temperature was 250°C. The RF signal was amplitude modulated (AM) by a square wave. The modulation frequency has been varied in a range of 1–400 kHz. The modulation depth was always 90%. The duty cycle was fixed at 50%. The pressure amounted to 0.2 mbar, and the *average* power was kept at 10 W. With a duty cycle of 50%, this leads to a power of 20 W during the plasma-on period.

The measured dependence of the deposition rate on the modulation frequency is shown in Figure 66a. It can be clearly seen that the deposition rate at the optimum of 100 kHz is about three times higher than the deposition rate of a cw discharge of 10 W (dotted line). In addition, there is a clear correlation between the deposition rate and the time-averaged optical emission intensity of the pulsed discharge. This average intensity is related to the electron temperature: a higher light intensity can be attributed to an increase in the electron temperature. This results in an increased production of radicals, which is reflected in an increased deposition rate.

The spectrally integrated (185–820 nm) emission of a 50-MHz discharge, modulated at 50 kHz, is shown in Figure 66b. Similar results are obtained for the SiH^* emission. The peak appearing at the onset is generated by the plasma. The rise time of about 1 μs and the FWHM of the peak of 6 μs are much longer than the 40-ns detection limit of the electronics. When the power is turned off, the light intensity decreases very fast and the plasma fully extinguishes.

Upon changing the modulation frequency it was observed that only the length of the plateau varied, whereas the height and FWHM of the peak at plasma onset remained almost constant. Therefore, the relative importance of the peak in the time-averaged light intensity is higher at higher modulation frequencies, and consequently, the average light intensity increases with increasing modulation frequency. This holds up to the observed optimum of 100 kHz. At this frequency the plasma-on period is 5 μs, i.e. about equal to the FWHM of the peak. A further increase in modulation frequency leads to a reduction of the peak and the average light intensity.

In order to be able to explain the observed results plasma modeling was applied. A one-dimensional fluid model was used, which solves the particle balances for both the charged and neutral species, using the drift–diffusion approximation for the particle fluxes, the Poisson equation for the electric field, and the energy balance for the electrons [191] (see also Section 1.4.1).

The calculated production of SiH_3 radicals during two modulation cycles at a modulation frequency of 50 kHz is also shown in Figure 66b. Good correlation with the behavior of average light intensity versus time is observed, including the sharp peak at plasma onset. The shape of the curve is determined by the instantaneous sharp rise of dissociation and by chemical reactions. The nonzero production of SiH_3 radicals during the plasma-off period is due to chemical reac-

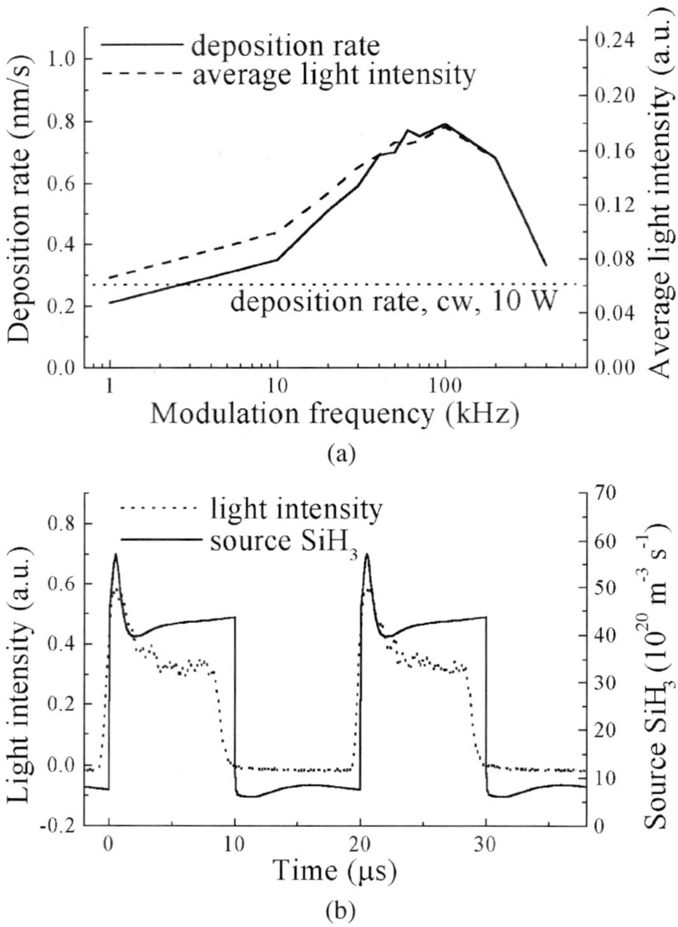

FIG. 66. Effects of modulating the RF excitation frequency: (a) deposition rate and average light intensity as a function of the modulation frequency, with the deposition rate at cw conditions indicated by the dotted line, (b) measured spectrally integrated emission and calculated production of SiH_3 radicals as a function of time, at a modulation frequency of 50 kHz and a 50% duty cycle. [From A. C. W. Biebericher, J. Bezemer, W. F. van der Weg, and W. J. Goedheer, *Appl. Phys. Lett.* 76, 2002 (2000), © 2000, American Institute of Physics, with permission.]

tions. SiH_3 is produced not only by dissociation of SiH_4 through electron impact, but also by an abstraction reaction of SiH_4 and H. These hydrogen atoms can originate from the dissociation of SiH_4 into SiH_2 and 2H [512].

The behavior of the electrons as a function of time can explain the observed results. Simulation results for a discharge operated at a pressure of 0.267 mbar,

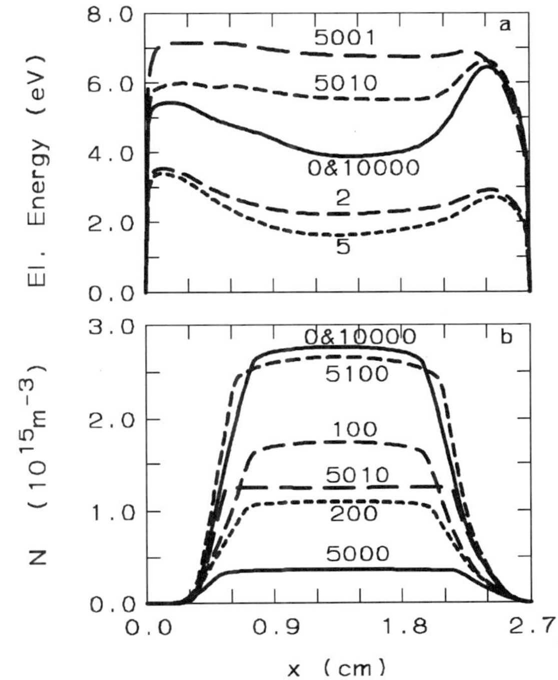

FIG. 67. The development in time of (a) the average electron energy and of (b) the density of electrons between the electrodes in the case of a 50-MHz plasma and a modulation frequency of 5 kHz. The RF electrode is located at $x = 0$, and the grounded electrode at $x = 2.7$ cm. The numbers near the curves in both graphs represent the number of RF cycles. One RF cycle $\equiv$ 20 ns. The plasma is turned off at cycle 0, turned on at cycle 5000, and turned off again at cycle 10,000. [From A. C. W. Biebericher, J. Bezemer, W. F. van der Weg, and W. J. Goedheer, *Appl. Phys. Lett.* 76, 2002 (2000), © 2000, American Institute of Physics, with permission.]

a frequency of 50 MHz, an electrode distance of 27 mm, a voltage modulated between 134 V (power 20 W) and 40 V (power <1 W), and a modulation frequency of 5 kHz are shown in Figure 67. In this figure the electron energy and density are shown as a function of the distance between the electrodes. The numbers indicate the numbers of passed RF cycles. A modulation frequency of 5 kHz corresponds to 10,000 RF cycles. The plasma is turned off at cycle 0 and is turned on at cycle 5000. It can be seen in Figure 67a that the average electron energy decreases very fast. This is consistent with the observed decay of the emission of the discharge (see Fig. 66b). The electron density also decreases, albeit with a longer decay time. The electron density is about halved after 100 cycles. Thus, very soon after shutoff, within a few cycles, no new radicals are produced, apart from those generated in neutral–neutral reactions. The electron (and ion) density decreases by the loss due to ambipolar diffusion to a value one order of magnitude lower than the value at cycle 0. When the power is switched on at cycle 5000, the electron energy increases to a value nearly two times larger than the value at cycle 0 (and 10,000). This fast increase is attributed to the low electron density at cycle 5000, which absorbs the power. The increase in electron energy results in an even steeper increase in reaction rates per electron, due to the energy dependence of the cross-sections. Thus, a peak in light intensity as well as a peak in SiH_3 radical production is observed. This leads to an increase of the average deposition rate to a value above the rate corresponding to the average deposited power (10 W); cf. Figure 66a. At a low modulation frequency (1 kHz) the decay of the profiles and the overshoot of the rates cover only a small fraction of the modulation period. In other words, the discharge is in a quasi steady state with respect to the applied power, and the deposition rate is only slightly above the corresponding deposition rate at cw conditions. As the modulation frequency increases, the overshoot becomes more important, which leads to an increase in deposition rate. When the period of the modulation becomes shorter than the decay time of the electron density, the power is distributed over too many electrons to yield considerable heating. The heating becomes less, and less of the radicals is produced, as is shown in Figure 66a. It is therefore concluded that an optimum in the modulation frequency exists, which corresponds to the decay time of the electron density. From Figure 67b it can be inferred that this optimum should be around 100 kHz, as at 200 cycles the electron density is reduced to $1/e$ of its value at cycle 0, i.e., the lifetime is 200 RF cycles, or 4 μs.

Another beneficial effect of modulation of the discharge is the observed improvement of uniformity of deposition. At cw discharge conditions that normally result in thickness variations of 70%, modulation lowers the variation to about 10% [519]. This was attributed to the fact that, as a result of modulation, reactions are not confined to the sheaths, and radicals are produced in the entire volume between the electrodes.

Madan et al. [515] have presented the effect of modulation on the properties of the material (dark conductivity and photoconductivity) and of solar cells. They also observe an increase in deposition rate as a function of modulation frequency (up to 100 kHz) at an excitation frequency of 13.56 MHz, in their PECVD system [159]. The optimum modulation frequency was 68 kHz, which they attribute to constraints in the matching networks. Increasing the deposition rate in cw operation of the plasma by increasing the RF power leads to worse material. Modulation with a frequency larger than 60 kHz results in improved material quality, for material deposited with equal deposition rates. This is also seen in the solar cell properties. The intrinsic *a*-Si:H produced by RF modulation was included in standard *p–i–n* solar cells, without buffer or graded interface layers. For comparison, solar cells employing layers that were deposited under cw conditions were also made. At a low deposition rate of about 0.2 nm/s, the cw solar cell parameters are: V_{oc} = 0.83 V, J_{sc} = 15.8 mA/cm^2, FF = 0.67, and η = 8.8% (at AM1.5 illumination conditions). Solar cells produced with the intrinsic layer deposited at 68-kHz modulation frequency are of the same quality: V_{oc} = 0.81 V, J_{sc} = 17.1 mA/cm^2, FF = 0.63, and η = 8.7%.

Bieberícher et al. [519] report that beyond a modulation frequency of about 40 kHz in their 50-MHz discharge the material is slightly worse than at standard cw conditions. However, they argue that due to the enhancement in deposition rate, lower partial pressures of silane can be used, which results in more efficient silane consumption. Indeed, they have shown that lowering the gas flows by a factor of 3 leads to a reduction in deposition rate of only a factor of 1.5 at 100-kHz modulation frequency [520]. Clearly, the efficiency of silane consumption is of great importance in production.

1.9. Hot Wire Chemical Vapor Deposition

1.9.1. General Description

The hot wire chemical vapor deposition (HWCVD) method was introduced as early as 1979 by Wiesmann et al. [521]. The principle of HWCVD is based on the thermal decomposition of silane at a heated tungsten (or tantalum) filament, foil, or grid. At temperatures of 1400–1600°C the silane is decomposed into a gaseous mixture of silicon and hydrogen atoms. Matsumura [522–524] further developed HWCVD, but termed it catalytical CVD (CTL-CVD); for he showed that the decomposition of silane at a heated tungsten filament is a catalytic process. The use of a much higher pressure (>0.1 mbar) than the one Wiesmann et al. used ($< 5 \times 10^{-4}$ mbar) led to a high deposition rate (0.5 nm/s) [521]. Doyle et al. showed that the silicon atoms deposited on the filament are then thermally evaporated onto the substrate, which is located within a few centimeters of the filament; hence the term *evaporative surface decomposition* [525]. At the same time, thermal and catalytic dissociation of silane and hydrogen lead to the generation of hydrogen atoms at the filament surface [526].

Device quality a-Si:H made by HWCVD (as they termed it) was first reported by Mahan et al. [19, 527]. They obtained a-Si:H with hydrogen concentrations as low as 1%. Deposition rates as high as 5 nm/s [528] and 7 nm/s [529] have been achieved for a-Si:H of high quality. In order to obtain *device quality* material it was shown by Doyle et al. [525] that the radicals that are generated at the filament (atomic Si and atomic H) must react in the gas phase to yield a precursor with high surface mobility. Hence, the mean free path of silane molecules should be smaller than the distance between filament and substrate, d_{fs}. Too many reactions between radicals and silane molecules, however, result in worse material. In fact, optimal film properties are found for values of pd_{fs} of about 0.06 mbar·cm [530, 531].

The combination of high deposition rate and the ability to produce *device quality* material is of particular interest for solar cell production and TFT fabrication [532–535]. Further, the low hydrogen content was expected to yield improved stability against light-induced degradation [527], as the Staebler–Wronski effect is related to the hydrogen content in the material (see also Section 1.1.2.5). This was demonstrated by Crandall et al. [536], who incorporated an HWCVD-deposited layer in solar cells, and observed reduced degradation upon light soaking as compared to devices with a conventional PECVD layer. In these *hybrid* solar cells only the intrinsic layer was made by HWCVD; all other (doped) layers were deposited by PECVD. Wang et al. [537] have reported solar cells that were completely made using HWCVD.

HWCVD-deposited *a*-Si:H layers have also successfully been used as the semiconductor layer in inverted–staggered TFTs [538, 539]. Moreover, it was demonstrated that these TFTs have excellent electrical properties, they do not suffer from a shift of the threshold voltage upon prolonged application of a gate voltage and these HWCVD TFTs are stable [538, 540]. In a TFT, the metastable character of *a*-Si:H manifests itself as a reversible threshold-voltage shift, upon prolonged application of gate voltage (see Section 1.11.2).

Dilution of silane with hydrogen leads to the formation of polycrystalline silicon films (poly-Si:H) that still contain a small amount ($<1\%$) of hydrogen [529, 541–544]. This is believed to occur as a result of increased etching by atomic hydrogen. Molecular hydrogen is decomposed at the filament to reactive atomic hydrogen, which is able to etch silicon atoms from disordered or strained bonding sites in the amorphous (sub)surface network. The amorphous network thus is transformed to a polycrystalline one, without amorphous tissue. Large grains are obtained (70-nm average grain size), and the crystalline fraction in these highly compact materials amounts to 95% [544]. The grain boundaries are passivated by the small amount of hydrogen.

Doping is achieved by adding dopant gases such as phosphine, diborane, or trimethylboron to the silane [545–547]. The doped material is of electronic quality comparable to PECVD-doped material.

1.9.2. EXPERIMENTAL SETUP

A cross-sectional overview of a HWCVD setup is shown in Figure 68 [548]. This particular reactor is part of a multichamber, ultrahigh-vacuum deposition system, PASTA, described by Madan et al. [159]. The background pressure of the reactor is 10^{-8} mbar. In this HWCVD reactor two coiled tungsten filaments are mounted parallel to each other in order to improve thickness uniformity. The filaments have a diameter of 0.5 mm and a length of 15 cm. The distance between the filaments is 4 cm, and the distance from the filaments to the substrate (d_{fs}) is 3 cm. The filaments are heated by sending a dc current through the wires. Calibration of the

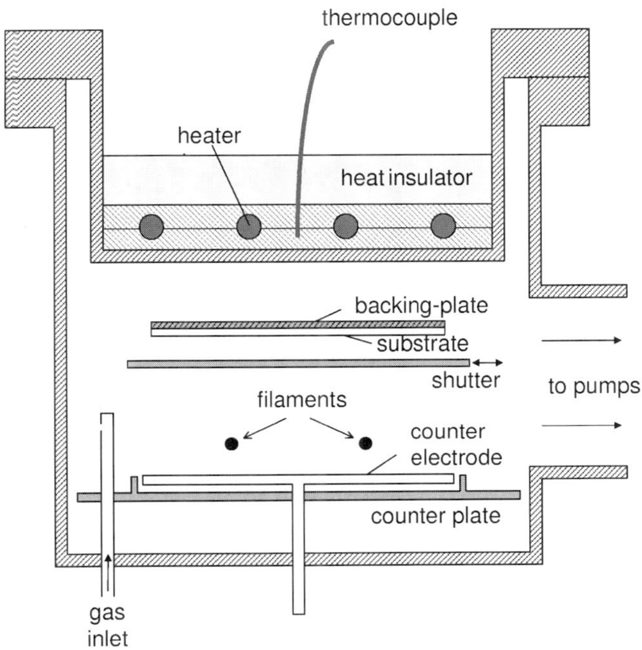

FIG. 68. Cross-sectional view of a HWCVD deposition reactor. (From K. F. Feenstra, Ph.D. Thesis, Universiteit Utrecht, Utrecht, the Netherlands, 1998, with permission.)

filament temperature as a function of dc current is needed, and can be done with a pyrometer. Temperatures exceeding about 1800°C should be used as to avoid silicide formation on the filament. On the other hand, the temperature should not be so high that tungsten is evaporated from the filament and is incorporated in the film [549]

A shutter can be placed in front of the substrate to prevent deposition during filament preheating. The reaction gases are injected on the lower-left side of the reactor, and directed towards the reactor wall in order to achieve more homogeneous gas flows. The (unreacted) gas is pumped out at the right side, so the gas flow is perpendicular to the filaments.

The substrate is radiatively heated by heaters that are placed outside the vacuum. A backing plate ensures a laterally homogenous temperature profile. In the same chamber also PECVD can be carried out. The backing plate then is the grounded electrode, and the RF voltage is applied to the counter electrode.

The high filament temperature used causes additional radiative heating of the substrate [530, 531]. Feenstra et al. [531, 548] have developed a heat transport model of their setup (Figure 68). All heat exchange is assumed to occur via radiation. They found, e.g., that at a heater temperature of 100°C and a filament temperature of 1900°C the substrate temperature reached a value of 350°C after a typical stabilization time of 25 min. With the shutter closed the stable substrate temperature was 50°C lower. This can be exploited to preheat the substrate, when the shutter is closed, to the desired temperature. Just before the deposition is started, the filament temperature is set to its correct value, and the shutter can be opened. This procedure avoids any temperature drift during deposition.

1.9.3. MATERIAL PROPERTIES AND DEPOSITION CONDITIONS

With the setup described, a series of depositions was carried out [531, 548], in which the substrate temperature was varied between 125 and 650°C, the pressure between 0.007 and 0.052 mbar, the gas flow rate between 15 and 120 sccm, and the dilution of the silane gas with hydrogen ($[SiH_4]/([SiH_4] + [H_2])$) between 0.1 and 1. Under these conditions the deposition rate varied between 1 and 2.5 nm/s [531]. Molenbroek et al. [530] reported a variation of the deposition rate between 1 and 9 nm/s for similar process conditions, at a filament temperature of 2000°C.

The variation of deposition temperature has similar effects on the material properties to those on PECVD-deposited material. With increasing temperature (125–650°C), the material becomes more dense (the refractive index extrapolated to 0 eV increases from 3.05 to 3.65), and the hydrogen content is decreased (15 to 0.3 at.%), as well as the microstructure factor (0.4 to 0). The activation energy is 0.83 eV up to a deposition temperature of 500°C. The dark conductivity and AM1.5 photoconductivity are about 5×10^{-11} and $5 \times 10^{-6}\ \Omega^{-1}\,\text{cm}^{-1}$, respectively, up to a deposition temperature of 500°C, i.e., the photoresponse is 10^5.

The deposition rate increases upon increasing the pressure. This is explained by noting that the impingement rate per unit area, r_i, of molecules on the filament is linearly dependent on the pressure as $r_i = p/\sqrt{2\pi k_B T_g}$, with T_g the gas temperature. However, as the pressure becomes higher, the collisional mean free path of the silane becomes smaller, and the silane supply to the filaments becomes restricted. Moreover, the transport of deposition precursors to the substrate is restricted as well. The mean free path of silane was estimated to be 2.5 cm at a pressure of 0.02 mbar [531], i.e., the mean free path about equals the distance between filament and substrate. Indeed, a maximum in deposition rate is observed at this pressure. This corresponds to a value of pd_{fs} of 0.06 (cf. [530]). The microstructure parameter plotted as a function of pd_{fs} has a minimum around $pd_{\text{fs}} = 0.06 \pm 0.02$ [530].

The silane flow rate determines to a large extent the possible deposition rate. If the supply of silane is too low, the environment in the immediate vicinity of the filament will become depleted of silane. The generation of deposition precursors (and also the deposition rate) is limited by the silane supply. Above a certain

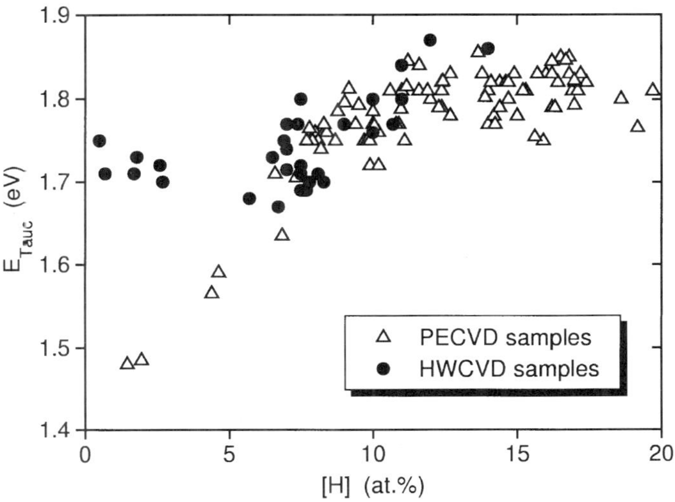

FIG. 69. Relation between the optical bandgap (Tauc convention) and the hydrogen content in *a*-Si:H deposited by HWCVD (closed circles) and PECVD (open triangles). (From K. F. Feenstra, Ph.D. Thesis, Universiteit Utrecht, Utrecht, the Netherlands, 1998, with permission.)

flow, the dissociation reaction rate is the limiting factor, and as a consequence the deposition rate will remain constant upon a further increase of the silane flow.

Dilution of silane with hydrogen to moderate amounts ($0.75 < [SiH]_4/([SiH_4] + [H_2]) < 1$) is not expected to have a large influence on material properties, as even in the pure silane case a large amount of atomic hydrogen is present. Every SiH_4 molecule dissociates to give four H atoms [525]. Also, the mean free path of the atomic H is larger than that of Si, being about 8 cm at 0.02 mbar [531], and consequently all hydrogen will reach the substrate. It is found that the deposition rate depends linearly on the dilution fraction, as it depends on the partial pressure of silane. Significant changes in the refractive index, the hydrogen content, the microstructure parameter, and electrical properties are observed for values of $[SiH_4]/([SiH_4] + [H_2])$ lower than about 0.3–0.4. A closer look at these data reveals that moderate hydrogen dilution slightly improves the dark conductivity and photoresponse, whereas dilution fractions smaller than 0.3–0.4 lead to deterioration of the material. Similar trends were also observed by Bauer et al. [550] and Molenbroek et al. [528, 530].

The common linear dependence of the bandgap on the hydrogen content as observed for PECVD-deposited material is not observed for HWCVD-deposited material. A large dataset is shown in Figure 69, in which the relation between bandgap and hydrogen content is compared for HWCVD- and PECVD-deposited material [531]. It is seen that the bandgap of HWCVD-deposited material follows a similar behavior, as a function of hydrogen content, to that of PECVD-deposited material, for values of the hydrogen content of 7% and higher: the bandgap varies between 1.7 and 1.85 eV [531]. The bandgap of PECVD-deposited materials decreases to 1.5 eV at a hydrogen concentration of 2%, whereas the bandgap of HWCVD-deposited material remains unchanged at 1.7 eV, even at hydrogen concentrations of 1%. This difference in behavior is due to the difference in the number of voids present in the material. A large number of voids reduces the effective density of the material, and increases the average Si−Si bond distance. As a result, the bandgap remains high. Crandall et al. [71] found that the number of voids increased with decreasing hydrogen content in HWCVD-deposited material. This observation is also supported by small-angle X-ray scattering (SAXS) [551] and nuclear magnetic resonance (NMR) [552] data.

1.9.4. Deposition Model

The deposition mechanism in HWCVD of *a*-Si:H can be divided into three spatially separated processes. First, silane is decomposed at the tungsten filament. Second, during the diffusion of the generated radicals (Si, H) from the filament to the substrate, these radicals react with other gas molecules and radicals, and new species will be formed. Third, these species arrive at the substrate and contribute to the deposition of *a*-Si:H.

The filament material acts as a catalyzer in the silane decomposition process. This is clear by comparing the Si−H bond energy (3–4 eV) in silane with the thermal energy at the filament temperature of 2000°C (0.25 eV): they differ by an order of magnitude. The dissociation of silane proceeds in two steps [525]. First, the silane reacts with the tungsten to form a tungsten silicide W_xSi_y at the surface of the filament. Hydrogen atoms leave the surface before they can recombine to form molecular hydrogen. Subsequently, the silicon atoms evaporate from the surface, which is an activated process with an activation energy of 3.6 eV [525]. At a filament temperature of about 1700°C the silane decomposition rate and the evaporation rate balance each other. For lower temperatures a tungsten–silicon alloy is formed on the filament.

A fraction of the generated radicals react with gas molecules before they are able to reach the substrate directly. This fraction depends on the process pressure. A number of reactions may take place, and only the exothermic ones are listed in Table IX, as endothermic reactions are not very likely to occur. As can be seen in Table IX, the Si and H radicals are highly reactive with silane. Most reactions will occur with the omnipresent silane. Quantum chemistry computations, confirming the findings of Molenbroek et al. [530], have shown that the reaction of Si with SiH_4 proceeds via insertion of Si into the SiH_4 to form Si_2H_4, followed by a rearrangement of the thus formed species and the elimination of H_2, which yields Si_2H_2 [553]. An increasing amount of SiH_3 will be created via the reaction of H with SiH_4 upon increasing the pressure. As a product of the reaction SiH_3

TABLE IX
EXOTHERMIC GAS PHASE REACTIONS IN THE HWCVD PROCESS, THEIR ENTHALPIES, AND THEIR RATES [192], ESTIMATED FOR TYPICAL OPERATING CONDITIONS

Chemical reaction		Enthalpy (eV)	Reaction rate ($cm^3\,s^{-1}$)
$H + SiH_4$	$\rightarrow SiH_3 + H_2$	-0.54	2.68×10^{-12}
$H + SiH_3$	$\rightarrow SiH_2 + H_2$	-1.46	2×10^{-11}
$SiH_3 - SiH_3$	$\rightarrow SiH_4 + SiH_2$	-0.91	1.5×10^{-10}
$Si + SiH_4$	$\rightarrow Si_2H_2 + H_2$	-0.34	3.5×10^{-10}
$Si_2H_2 + SiH_4$	$\rightarrow Si_3H_4 + H_2$	$-0.4/-0.05$	2×10^{-10}

+ SiH_3, SiH_2 is created which at higher pressures may polymerize further to Si_3H_4. Holt et al. [554] have performed Monte Carlo simulations to study gas phase and kinetic processes in HWCVD. They showed that under the conditions for obtaining *device quality* material SiH_3 is the most abundant species at the substrate. At a pressure of about 0.08 mbar the fluxes of SiH_3, Si_2H_2, and Si to the substrate were calculated to be 5×10^{17}, 1×10^{16}, and 5×10^{13} $cm^{-2}\,s^{-1}$, respectively.

The HWCVD deposition process is more or less the same as for PECVD, and was described in Section 1.7. Important differences between the two is the absence of ions, and the limited number of different species present in the gas phase, in the former. At low pressure atomic Si is the main precursor. This yields void-rich material with a high microstructure factor. Increasing the pressure allows gas phase reactions with Si and H to create more mobile deposition precursors (SiH_3), which improves the material quality. A further increase leads to the formation of higher silanes, and consequently to a less dense film.

Using threshold ionization mass spectrometry and *in situ* ellipsometry, Schröder and Bauer [555] have shown that the Si_2H_4 radical may well be the species responsible for deposition, rather than SiH_3 as in PECVD. This larger and less mobile precursor is thought to be the cause of the observed differences in the deposition conditions required in HWCVD and PECVD to obtain *device quality* material.

1.10. Expanding Thermal Plasma Chemical Vapor Deposition

1.10.1. GENERAL DESCRIPTION

The expanding thermal plasma chemical vapor deposition (ETP CVD) technique has been developed in the group of Schram [20, 556] and has been used for the deposition of several materials, such as hydrogenated amorphous silicon [21] and carbon [557], and even diamond [558]. The technique is a remote-plasma deposition method. The generation of the plasma, the transport of the plasma, and the deposition are geometrically separated. In remote-plasma CVD the substrate holder is not a necessary electrode for the plasma. The absence of direct plasma contact with the substrate allows for better control of ion bombardment, which is advantageous. Also, the properties of the plasma can be varied independently, which makes optimization of the whole process simpler.

An argon–hydrogen plasma is created in a dc thermal arc (cascaded arc) operated at high pressure ($\approx$ 0.5 bar) [556, 559, 560] (the cascaded arc is also employed in IR ellipsometry, providing a well-defined source of intense IR radiation; see Section 1.5.4 [343]). As the deposition chamber is at much lower pressure (0.1–0.3 mbar), a plasma jet is created, expanding into the deposition chamber. Near the plasma source silane is injected, and the active plasma species dissociate the silane into radicals and ions. These species can deposit on the substrate, which is positioned further downstream.

The main advantage of this technique is the very high deposition rate that can be obtained for the different materials. However, this high deposition rate may not always be compatible with good material quality. A large effort has been made and has resulted in the deposition of good-quality *a*-Si:H at deposition rates as high as 10 nm/s [432, 561, 562].

1.10.2. Experimental Setup

The ETP CVD setup is schematically shown in Figure 70. It consists of a thermal plasma source, a cascaded arc, and a low-pressure deposition chamber [563, 564]. The cascaded arc consists of three cathodes, a stack of ten copper plates with a central bore of 4 mm, and an anode plate with a conical nozzle [556]. The length of the cascade is about 10 cm. All parts are water-cooled. Pure argon or an argon–hydrogen mixture is introduced into the arc. The argon flow is varied between 1800 and 6000 sccm, and the hydrogen flow between 0 and 1200 sccm. The arc pressure is between 0.4 and 0.6 bar. The dc discharge is current-controlled. The arc current and voltage are 30–75 A and 45–120 V, respectively. Typical arc plasma parameters are a electron density and temperature of 10^{16} cm^{-3} and 1 eV [565], respectively.

The deposition chamber is a cylindrical vessel with a diameter of 50 cm and a length of 80 cm. At about 5 cm from the arc outlet, silane can be introduced via an injection ring (7.5-cm diameter) that contains eight holes of 1-mm diameter each. The distance between arc outlet and substrate is 38 cm. The substrates are heated via the substrate holder, of which the temperature can be controlled between 100 and 500°C. Samples can be loaded via a load lock equipped with a magnetic transfer arm. The substrate can be optionally RF-biased. A residual gas analyzer

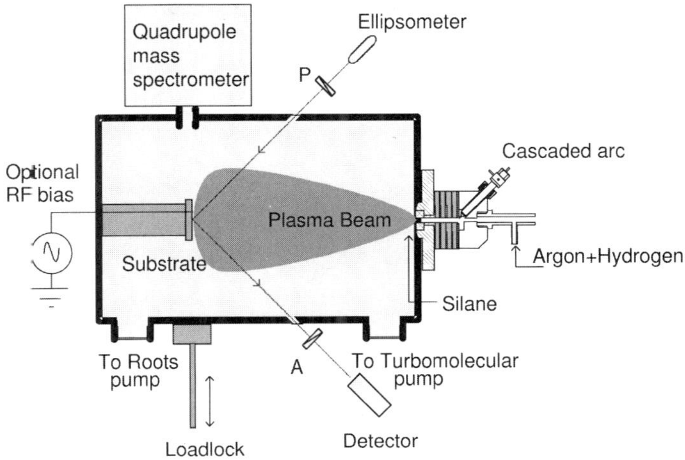

FIG. 70. Cross-sectional view of a expanding thermal plasma deposition reactor. (Courtesy of M. C. M. van de Sanden, Eindhoven University of Technology, Eindhoven, The Netherlands.)

(QMS) and an ellipsometer complete the setup. The typical pressure is in the range of 0.15–0.5 mbar. The deposition chamber has a volume of 180 l. During processing it is pumped by a stack of two Roots blowers and one forepump (total pumping capacity is about 1500 m^3/h); otherwise it is pumped by a turbo pump (450 l/s), with which a base pressure of 10^{-6} mbar is reached.

As a result of the large pressure gradient between arc and deposition chamber the plasma expands supersonically from the nozzle into the deposition chamber. After a stationary shock front at about 5 cm from the nozzle, the plasma expands subsonically. As a consequence, the electron density and temperature are drastically reduced [566]. In this downstream region the plasma is recombining, much as in an afterglow plasma. The low electron temperature in the region where the silane is introduced implies that electron-induced reactions are ineffective for the ionization and dissocation of silane. It is found that the admixture of H_2 to Ar in the arc is the cause of the strong reduction in the ion and electron density [567]. For pure Ar, the ion density is 10^{13} cm^{-3}, and consists mainly of Ar^+. At high H_2 flows (1800 sccm), the ion density is about 4×10^{10} cm^{-3}, and is mainly H^+. At low hydrogen flows, ArH^+ is the most aboundant ion. The reason for this drastic change in density is that fast molecular processes become an important recombination channel for the ions [560, 568]. The atomic H concentration increases from about 2×10^{11} cm^{-3} at low (240 sccm) H_2 flow to about 2×10^{12} cm^{-3} at high (1800 sccm) H_2 flow. The arc changes basically from an argon ion source to an atomic hydrogen source on going from pure argon to admixture of a large amount of hydrogen.

1.10.3. Material Properties and Deposition Conditions

The deposition rate is found to be independent of the deposition temperature, for low (0.3 nm/s) to high (30 nm/s) rates [472, 473]. The deposition rate and the silane consumption increase with increasing partial pressure of silane [569]. A higher arc current yields a higher silane depletion and deposition rate [563]. The bond-angle distortion, as determined from Raman spectroscopy measurements, varied between 8.3° at low deposition rate and 9.1° at high deposition rate [570], indicating a slightly increased disorder.

The incorporated amount of hydrogen depends on the deposition temperature. The hydrogen content decreases with increasing temperature. For films deposited at a high rate, the hydrogen content always is higher than for the films that were deposited at a low rate. The hydrogen content varies from 60 at.% at a deposition temperature of 100°C to 10 at.% at a deposition temperature of 400°C for high-rate deposited films. The corresponding values for low-rate deposited films are 22 at.% (100°C) and 5 at.% (400°C). The microstructure factor also decreases with increasing temperature. The refractive index increases with increasing temperature [468]. Both dark conductivity and photoconductivity show an increase with increasing temperature, and have a maximum at 400–450°C [571] of 4×10^{-6} and $10^{-9}\ \Omega^{-1}\,\mathrm{cm}^{-1}$, respectively.

The deposition rate and the silane consumption decrease sharply by a factor of about four upon adding 60–120 sccm hydrogen to pure argon in the arc, as in shown in Figure 71a. Further addition of hydrogen does not influence these parameters much [569, 572]. Similar effects are observed for the refractive index, microstructure factor, activation energy, and dark conductivity and photoconductivity: adding a small amount of hydrogen leads to a large change, which is saturated upon further hydrogen admixture.

Using ion and threshold ionization mass spectrometry (TIMS), Langmuir probe measurements, and cavity ring down absorption spectroscopy (CRDS), the influence of hydrogen admixture on the film growth precursors has been investigated [567, 569, 573]. For the argon–hydrogen plasma, it was found that electron temperature and ion fluence decrease with increasing hydrogen flow. At low hydrogen flow the electron temperature is 0.3 eV, whereas at high hydrogen flow it is 0.1 eV. The ion fluence is decreased from about 120 sccm at low hydrogen flow to about 4 sccm at high flow.

Adding silane to the argon–hydrogen plasma leads to the formation of large, hydrogen-poor positive silicon ions [567]. With increasing flow the cluster size increases; in fact, the higher silane clusters $\mathrm{Si}_n\mathrm{H}_m^+$ ($n > 5$) are much more abundant than the lower silane clusters. The total ion flux towards the substrate scales linearly with the silane flow, while it decreases with increasing hydrogen flow. The maximum contribution of the cationic clusters to film deposition ranges from 4% to 7% for varying silane flows, and from 5% to 9% for varying hydrogen

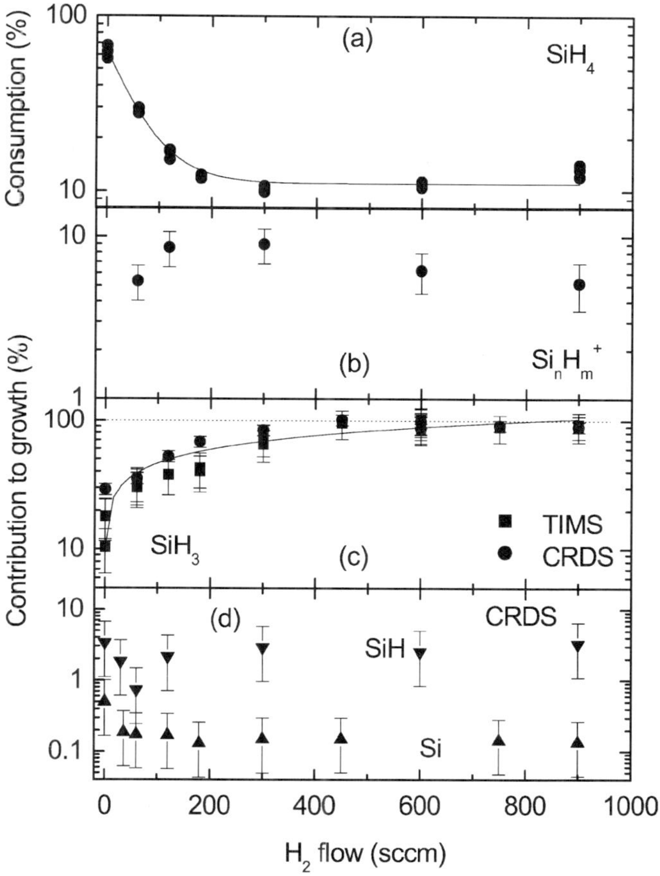

FIG. 71. The influence of hydrogen admixture on (a) the silane consumption, and the contribution of (b) $Si_nH_m^+$, (c) SiH_3, (d) SiH and Si to the deposition, as measured by TIMS and CRDS. (From W. M. M. Kessels, Ph.D. Thesis, Technische Universiteit Eindhoven, Eindhoven, the Netherlands, 2000, with permission.)

flows (Fig. 71b), and is nearly independent of both the silane and the hydrogen flow [567].

As a consequence, the contribution of neutral species to the deposition is more than 90%, and consists mainly of SiH_3. This contribution increases from about 20% at low hydrogen flow to about 90% at intermediate hydrogen flow, and remains constant with further increase (Fig. 71c). A contribution of SiH_2 larger than 0.1% could only be measured at zero hydrogen flow, and amounted to about 5%. Other radicals, viz., SiH and Si, contribute about 2% and 0.2%, respectively, to the deposition (Fig. 71d), irrespective of hydrogen flow.

The optimized conditions for the deposition of good-quality a-Si:H have been found to be: an argon flow of 3300 sccm, a hydrogen flow of 600 sccm, a silane flow of 600 sccm, an arc current of 45 A, an arc voltage of 140 V, an arc pressure of 0.4 bar, a process pressure of 0.2 mbar in the downstream region, and a substrate temperature of 400°C [571, 572]. At this high deposition temperature the hydrogen content is 6–7 at.%. Consequently the (cubic) bandgap is low, 1.51 eV. Also, the microstructure factor still is nonzero, viz. 0.2, and what is more, the dark conductivity is a factor of 10 higher than needed for *device quality* material. The electron drift mobility is somewhat smaller than usual for device quality films, but the hole drift mobility is one order of magnitude larger. The activation energy, Urbach energy, and defect density are 0.75 eV, 50–55 meV, and $10^{16}\,\mathrm{cm}^{-3}$.

1.10.4. Deposition Model

Kessels has summarized the information on the reactive species emanating from the Ar–H_2-operated cascaded arc [571], and has formulated a global reaction scheme. For a hydrogen flow between 0 and 120 sccm, the dissociation of silane is governed by dissociative charge transfer of Ar^+ with SiH_4, which generates SiH_n^+ ($n \leq 3$) ions: $Ar^+ + SiH_4 \rightarrow Ar + SiH_n^+ + \cdots$. Then, as the electron density is high, these reactions are quickly followed by dissociative recombination reactions with electrons, which generates SiH_m^+ ($m \leq 2$) radicals: $SiH_n^+ + e^- \rightarrow SiH_m + \cdots$. These highly reactive radicals react with silane molecules, forming polysilane radicals: $SiH_m + SiH_4 \rightarrow Si_2H_p + \cdots$. When the electron density is decreased as a result of the dissociative recombination reactions, ion–molecule reactions between silane ions and silane will become significant as well. These reactions ($SiH_n^+ + SiH_4 \rightarrow Si_2H_m^+ + \cdots$) lead to sequential ion–SiH_4 reactions ($Si_pH_q^+ + SiH_4 \rightarrow Si_{p+1}H_r^+ + \cdots$), which generate cationic clusters (up to $Si_{10}H_r^+$ have been observed).

At a hydrogen flow larger than 300 sccm the ion fluence from the arc is greatly reduced. Reactions with atomic H emanating from the arc now are more effective. The dissociation of silane is governed by hydrogen abstraction, which generates the SiH_3 radical ($H + SiH_4 \rightarrow H_2 + SiH_3$). The small flow of H^+ from the arc is responsible for a charge exchange reaction with silane, which creates SiH_n^+ ($n \leq 3$), which initiates sequential ion–molecule reactions as above.

The deposition of good-quality a-Si:H using ETP CVD is a result of the dominant presence of the SiH_3 radical under conditions of high hydrogen flow in the cascaded arc. The deposition process itself is more or less the same as for PECVD, and was described in Section 1.7. An important difference between the ETP CVD and the PECVD process is the absence of ions with considerable energy (>1 eV) and the larger number of cationic clusters in the former. Nevertheless, good-quality a-Si:H is deposited at high rates.

1.11. Applications

The first practical device demonstrating the use of a-Si:H as a photovoltaic material was the 2.4%-efficient solar cell reported by Carlson and Wronski [34]. Since then, interest in a-Si:H has been growing rapidly, more or less prompted by the many possible applications. It is used in solar cells and optical sensors based on the photovoltaic effect. Especially the possibility of uniform large-area deposition has been exploited in a-Si:H thin film transistors (TFTs) that are used in controlling active-matrix liquid crystal displays (LCDs). The high photoconductivity and fast photoresponse of a-Si:H is of great importance for use in large-size linear image sensors.

A selection of applications is presented in the following subsections: solar cells, TFTs, light sensors (visible, IR, X-ray), and chemical sensors. Also, light-emitting devices, in particular utilizing erbium incorporation in a-Si:H, are presented. Finally, electrostatic loudspeakers in which an a-Si:H film is incorporated are described. Details of various applications described here, as well as many other applications, can be found in the excellent edited books [4, 5, 11, 13, 574].

1.11.1. SOLAR CELLS

1.11.1.1. Operation Principle

The operation principle of a solar cell is based on charge separation. Photogenerated electrons and holes must be spatially separated in order to contribute to the net current of the device. Charge separation is done by an internal electric field. In crystalline semiconductors this can be achieved by stacking a p-type doped material on an n-type doped material. The internal field at the p–n junction prevents the recombination of photogenerated electrons and holes. Once separated, however, recombination may occur, and is an important loss process in a solar cell [575]. In amorphous semiconductors a p–n junction hardly shows photovoltaic action. Photogenerated electrons and holes cannot diffuse over long distances, as the defect density in p- and n-type doped material is high. Therefore an undoped (intrinsic) layer has to be introduced between the p- and the n-layer, with a thickness smaller than the mean free path of the slower carriers, viz. the holes.

In the presence of an electric field the drift length is the mobility–lifetime product times the electric field: $\lambda_{\mathrm{mfp}} = \mu\tau E$ [576]. With typical values of $\mu\tau$ and E the mean free path usually exceeds by far the thickness of the solar cell, and virtually all photogenerated carriers can be collected. However, under certain operating conditions, field-free regions in the i-layer may exist, and the collection efficiency is decreased because the diffusion lengths of the carriers are much smaller than the thickness of the solar cell [11, 577].

The p and n layers provide the built-in potential but do not contribute to the collection of carriers. Therefore these layers need to be only as thick as the de-

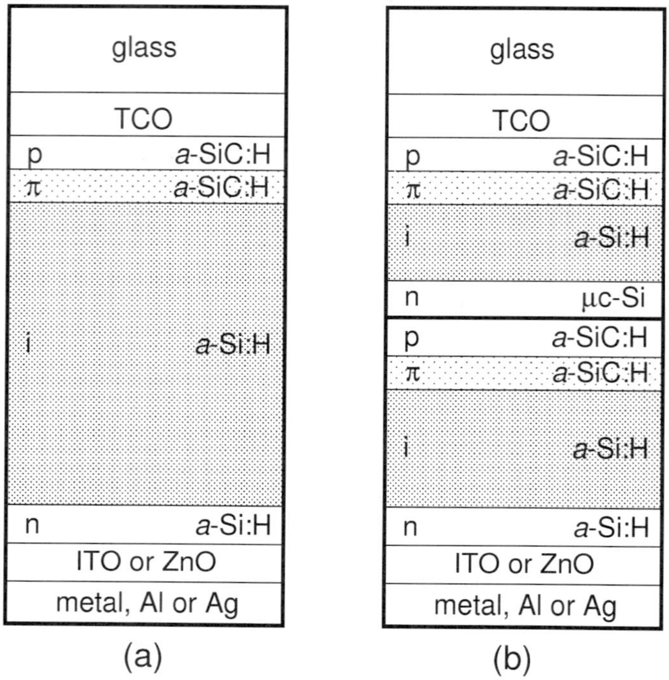

FIG. 72. Schematic cross-section of (a) a single junction *p–i–n a*-Si:H superstrate solar cell and (b) a tandem solar cell structure. (From R. E. I. Schropp and M. Zeman, "Amorphous and Microcrystalline Silicon Solar Cells—Modeling, Materials and Device Technology," Kluwer Academic Publishers, Boston, 1998, with permission.)

pletion layer (5–20 nm), as any additional thickness will unnecessarily decrease the collection efficiency by absorbing light. The intrinsic layer should be as thick as possible to absorb the maximum fraction of photons, but the depletion width at operating conditions of about 1 μm sets an upper limit. Typically, one uses a thickness of 500 nm.

A schematic cross-section of a *p–i–n a*-Si:H solar cell [11] is shown in Figure 72a. In this so-called superstrate configuration (the light is incident from above), the material onto which the solar cell structure is deposited, usually glass, also serves as a window to the cell. In a substrate configuration the carrier onto which the solar cell structure is deposited forms the back side of the solar cell. The carrier usually is stainless steel, but flexible materials such as metal-coated polymer foil (e.g. polyimid) or a very thin metal make the whole structure flexible [11].

The current–voltage (*I–V*) characteristic of a solar cell is given by

$$I(V) = I_{\mathrm{ph}}(V) - I_0\left[\exp\left(\frac{e(V + I(V)R_{\mathrm{s}})}{Ak_BT}\right) - 1\right] - \frac{V + I(V)R_{\mathrm{s}}}{R_{\mathrm{sh}}} \tag{69}$$

where $I(V)$ is the current I at voltage V, $I_{ph}(V)$ the photogenerated current, I_0 the reverse saturation current, A the diode ideality factor, R_s the series resistance, and R_{sh} the shunt resistance parallel to the junction. Usually the current density $J(V)$ rather than the current $I(V)$ is shown, e.g., see Figure 60. An I–V measurement is performed by connecting a variable load or by applying an external voltage [578], in the dark or under illumination with an AM1.5 spectrum. Four parameters determine the performance of a solar cell: the efficiency η, the fill factor FF, the open circuit voltage V_{oc}, and the short circuit current I_{sc}. The efficiency is defined as the ratio of the maximum delivered power to the power of the incident light. The maximum-power point is the point where $I_{max}V_{max}$ has the maximum value. The open circuit voltage V_{oc} and the short circuit current I_{sc} are defined as the values of the voltage and current at $I(V) = 0$ and $V = 0$, respectively. The fill factor FF is defined as the ratio of the maximum power to the product of I_{sc} and V_{oc}. Typical values of these parameters were shown in Figure 60. A characteristic resistance R_{ch} is sometimes defined as V_{oc}/I_{sc} in order to assess the effect of R_s and R_{sh} on the fill factor. When $R_s \ll R_{ch}$ or $R_{sh} \gg R_{ch}$ little effect is seen on the fill factor [575].

The short circuit current is the product of the photon flux $\Phi(\lambda)$ of the incident solar spectrum and the wavelength-dependent spectral response or collection efficiency $Q(\lambda)$ integrated over all wavelengths: $J_{sc} = \int \Phi(\lambda)Q(\lambda)\,d\lambda$ (see Fig. 61b). The collection efficiency is about 80% between 450 and 600 nm, demonstrating that there is little loss due to recombination (the i-layer is of *device quality*). The decreasing collection efficiency at the red side is due to the decreasing absorption coefficient of a-Si:H. In the blue, the decreasing collection efficiency is due to absorption in the p-layer and/or buffer layer.

1.11.1.2. Single-Junction Solar Cells

A schematic cross section of a single-junction p–i–n a-Si:H solar cell in the superstrate configuration was shown in Figure 72a. The front electrode consists of a transparent conductive oxide (TCO). Examples are tin oxide (SnO_2:F), indium tin oxide (ITO, In_2O_3:Sn), and doped zinc oxide (ZnO:Al) [579]. A high transparency in the spectral range of the solar cell is required, as well as a low sheet resistance, to minimize series resistance losses. Further, the contact resistance with the p-layer should be low. The surface of the TCO should be textured to reduce optical reflection and to increase the optical path length by scattering, thereby increasing the absorption in the active i-layer [577, 580, 581]. Shunting paths and pinholes should be avoided by a proper morphology. The TCO should also be PECVD-compatible, i.e., it should be highly resistant to the reducing H ambient during PECVD deposition.

ITO, SnO_2:F, and ZnO:Al share high transparency (>90%), large bandgap (>3.5 eV), and low sheet resistance (<10 $\Omega/\square$). Tin oxide is made by atmospheric CVD, which inherently produces a native texture. Zinc oxide and ITO, both made by sputtering or evaporation, usually are flat. Zinc oxide can be etched by HCl to produce a rough surface [11]. Both tin oxide and zinc oxide are chemically resistant. ITO can be made chemically resistant by adding a thin titanium oxide layer as a protective coating [582]. Zinc oxide has a high contact resistance with doped *a*-Si:H, which can be circumvented by using a microcrystalline doped layer [583].

Absorption in the *p*-layer can be reduced by using an *a*-SiC:H alloy with a bandgap of about 2 eV [584]. Carbon profiling within the *p*-layer further improves the window properties [585]. An intentionally graded *p–i* interface (*buffer layer*) 10 nm in thickness enhances the spectral response in the blue [125, 494, 586], which can be attributed to a reduced interface recombination.

The properties of the *i–n* interface and the *n*-layer are not critical, as long as the conductivity is high. Introducing a TCO layer between *n*-layer and metal back contact enhances the light trapping even further. Absorption mainly is enhanced in the long-wavelength region [587]. The optimum thickness is found to be 70 nm [588].

In the substrate configuration the stainless steel carrier is coated with a Ag–ZnO bilayer in order to enhance the back reflection of the back contact; see Figure 73 [11]. An increase in J_{sc} of about 50% was achieved by Banerjee and Guha [589] by using a textured Ag–ZnO bilayer, which further enhances the optical path length and consequently the absorption. As at this stage no *a*-Si:H has been deposited, there are virtually no restrictions on the process temperature.

After depositing the *n*-layer, a similar *i–n* buffer layer can be deposited as in *p–i–n* cells. Bandgap profiling, i.e., increasing the bandgap towards the top *p*-layer, is much easier for *n–i–p* structures, as here simply decreasing the deposition temperature during deposition automatically yields a larger bandgap [11]. In superstrate *p–i–n* cells, deposition of the *i*-layer is from the *p*- to the *n*-layer. In this case, a decrease in the bandgap is achieved by increasing the temperature during deposition, which may not always be desirable for the already deposited *p*-layer and *p–i* interface. Another advantage is that the most critical top layers are deposited last.

The transparent top contact is deposited last of all, which imposes restrictions on the process temperature. Thermally evaporated ITO and ZnO deposited by metal–organic CVD (MOCVD) are most suitable. At a typical thickness of 70 nm the ITO serves as a good antireflection coating as well. Due to the somewhat high sheet resistance, a metal (Ag) grid is necessary to reduce the series resistance [11].

1.11.1.3. Multiple-Junction Solar Cells

The performance of *a*-Si:H can be improved by stacking two (or more) cells with different optical bandgaps on top of each other; see Figure 72b [11]. In

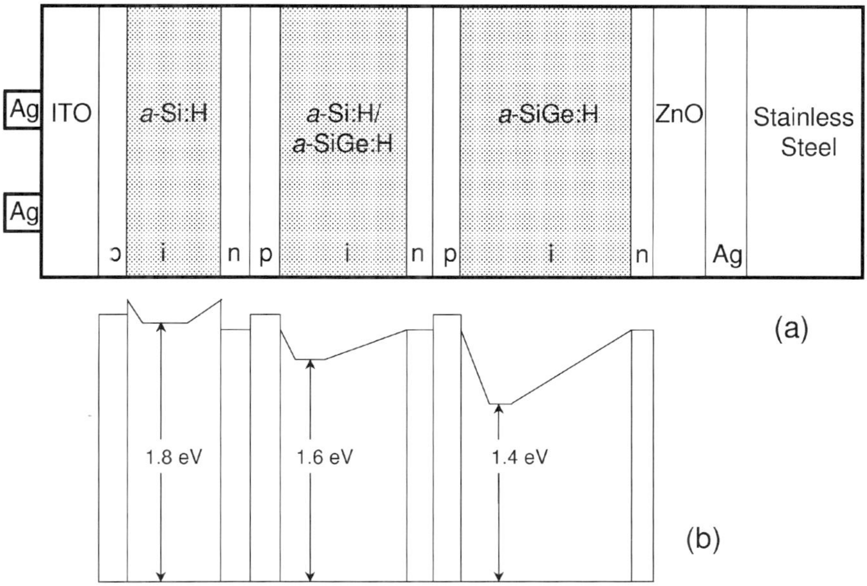

FIG. 73. Schematic cross section of a triple-junction *a*-Si:H substrate solar cell on stainless steel (a), and the corresponding schematic band diagram (b). (From R. E. I. Schropp and M. Zeman, "Amorphous and Microcrystalline Silicon Solar Cells—Modeling, Materials and Device Technology," Kluwer Academic Publishers, Boston, 1998, with permission.)

such a *tandem* solar cell, the larger-bandgap material, of which the top cell is made, collects the high-energy photons, while the lower-energy photons are collected in the lower cell. While theoretical efficiencies of single-junction *a*-Si:H solar cells are around 15%, values above 20% are calculated for multijunction cells. A two-terminal (top and bottom contact) double-junction cell consisting of a top cell with a bandgap of 1.75 eV and a bottom cell with a bandgap of 1.15 eV is predicted to have a conversion efficiency of 21%, while a triple stack of 2.0-, to 1.7-, 1.45-eV cells is predicted to have an efficiency of 24% [590]. In the latter cell carbon alloying of *a*-Si:H would make the top cell, and germanium alloying the bottom cell. However, due to high defect densities present in *a*-SiC:H and *a*-SiGe:H, other ways of obtaining good-quality high- and low-bandgap material have been developed. Dilution of SiH_4 with hydrogen in the regime $0.05 \leq [SiH_4]/([SiH_4] + [H_2]) \leq 0.15$ yields material of better optoelectronic properties than *a*-SiC:H [369], and is therefore used for the top cell (Fig. 72b).

Hydrogen dilution also improves the quality of *a*-SiGe:H alloys [369]. An *a*-SiGe:H bottom cell must be designed to operate at long wavelengths, with relatively weakly absorbed light. The concept of bandgap grading at the *p*–*i* interface, as in *a*-Si:H cells, has been demonstrated to be inappropriate for *a*-SiGe:H cells [591]. Several bandgap profiles were investigated. The best solar cell performance was obtained for a bandgap that first decreased but then increased; see also Zimmer et al. [592].

A triple junction structure based on 1.79-, 1.55-, and 1.39-eV materials [593] was reported to have an initial efficiency of 14.6% and a stabilized (see Section 1.11.1.4) efficiency of 12.1% [154]. In these series-connected structures the open circuit voltage is the sum of the V_{oc}'s of the individual cells (here $V_{oc} = 2.3$ V), while the currents are matched to be equal in all cells: $J_{sc} = 7.56$ mA/cm^2. The fill factor is 0.7 [154]. A schematic cross section of such a triple-junction *p*–*i*–*n* *a*-Si(Ge):H solar cell structure in the substrate configuration is shown in Figure 73 [11], together with the corresponding band diagram. This band diagram shows the intricate bandgap profiling scheme that is needed to obtain these high-efficiency solar cells.

A critical part in the structure of multijunction solar cells is the *n*–*p* junction that separates the individual cells. Here, electrons that are generated in the top cell (Fig. 72b) flow to the *n*–*p* junction, where they must recombine with holes from the bottom cell. As a result of the very high doping levels used, recombination takes place via tunneling. When the recombination rate is not balanced with the supply of carriers, space-charge accumulation will occur, which negatively influences the electric field in the adjacent cell that has the highest generation rate [11]. Often one of the layers of the tunnel junction is microcrystalline [594].

Another important issue in multijunction cells is current matching. The individual currents must exactly balance; otherwise a loss in efficiency occurs. A current mismatch can be easily revealed by measuring the spectral response [595]. If the currents are matched, then the quantum efficiency is flat over a wide range of wavelengths. If one of the cells is limiting the current, then the observed quantum efficiency is not flat, and in fact is the quantum efficiency of the current-limiting cell.

Empirical optimization of the thicknesses of the individual cells within the structure is to be combined with computer modeling. A comprehensive model is required in which an optical and an electrical model of *a*-Si:H are integrated [11]. Since such a model contains about 100 input parameters for a single-junction solar cell, a careful calibration procedure is needed to extract input parameters from measured layer properties [596, 597]. Current matching in tandem cells has been investigated using a model that not only well describes recombination via tunneling but also takes into account the optical enhancement due to the textured TCO [598]. It was found that for an *a*-Si:H–*a*-Si:H stack the optimum thickness of the bottom cell amounted to 300 nm, while for an *a*-Si:H–*a*-SiGe:H stack it amounted to 150 nm. The optimum thickness of the top cell was 50 and 60 nm for the *a*-Si:H–*a*-Si:H and the *a*-Si:H–*a*-SiGe:H stack, respectively [11, 598].

1.11.1.4. Stability

Illumination of solar cells causes a reduction of efficiency and fill factor, as a result of light-induced creation of defects (Staebler–Wronski effect, Section 1.1.2.5). This reduction is halted after several hundred hours of illumination. The reduction is correlated with solar cell thickness. A large intrinsic layer thickness leads to a large reduction of efficiency and fill factor compared to a small intrinsic layer thickness. The solar cell properties can be completely recovered by annealing at about 150°C. The open circuit voltage and short circuit current decrease only slightly.

Material properties cannot always be correlated with degradation behavior [493, 495, 599, 600]. Lee et al. [600] have shown the degradation kinetics of cells in which the intrinsic layer was deposited by highly diluting the silane. A clear correlation was observed between the decrease in efficiency and the increase of $\mu\tau$. Another example that demonstrates a correlation between degradation behavior and material quality was shown in Figure 62. Here the normalized cell efficiency as a function of illumination time was depicted for solar cells, where the intrinsic layer is deposited at different discharge power levels [493]. At the lower power levels the efficiency stabilizes at about 60% of its initial value, whereas at the higher power levels it stabilizes at 40%. This has been attributed to the fact that at the higher power densities the density of the material is lower and the hydrogen content is higher.

As a result of the creation of defects, trapping of electrons or holes is enhanced. Thus, the created defects reduce the product $\mu\tau$, which depresses charge collection and solar cell efficiency. The charge collection length, i.e., the average distance that a carrier travels before it is trapped, is defined as $d_c = \mu\tau V_{\mathrm{bi}}/d$, with V_{bi} the built-in voltage (typically around 1.2 V [3]) and d the thickness of the cell. An empirical relationship between fill factor and charge collection length d_c was reported by Faughnan and Crandall [601]:

$$\mathrm{FF} = 0.35 + 0.15\ln\left(\frac{d_c}{d}\right) \tag{70}$$

Smith et al. [602] have derived the dependence of FF on illumination time. They combined the empirical relationship between FF and d_c [Eq. (70)] with the time dependence of the defect density ($N_{\mathrm{db}}(t) \propto G^{0.6}t^{1/3}$ [89]) and the relation between $\mu\tau$ and defect density ($N_{\mathrm{db}} \propto 1/\mu\tau$ [3]). They arrived at

$$\mathrm{FF}(t) = \mathrm{FF}_i - \frac{k}{3}\log\left(\frac{t}{t_i}\right) \tag{71}$$

where FF_i is the initial FF, k a kinetic constant for degradation, and t_i the initial time. Their data were fitted with $\mathrm{FF}_i = 0.68$ and $k = 0.25$.

Catalano et al. [603] have introduced the device thickness into an empirical time dependence of the efficiency:

$$\eta = \eta_i \left[1.1 - K \log\left(\frac{t}{t_i} \right) \right] \tag{72}$$

where η_i is the initial efficiency, and K a rate constant. The removal of defects was found to be a temperature-activated process [604] represented by $K = K' \exp(E_A/k_B T)$, where K' is a constant increasing with the thickness ($\propto d^{0.54}$), and E_A an activation energy (0.2 eV [603]). At 35°C Eq. (72) can be rewritten as [603]

$$\eta = \eta_i \left[1.1 - 0.165 d^{0.54} \log\left(\frac{t}{t_i} \right) \right] \tag{73}$$

From the increase of the rate constant with thickness it is clear that thick cells will degrade deeper than thin cells. However, because the initial efficiency increases with thickness, an optimum thickness of 200–300 nm is found [577].

The degradation of solar cell properties can be circumvented by proper device design [11]. The parameter of interest here is the electric field profile after degradation, which should be optimized for carrier collection. A thinner intrinsic layer combined with enhanced light confinement leads to a higher electric field after degradation. Also, bandgap profiling may assist the carrier transport in the low-field region that is present in the intrinsic layer after degradation [605]. Further hydrogen dilution during deposition of the intrinsic layer has been reported to improve the stability [606].

Individual cells in multijunction cells are more stable, due to their reduced thickness compared to single-junction cells. Moreover, as the amount of light that is absorbed in the bottom cell is reduced and as the degradation rate is reduced at lower light levels, the stability is improved [606]. Therefore the multijunction cell is expected to be more stable than a single-junction cell of the same thickness, which indeed has been demonstrated [577, 606–608].

Outdoor operating temperatures of solar cells are around 60°C. At this temperature significant annealing of defects occurs [606], and the stability is better than at 25°C. Interestingly, seasonal measurements of the performance of solar cells show that the efficiency is higher in summer than in winter [609], which was attributed to annealing of defects [610]. However, it was subsequently reported that these seasonal effects are due to the spectral differences in summer and winter, rather than the increased operating temperature in summer [611, 612].

The research effort in many laboratories around the world on the optimization of laboratory-scale (1-cm^2 area) cells has led to the development of large-area (1 m^2) solar cell modules with an efficiency exceeding 10% [613, 614]. In a module individual cells can be connected in series or parallel, depending on the desired output voltage and current. Details on this topic can be found elsewhere, e.g., [11, 157, 577, 607, 615, 616].

1.11.2. THIN FILM TRANSISTORS

1.11.2.1. Operation Principle

The most common *a*-Si:H TFT structure is the so-called inverted staggered transistor structure [40], in which silicon nitride is used as the gate insulator. A schematic cross section is shown in Figure 74. The structure comprises an *a*-Si:H channel, a gate dielectric (SiN_x), and source, drain, and gate contacts.

The operation principle of these TFTs is identical to that of the metal–oxide–semiconductor field-effect transistor (MOSFET) [617, 618]. When a positive voltage V_G is applied to the gate, electrons are accumulated in the *a*-Si:H. At small voltages these electrons will be localized in the deep states of the *a*-Si:H. The conduction and valence bands at the SiN_x–*a*-Si:H interface bend down, and the Fermi level shifts upward. Above a certain threshold voltage V_{th} a constant proportion of the electrons will be mobile, and the conductivity is increased linearly with $V_G - V_{th}$. As a result the transistor switches on, and a current flows from source to drain. The source–drain current I_{SD} can be expressed as [619]

$$I_{SD} = \mu_{FE} C_G \frac{W}{L}\left[(V_G - V_{th}) - \frac{V_D}{2}\right] V_D \tag{74}$$

where μ_{FE} is the effective carrier or field-effect mobility, C_G the gate capacitance, V_D the drain voltage, and W and L the channel width and length, respectively. Saturation of the source–drain current occurs when $dJ_{SD}/dV_D = 0$, and it is then expressed as

$$I_{SD,sat} = \mu_{FE} C_G \frac{W}{2L}(V_G - V_{th})^2 \tag{75}$$

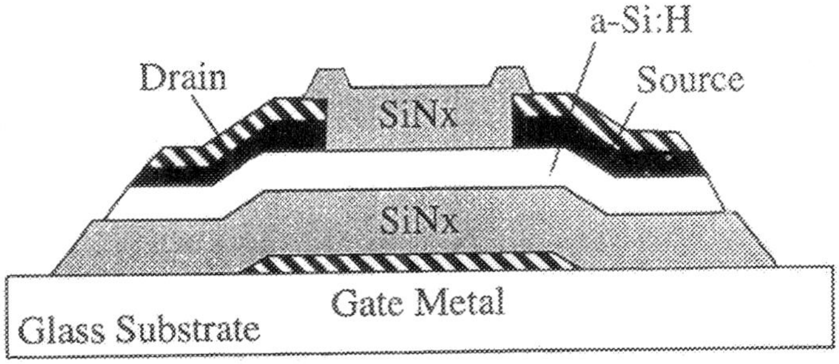

FIG. 74. Schematic cross section of an inverted staggered TFT structure.

The maximum current in the on state is about 10^{-4} A, while the current at $V_G = 0$ is 10^{-11} A, which shows the large dynamic range. The threshold voltage is around 1 V, typically, and μ_{FE} can amount to 1 $cm^2\ V^{-1}\ s^{-1}$ [618].

At the interface of the nitride ($E_g = 5.3$ eV) and the a-Si:H the conduction and valence band line up. This results in band offsets. These offsets have been determined experimentally: the conduction band offset is 2.2 eV, and the valence band offset 1.2 eV [620]. At the interface a small electron accumulation layer is present under zero gate voltage, due to the presence of interface states. As a result, band bending occurs. The voltage at which the bands are flat (the flat-band voltage V_{FB}) is slightly negative.

1.11.2.2. Stability

TFTs are not perfectly stable. After prolonged application of a gate voltage, the threshold voltage starts to shift. Two mechanisms can account for this [621]. One is the charge trapping of electrons in the nitride, where the silicon dangling bond is the dominant electron trap. When electrons are transferred from the a-Si:H to the nitride, they can be trapped close to the interface, and a layer of fixed charge results. This fixed charge Q_F reduces the charge induced in the a-Si:H, and the transfer characteristic shifts by an amount Q_F/C_G. In the second mechanism deep defect states are created by the breaking of strained or weak Si—Si bonds. Electrons are trapped in these states and act as fixed negative charges, which results in band bending. This gives a decrease in source–drain current or a positive shift of V_{th} of several volts. It has been found that the time dependence of the threshold current follows a stretched exponential behavior [622]:

$$\Delta V_{th} = (V_{th}(\infty) - V_{th}(0))\left[\left(\exp\left(-\frac{t}{t_0}\right)^{\beta}\right) - 1\right] \tag{76}$$

where $V_{th}(\infty)$ and $V_{th}(0)$ are the threshold voltages at infinity and at $t = 0$, respectively. The factor β has been related to the time-dependent ($\propto t^{\beta-1}$) hydrogen diffusion coefficient in a-Si:H [623, 624].

HWCVD-deposited a-Si:H layers incorporated in inverted–staggered TFTs have been reported to result in stable behavior of the TFT [538, 539]. The threshold voltage does not shift upon prolonged application of a gate voltage. It has been suggested that not only the H content but also the H bonding structure is responsible for the difference in metastable character for PECVD and HWCVD layers [94, 96, 97, 625].

Another disadvantageous phenomenon in TFTs is the photoconductivity of a-Si:H [626]. Electrons and holes are photogenerated and recombine at the back surface (gate insulator). The photocurrent reduces the on/off ratio of the TFT. Illumination, however, cannot always be avoided, e.g., in active matrix displays. A way of circumventing this is to make the a-Si:H as thin as possible.

1.11.2.3. Fabrication

The TFTs are made on transparent glass substrates, onto which gate electrodes are patterned. Typically, the gate electrode is made of chromium. This substrate is introduced in a PECVD reactor, in which silane and ammonia are used for plasma deposition of SiN_x as the gate material. After subsequent deposition of the *a*-Si:H active layer and the heavily doped *n*-type *a*-Si:H for the contacts, the devices are taken out of the reactor. Cr contacts are evaporated on top of the structure. The transistor channel is then defined by etching away the top metal and *n*-type *a*-Si:H. Special care must be taken in that the etchant used for the *n*-type *a*-Si:H also etches the intrinsic *a*-Si:H. Finally the top passivation SiN_x is deposited in a separate run. This passivation layer is needed to protect the TFT during additional processing steps.

The thickness of the active layer is about 100–300 nm, while the source–drain distance (channel length) amounts to a few micrometers. The channel length is determined by the current requirements and usually exceeds 10 μm. Other manufacturing schemes as well as alternative structures are described elsewhere [619, 621]. Technology developments for the next generation TFTs that are to be used for high-resolution displays have been summarized by Katayama [627].

1.11.2.4. Application in Active Matrix Displays

One of the most important applications of TFTs is in active matrix addressing of liquid crystal displays (AMLCD) [619, 621]. The light transmission of a liquid crystal between two glass plates is controlled by the voltage on the pixel electrode. Nematic liquid crystals consist of rod-shaped molecules that align themselves parallel to an applied electrical field. When no field is present, they align themselves according to the surfaces in between which they are contained. By orienting the crystals, a change in polarization of the light passing through the crystals can be achieved. In this way, the crystals act as light switches. The voltage on the pixel electrode is controlled by the transistors. These act as switches that pass the voltage on their source to the drain, which is, the pixel electrode. When the gate is switched off, the pixel electrode is electrically isolated from the source, and the information remains on the pixel electrode. An $n \times m$ matrix of source and gate lines is able to address $n \times m$ pixels. This allows for high-definition large-area television displays. As an example, a 640×480 pixel display with three subpixels for each color consists of over one million transistors.

Active matrix addressing can also be used in printer heads [628] and linear sensor arrays [629]. *a*-Si:H TFTs are used to address a linear array of output or input transducers, respectively. In the linear sensor array, *a*-Si:H photodiodes are used.

1.11.3. LIGHT SENSORS

1.11.3.1. Linear Arrays

Linear arrays of a-Si:H photodiodes are widely used in optical page scanning applications such as fax machines and document scanners [630]. The large linear dimension of the array (as wide as the page to be scanned) allows a much simpler design. No optics are needed for image size reduction, which is required when a CCD camera is used as the detecting element. The matrix addressing is similar to that in AMLCD technology, but confined to one dimension. For each pixel a p–i–n photodiode is connected to a transistor. During illumination of the pixel, charge is transferred to the bottom electrode, and accumulates when the TFT is off. Upon switching on of the TFT, the charge will flow out and can be read by external electronic circuitry. In this way, the sensor integrates the signal during the time that it is not addressed. Readout times are very short in comparison with accumulation times.

1.11.3.2. Photoreceptors

a-Si:H photoconductors can be used as the light-sensitive components in electrophotographic or xerographic processes [39, 631, 632]. In contrast with conventional materials such as As_2Se_3, CdS, or ZnO, a-Si:H is nontoxic and provides a hard surface. Typically, a three-layer structure is used on top of a glass substrate: n-a-Si:H/i-a-Si:H/a-SiC:H.

The electrophotographic process consists of a number of steps [633]. First, the photoconductor is charged in the dark via a corona discharge. Then the photoconductor is illuminated by projecting the image to be copied onto the surface. In the areas that are exposed to light the material becomes conductive and charge flows to the substrate. The nonilluminated parts still contain charge. The photoconductor then is moved under a bias electrode, and an electric field is present in the gap between electrode and charged surface. Now, liquid toner (ink) is introduced in this gap, and the pigment particles from the toner drift to the surface of the photoconductor due to the electrostatic force. The last step is the transfer of the toner to paper, where it is fused by heat or pressure to make the image permanent.

A charged photoconductor discharges slowly in the dark, as is evident from the small decrease in time of the surface voltage. The discharge during illumination is much faster. When the discharge is not complete, a small residual voltage remains on the surface. A low dark current is required, which limits the dark photoconductivity of a-Si:H to $10^{-12}\ \Omega^{-1}\,\mathrm{cm}^{-1}$. The residual voltage can be minimized by optimizing the $\mu\tau$ of the material.

The purpose of the n-type doped a-Si:H layer is to prevent injection of charge from the substrate into the photoconductor. Thus it serves as a blocking layer. Injection of surface charge into the photoconductor is prevented by the surface blocking layer. The a-SiC:H is of low quality, with a high midgap density of states. Surface charge will be trapped in these midgap states.

The corona discharges produces oxygen ions and ozone, which may react with the photoconductor [634]. As a means to circumvent possible degradation of the surface layer, an extra, protective thin layer was proposed, with high carbon content [101, 635, 636]. This would reduce silicon–oxygen reactions at the surface. Excellent electrophotographic characteristics have been obtained with a thin device comprising a 0.1-μm-thick n-type a-Si:H layer, a 1.0-μm intrinsic a-Si:H layer, a 0.1-μm undoped a-$SiC_{0.1}$:H layer, and a 0.014-μm undoped a-$SiC_{0.3}$:H layer [101].

1.11.3.3. Position Sensor

An a-Si:H-based position sensor consists of an intrinsic film sandwiched between two transparent conductive electrodes [637]. Two line contacts on the top are perpendicular to two on the bottom. When a light spot is incident on the device, carriers are generated, and a photocurrent flows to the contacts. The contacts form resistive dividers, so that from the ratio of the photocurrents the lateral position relative to the top or bottom contacts can be determined. The top contacts give the x-position, and the bottom contacts the y-position.

A p–i–n diode has been used on glass and on polyimid as a position-sensitive detector [638, 639]. The position of an incident light spot is measured by means of the lateral photovoltage.

1.11.3.4. Color Sensors

Similar to multijunction solar cells are color detectors, which have been designed to detect two [640, 641] or three [642–644] colors. A so-called adjustable threshold color detector (ATCD) consists of an a-Si:H–a-SiC:H p–i–n–i–p–i–n stack [642]. This ATCD discriminates between the three fundamental colors, red, green, and blue. As a result of proper thickness and bandgap design, the blue is absorbed in the top p–i–n junction, the green in the middle n–i–n junction, and the red in the bottom n–i–p junction. A change in polarity and value of the voltage selects the color that can be detected. The operation of the device is based on adjusting the photocurrent detection in the middle n–i–n junction.

A TCO/p–i–n–i–p/TCO/p–i–n/metal stack has been designed as a three-color sensor [643, 644]. An extra contact is made to the middle TCO. With appropriate bandgaps the peak detection is at 450, 530, and 635 nm for the blue, green, and red, respectively.

An active matrix of two-color a-Si:H photodetectors has been reported [645], where a n–i–p–i–n switching device is stacked on a two-color p–i–n–i–p structure.

1.11.3.5. IR Sensor

The absorption values of *a*-Si:H in the IR are much higher than of *c*-Si, due to transitions between extended states and deep levels. The IR sensitivity increases with increasing defect density. However, as the collection efficiency decreases drastically with increasing defect density, photodiodes cannot be used. Another approach is based on the change in photocapacitance due to IR radiation. A structure similar to a *p–i–n* *a*-Si:H solar cell has been used as an infrared detector [646]. The intrinsic layer in fact is a so-called *microcompensated* layer with very low concentrations of boron and phosphorus, but with a high defect density. In this layer the IR sensitivity is enhanced, due to transitions between extended states and hole traps. The increase in trapped charge modifies the electric field in the device, which is observed as a change in capacitance. A high responsivity is observed between 800 and 1400 nm, as well as between 3500 and 4500 nm.

1.11.3.6. X-Ray Sensor

High-energy radation can be imaged with *a*-Si:H, either directly or via a converter [3]. A thick film is required for direct detection, due to the weak interaction of the radiation with the material. A converter usually is a phosphor, which emits in the visible, and thin *a*-Si:H films are needed. X-rays with an energy up to 100 keV eject the electrons from the inner atomic core levels to high levels in the conduction band. The emitted electrons create electron–hole pairs due to ionization. These pairs can be detected in the same way as in *p–i–n* photodiodes.

X-ray imagers consist of a phosphor in direct contact with the surface of an array of *a*-Si:H photodiodes. The device is a matrix-addressed array, in which each imaging pixel consists of a photodiode and a TFT [647–649]. These X-ray imagers are very suitable for static and dynamic imaging in medical diagnosis.

1.11.4. CHEMICAL SENSORS

1.11.4.1. pH ISFET

An ion-sensitive field effect transistor (ISFET) with a reference electrode is similar to a MOSFET, with the gate exposed to an electrolyte. Selected ion concentrations can be measured. The threshold voltage of an ISFET depends on the potential of the electrolyte–insulator surface, which in turn is dependent on temperature and on the pH of the electrolyte. The threshold voltage has been found to shift down linearly by about 0.5 V on going from pH 7 to pH 1 [650]. Also, a drift of the gate voltage in time is found; it is small (1 mV/h) for low pH and large (7 mV/h) for high pH [651]. This is related to a small temporal change in capacitance of the insulator [652].

1.11.4.2. Hydrogen Sensor

A metal–insulator–semiconductor (MIS) device where palladium is used as the metal is sensitive to hydrogen. This sensitivity is a result of dissociation of hydrogen on the Pd surface, which induces an interface dipole layer that changes the barrier height and dark diode current (reverse bias current) [653]. Using a MIS structure with *a*-Si:H, i.e., glass/Cr/*n*-*a*-Si:H/*i*-*a*-Si:H/SiO_x/Pd, Fortunato et al. [654] reported that the presence of 400 ppm of hydrogen leads to an increase of the reverse current of this device by three orders of magnitude. When these devices are illuminated during operation, 400 ppm of hydrogen induces a shift of the open circuit voltage to lower values. Hence, these photochemical sensors can operate as a gas sensor either in the dark or under illumination.

1.11.5. OTHER APPLICATIONS

1.11.5.1. Loudspeakers

The capability of depositing *a*-Si:H uniformly over large areas has been of special interest for application in electrostatic loudspeakers, where the *a*-Si:H layer is suitable for the retention layer on the vibrating foils in the loudspeaker. One aims at obtaining a large-area thin film that can be either positivily or negatively charged by applying a high voltage, but at the same time can hold large surface charge densities that do not displace laterally. Here, the high photosensitivity is not important, and in fact should be quenched for proper operation.

Electroacoustic transducers can be divided into electrodynamic and electrostatic transducers. The most commonly used loudspeakers convert electrical energy to acoustical energy by electrodynamical means: driving an electromagnetic coil that activates a diaphragm. At low frequencies, however, the diaphragm acts as a high-pass filter, with a cutoff frequency that depends on the diameter of the diaphragm. The mechanical construction of large-area loudspeakers adversely influences the amplitude and phase behavior, especially at high frequencies. These drawbacks usually are overcome by using different-size loudspeakers for different frequency ranges.

The electrostatic loudspeaker is used for high-end sound reproduction. The operation principle of an electrostatic loudspeaker is illustrated in Figure 75. Essentially, the electrostatic loudspeaker is a capacitor with air as the dielectric. One of the electrodes is a thin, electrically conducting foil (a in Fig. 75), and the other is a perforated flat fixed plate (b in Fig. 75) that also is conductive. The electrodes are oppositely charged by a high voltage, 5 kV. Sound is produced by superimposing audio-frequency voltages that drive the movement of the foil [655].

The following issues need to be addressed. The amplitude of the foil motion should be very small to avoid nonlinear distortion. A large-area foil therefore is used. Further, the electrical force between the electrodes can be as large as 20

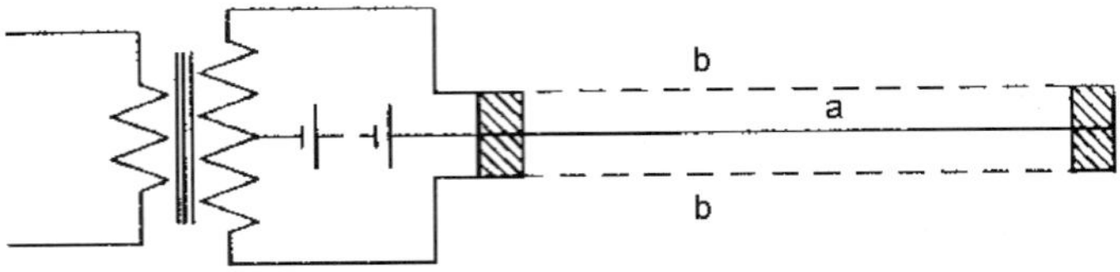

FIG. 75. Schematic drawing of an electrostatic loudspeaker: a, thin flexible foil coated with a-Si:H (or graphite); b, perforated flat fixed plates.

kV/cm, which may lead to electrical breakthrough. Another issue is the presence of an electric field component over the surface of the foil, as a result of clamping the foil to the edges of the support. This field will vary with the amplitude of the foils, which leads to charge displacements and associated power dissipation. This effect especially may be present at low frequencies, where amplitudes are large.

The thin film should meet the following requirements. Areas in excess of 1 m^2 need to be deposited continuously, reproducibly, and homogeneously. The use of plastic foils sets a limit to the deposition temperature. The thin film should be capable of holding large surface charge concentrations, and have a large lateral resistivity. It should be mechanically stable, capable of accommodating mechanical deformations, and resistant to humidity. Usually, the foils are coated with a graphite layer as the semiconducting thin film. However, these layers are frequently unstable, and suffer from charge discplacement effects, which eventually lead to electrical breakdown [656].

Prototype electrostatic loudspeakers where the graphite is replaced by a-Si:H have been made, where a Mylar foil (area 10×10 cm^2, thickness 6 μm) is used [657]. Deposition of the a-Si:H layer was carried out in the ASTER deposition system. Uniform deposition (standard deviation of thickness, 1.5%) was achieved by diluting the SiH_4 with H_2 with $SiH_4 : H_2 = 1 : 2$ [370]. The deposition was at room temperature. The hydrogen content amounted to 18 at.%, and the bandgap was 1.81 eV. The dark conductivity and AM1.5 photoconductivity were 7.5×10^{-9} and 1.8×10^{-8} Ω^{-1} cm^{-1}, respectively. In practice the film would not be exposed to light.

The frequency reponse characteristics of the foil with a-Si:H were measured in the range from 2.9 Hz to 5.46 kHz, by making use of generated pink noise with a continuous energy distribution. In Figure 76 the frequency response characteristics of the a-Si:H-coated foil are compared with those of a conventional graphite-coated foil of the same area. It can be concluded from this figure that the output of the a-Si:H-coated loudspeaker is lower than that of the graphite-coated one from 60 Hz upwards. In spite of the lower efficiency in this range, the a-Si:H-coated loadspeaker outperforms the graphite-coated one in the low range (10–60 Hz). The response is more extended towards low frequencies, and the cut-

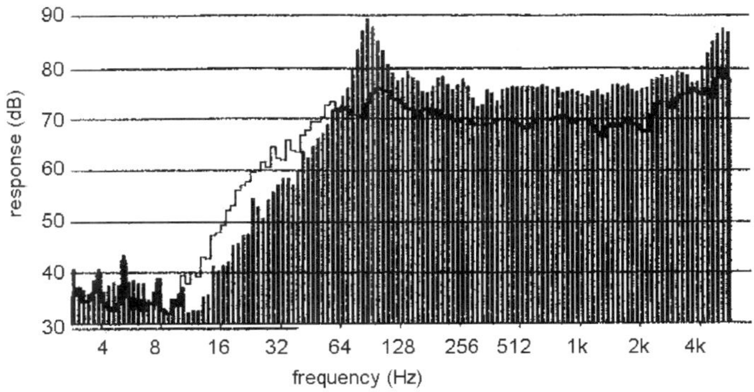

FIG. 76. Frequency response characteristics of a conventional, graphite-coated diaphragm (bar graph) and an *a*-Si:H-coated diaphragm (line graph).

off behavior (dB/octave) also seems improved. The frequency response is stable, even after prolonged application of low-frequency signals. The improvement was attributed to the reduced power dissipation at low frequencies due to the absence of charge displacement currents at the surface of the *a*-Si:H coated foil [657].

1.11.5.2. Erbium in a-Si:H

Erbium, one of the rare-earth materials, has been very important in the development of optical communication technology in the past decades. The trivalent erbium ion (Er^{3+}) has an incomplete $4f$ shell, which is shielded from the outside by $5s$ and $5p$ shells. Due to this, in erbium-doped materials a sharp optical transition from the first excited state to the ground state of Er^{3+} is observed, which occurs at an energy of 0.8 eV. The corresponding wavelength is 1.54 μm, which is of great importance in that standard silica-based optical fibers have the highest transparency at this wavelength. Indeed, Er-doped silica optical fibers have been demonstrated to operate as optical amplifiers at 1.54 μm [658, 659]. Planar amplifiers in which Er-doped channel waveguides are manufactured on a planar substrate have also been demonstrated [660]. These are of particular importance in that they can be integrated with other waveguide devices on a single chip. If efficient light emission from Er-doped silicon were possible, integration of optical and electrical functions on a single silicon chip would be within reach.

Due to the small emission and absorption cross sections of Er^{3+}, a high Er density is needed to reach reasonable values of optical gain. Typically Er densities are between 0.1% and 1.0% (10^{19}–10^{20} Er/cm^3). These values are far beyond the equilibrium solubility limits of Er in silicon. Therefore, nonequilibrium methods have to be used, such as ion implantation. Er implantation in crystalline silicon leads to amorphization, and additional annealing (600°C) is required to recrystallize the silicon. Optical activation of the Er may even require subsequent annealing at higher temperatures. Impurities such as oxygen or carbon have been found to enhance the luminescence. More can be found in the reviews on Er in silicon that have been published by Polman [661] and Priolo et al. [662, 663].

In this sub-subsection, the Er doping of amorphous silicon is discussed. The problem of limited solubility of Er in crystalline silicon has been circumvented. However, the electrical properties of pure *a*-Si are poor compared to *c*-Si. Therefore, hydrogenated amorphous silicon is much more interesting. Besides, the possibility of depositing *a*-Si:H directly on substrates, i.e., optical materials, would make integration possible. Both low-pressure chemical vapor deposition (LPCVD) [664] and PECVD [665, 666] have been used to make the *a*-Si:H into which Er is implanted. In both methods oxygen is intentionally added to the material, to enhance the luminescence.

The *a*-Si:H is deposited on Si(100) by LPCVD from SiH_4 and N_2O at 620°C, with hydrogen and oxygen contents of 10% and 31%, respectively. The *a*-Si:H layer was 340 nm thick. Erbium was implanted at 500 keV to a dose of 1×10^{15} Er/cm^2. The peak concentration of Er at a depth of 150 nm was 0.2%. Upon annealing at 400°C, room-temperature photoluminescence is observed [664], with the characteristic Er^{3+} peak at 1.54 μm. Temperature-dependent luminescence measurements in the range from 77 to 300 K show quenching of the peak luminescence by only a factor of 3. In the case of to Er-implanted crystalline silicon, this quenching is 10–100 times larger. Moreover, the luminescence intensity for the amorphous material is higher over the whole range of measured temperatures.

Luminescence lifetime measurements reveal a double-exponential decay, with lifetimes around 160 and 800 μs, independent of temperature in the range of 77–300 K. This behavior differs from the lifetime quenching as observed for oxygen- or nitrogen-codoped *c*-Si, where the lifetime is quenched by one to nearly three orders of magnitude [667, 668]. This has been attributed to nonradiative de-excitation of excited Er^{3+} that occurs at higher temperatures and depends on the bandgap of the material [669]: the larger the bandgap, the less quenching should be observed. Indeed, the large bandgap of the LPCVD *a*-Si:H (about 2 eV) seems to inhibit quenching.

With PECVD *a*-Si:H is deposited on a Corning glass substrate from SiH_4 at 230°C, employing the PASTA deposition system described by Madan et al. [159]. The hydrogen and oxygen contents were of 10% and 0.3%, respectively. The presence of this background concentration of oxygen probably was due to postponed maintenance; however, it was advantageous for this particular experiment. The *a*-Si:H layer was 250 nm thick. Erbium was implanted at 125 keV to a dose of 4×10^{14} Er/cm^2. The peak concentration of Er at a depth of 35 nm was 0.2%. In some samples additional oxygen was implanted at 25 keV to a dose of 7×10^{15} Er/cm^2, which resulted in an oxygen peak concentration of 1.7%, which over-

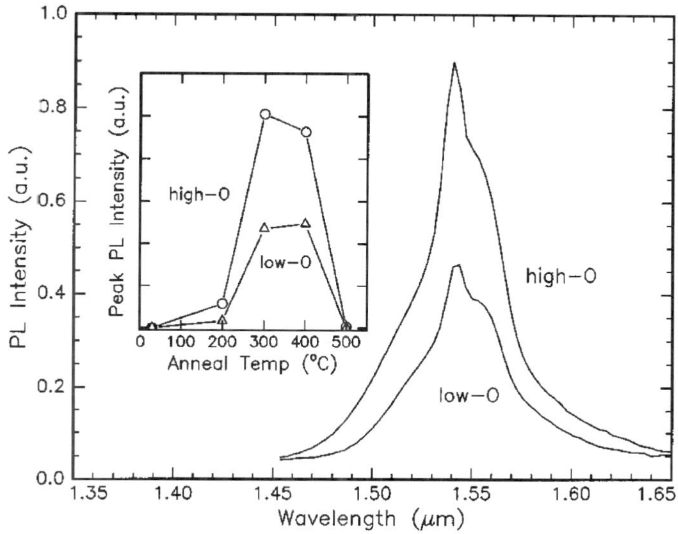

FIG. 77. Room-temperature photoluminescence spectra of Er-implanted PECVD a-Si:H, annealed at 400°C. The implantation energy and dose were 125 keV and 4×10^{14} Er/cm^2, respectively, which resulted in peak concentration of 0.2 at.%. "Low-O" and "high-O" denote a peak oxygen concentration in a-Si:H of 0.3 and 1.3 at.%, respectively. The inset shows the 1.54-μm peak intensity as a function of annealing temperature, for both oxygen concentrations. [From J. H. Shin, R. Serna, G. N. van den Hoven, A. Polman, W. G. J. H. M. van Sark, and A. M. Vredenberg, *Appl. Phys. Lett.* 68, 697 (1996), © 1996, American Institute of Physics, with permission.]

lapped the Er depth profile. Upon annealing at 400°C, room-temperature photoluminescence is observed [665, 666], with the characteristic Er^{3+} peak at 1.54 μm, as shown in Figure 77. The inset shows the behavior of the peak intensity as a function of annealing temperature. Temperatures between 300 and 400°C clearly are optimum annealing temperatures. Both the as-implanted samples and the samples annealed at 500°C show no erbium-related luminescence. In order to remove irradiation-induced effects, high annealing temperatures are needed; however, that increases the outdiffusion of hydrogen.

Temperature-dependent luminescence measurements in the range from 77 to 300 K show quenching of the peak luminescence by a factor of about 15. Similar behavior is observed in the lifetime quenching [665, 666]. As the band gap of the PECVD a-Si:H is about 1.6 eV, nonradiative deexcitation of Er may occur at elevated temperatures. The amount of quenching lies in between that of c-Si and LPCVD a-Si:H, just like the bandgap.

Electroluminescent devices were made to demonstrate the possible application of these a-Si:H materials. Er-doped p–n diodes in c-Si show electroluminescence, both in forward and reverse bias [670–672]. Under forward bias the electroluminescence signal is attributed to electron–hole recombination, which results in excitation of the Er^{3+}. Under reverse bias, in particular beyond the Zener breakdown, the erbium is excited by impact excitation by hot electrons that are accelerated across the junction.

Incorporated in a device, the LPCVD *a*-Si:H material shows electroluminescence only in reverse bias [673]. The mechanism is similar to the one described for *c*-Si. The PECVD *a*-Si:H material was incorporated in a *p–i–n* solar cell structure, with a thickness of the intrinsic layer of 500 nm (see Section 1.11.1). Oxygen was coimplanted at 80 keV (3.2×10^{15} O/cm^2) and at 120 keV (5.5×10^{15} O/cm^2), which resulted in a roughly constant oxygen concentration of 1.0% in the Er projected range in the middle of the intrinsic *a*-Si:H layer. Electroluminescence is observed under forward bias [674].

Incorporation of Er *during* deposition of *a*-Si:H has also been achieved using molecular beam epitaxy [675], magnetron-assisted decomposition of silane [676], PECVD with a metal–organic source of solid erbium (tris(2,2,6,6-tetramethyl-2,5-heptadionato)Er(III) [677, 678] and tris(hexafluoroacetylacetate)Er(III)·1,2-dimethoxyethane [679]), electron cyclotron resonance PECVD in combination with sputtering of erbium [680], and catalytic CVD combined with pulsed laser ablation [681]. Explanations of the quenching behavior seem to be dependent on the sample preparation technique; recently, however, a comprehensive energy transfer model based on a Förster mechanism (resonant dipole coupling [682]) has been proposed [683].

1.11.5.3. Miscellaneous

Photodiode arrays have been used as retinal implants [684]. These arrays of *p–i–n* diodes are fabricated on a thin titanium layer bonded to a glass plate. The total thickness of this flexible structure is 1.5 μm. The microphotodiode array (MPDA) is used to replace photoreceptors (rods and cones) that have become defective due to disease.

Near-field optical microscopy (NSOM) [685, 686] has been used as a tool for nanolithography on the submicron length scale. Here, the fiber optic probe is used as a light source to expose a resist by scanning the surface. Dithering (oscillating the probe tip parallel to the surface) is needed to keep a constant probe–sample distance. As NSOM is not diffraction-limited, it is in principle possible to generate features with smaller dimensions (<100 nm [687, 688]) than those that are achievable with far-field optical lithography. Patterning of *a*-Si:H involves oxidizing the surface by local removal of hydrogen passivation. The thus formed oxide layer is the mask for subsequent etching. Herndon et al. [689] have reported on NSOM patterning of PECVD *a*-Si:H in open air, by using an Ar ion laser as the light source. Subsequent etching in a hydrogen plasma revealed an etch selectivity of 500 : 1. The feature widths are found to be dependent on the dither amplitude.

A monolithic, amorphous-silicon-based, photovoltaic-powered electrochromic window has been reported by Gao et al. [690]. A thin, wide-bandgap *a*-SiC:H *n–i–p* solar cell on top of a glass substrate is used as a semitransparent power supply. An electrochromic device consisting of $Li_yWO_3/LiAlF_4/V_2O_5$ is deposited directly on top of the solar cell. The operation principle is based on the change of optical absorption by inserting and extracting Li^+ in and out of the WO_3 film. A prototype 16-cm^2 device is able to modulate the transmittance of the stack by more than 60% over a large part of the visible spectrum.

1.12. Conclusion

In this chapter the deposition of *a*-Si:H by PECVD has been described. The chapter covers material as well as discharge issues. It tries to relate material and discharge properties in various ways. Plasma modeling provides a means to study in detail the physical and chemical processes that occur in the plasma. The presented models show a high degree of sophistication, but from the comparison with experimental data it is clear that especially the deposition model needs improvement. Also, a full 2D model most probably is not needed, as differences between 1D and 2D modeling results are not very large.

Plasma analysis reveals information on the products of chemical processes, and can be used to good effect as a feedback to plasma modeling. The role of ions has been thoroughly illustrated, and the important result that ion bombardment with moderate energy is beneficial for material quality has been quantified.

The many modifications to the conventional RF PECVD method show that one still is trying to find methods that will in the end lead to improved material properties. This is especially the case for the intrinsic metastability of *a*-Si:H. In this respect, the stable material that is obtained at discharge conditions "at the edge of crystallinity" is very promising. Also, the quest for higher deposition rates while at least maintaining *device quality* material properties shows the industrial drive behind the research. Faster deposition allows for more solar cells to be produced in the same time.

In the three deposition methods PECVD, HWCVD, and ETP CVD one can see certain similarities. In one region in the reactor system, the generation of growth precursors takes place. For PECVD this occurs in the discharge bulk, for HWCVD at and around the wire, and for ETP CVD in the vicinity of the expanding beam. A transport region exists where precursors are transported to the substrate. For HWCVD and ETP CVD this region is between wire/source and substrate. For PECVD this region is the sheath. The third region is the deposition region, in all cases the substrate. Looking at these techniques from this perspective, one can conclude that the deposition mechanism of *a*-Si:H is universal. The best material quality is obtained when the SiH_3 radical is abundantly present.

In the last part of this chapter a summary was given of some applications. These are the driving force for the large research effort that continues. From these applications and the many more that were not treated here, it will be clear that a-Si:H is a material that is omnipresent and contributes greatly to our well-being.

Acknowledgments

It goes without saying that the work presented here could not have been done without the help and support of many people. First, many thanks go to John Bezemer, my supervisor, who taught me all there is to know about plasma deposition of a-Si:H, and now enjoys his retirement. Second, I gratefully acknowledge the colleagues with whom I performed and discussed numerous deposition and plasma analysis experiments: Hans Meiling, Edward Hamers, Stefania Acco, Martin van de Boogaard, Arjan Berntsen, Ernst Ullersma, René van Swaaij, Meine Kars, Gerard van der Mark, and students Renée Heller, Remco van der Heiden, Pieter van de Vliet, Micha Kuiper, Tjitske Kooistra, Jan-Wijtse Smit, Edward Prendergast, Tom Thomas, and Lenneke Slooff. Further, I would gratefully like to thank the following persons for various types of continuous support during my active time of research on a-Si:H deposition: Wim Arnold Bik, Frans Habraken, Arjen Vredenberg, Dirk Knoesen, Wim de Kruif, Albert Polman, Gerhard Landweer, Jeroen Daey Ouwens, Kees Feenstra, Maarten van Cleef, Jatindra Rath, Wim Sinke, Bert Slomp, Jeike Wallinga, Karine van der Werf, Ruud Schropp, and Werner van de Weg, who headed the group for over a decade.

I very much enjoyed the fruitful collaboration with Wim Goedheer and his group—Gert Jan Nienhuis, Peter Meijer, and Diederick Passchier—at the FOM Institute for Plasmaphysics "Rijnhuizen," and thank him for supplying me with published and additional data used in the section on plasma modeling.

Also, the biannual Dutch meetings on solar cell research always allowed for many discussions with colleagues from Eindhoven Technical University (Daan Schram, Gijs Meeusen, Richard van de Sanden) and Delft Technical University (Miro Zeman, Wim Metselaar).

Further I thank Hans Gerritsen and Gijs van Ginkel for allowing and stimulating me to write this chapter.

And last but not least, I thank my wife Joni, son Guido, and daughter Sophie for the enormous joy they give me every day.

Parts of the work presented here were financially supported by the Netherlands Agency for Energy and the Environment (NOVEM), the Foundation for Fundamental Research on Matter (FOM), the Royal Netherlands Academy of Art and Sciences (KNAW), the Netherlands Technology Foundation (STW), and the Netherlands Organization for Scientific Research (NWO).

References

1. J. I. Pankove, Ed., "Hydrogenated Amorphous Silicon, Semiconductors and Semimetals, Vol. 21, Parts A-D." Academic Press, Orlando, FL, 1984.
2. K. Tanaka, Ed., "Glow-Discharge Hydrogenated Amorphous Silicon." KTK Scientific Publishers, Tokyo, 1989.
3. R. A. Street, "Hydrogenated Amorphous Silicon" Cambridge Solid State Science Series. Cambridge University Press, Cambridge, U.K., 1991.
4. J. Kanicki, Ed., "Amorphous and Microcrystalline Semiconductor Devices—Optoelectronic Devices." Artech House, Norwood, MA, 1991.
5. J. Kanicki, Ed., "Amorphous and Microcrystalline Semiconductor Devices—Materials and Device Physics." Artech House, Norwood, MA, 1992.
6. W. Luft and Y. S. Tsuo, "Hydrogenated Amorphous Silicon Alloy Deposition Processes." Marcel Dekker, New York, 1993.
7. R. F. Bunshah, Ed., "Handbook of Deposition Technologies for Films and Coatings—Science, Technology and Applications." Second Edition. Noyes Publications, Park Ridge, NJ, 1994.
8. G. Bruno, P. Capezzuto, and A. Madan, Eds., "Plasma Deposition of Amorphous Silicon-Based Materials." Academic Press, Boston, 1995.
9. E. S. Machlin, "Materials Science in Microelectronics—the Relationships between Thin Film Processing and Structure." Giro Press, Croton-on-Hudson, NY, 1995.
10. E. S. Machlin, "Materials Science in Microelectronics—Volume 2, The Effects of Structure on Properties in Thin Films." Giro Press, Croton-on-Hudson, NY, 1998.
11. R. E. I. Schropp and M. Zeman, "Amorphous and Microcrystalline Silicon Solar Cells—Modeling, Materials and Device Technology." Kluwer Academic Publishers, Boston, 1998.
12. T. M. Searle, Ed., "Properties of Amorphous Silicon and its Alloys, EMIS Datareview Series, No. 19." IEE, London, 1998.
13. R. A. Street, Ed., "Technology and Applications of Amorphous Silicon." Springer-Verlag, New York, 2000.
14. C. A. M. Stap, H. Meiling, G. Landweer, J. Bezemer, and W. F. van der Weg, "Proceedings of the Ninth E.C. Photovoltaic Solar Energy Conference, Freiburg, F.R.G., 1989" (W. Palz, G. T. Wrixon, and P. Helm, Eds.), p. 74. Kluwer Academic, Dordrecht, 1989.
15. H. Meiling, Ph.D. Thesis, Universiteit Utrecht, Utrecht, 1991.
16. H. Curtins, N. Wyrsch, and A. V. Shah, *Electron. Lett.* 23, 228 (1987).
17. H. Shirai, J. Hanna, and I. Shimizu, *Japan. J. Appl. Phys.* 30, L881 (1991).
18. J. T. Verdeyen, L. Beberman, and L. Overzet, *J. Vac. Sci. Technol. A* 8, 1851 (1990).
19. A. H. Mahan, J. Carapella, B. P. Nelson, R. S. Crandall, and I. Balberg, *J. Appl. Phys.* 69, 6728 (1991).
20. G. M. W. Kroesen, D. C. Schram, and M. J. F. van de Sande, *Plasma Chem. Plasma Proc.* 10, 49 (1990).
21. G. J. Meeusen, E. A. Ershov-Pavlov, R. F. G. Meulenbroeks, M. C. M. van de Sanden, and D. C. Schram, *J. Appl. Phys.* 71, 4156 (1992).
22. H. Overhof, *J. Non-cryst. Solids* 227–230, 15 (1998).

23. J. Tauc, Ed., "Amorphous and Liquid Semiconductors." Plenum, New York, 1974.
24. A. J. Lewis, G. A. N. Connell, W. Paul, J. R. Pawlik, and R. J. Temkin, "Tetrahedrally Bonded Amorphous Semiconductors," AIP Conference Proceedings, Vol. 20 (M. H. Brodsky, S. Kirkpatrick, and D. Weaire, Eds.), p. 27. American Institute of Physics, New York, 1974.
25. E. C. Freeman and W. Paul, *Phys. Rev. B* 18, 4288 (1978).
26. W. Paul and H. Ehrenreich, *Phys. Today* 38 (8), 13 (1985).
27. H. F. Sterling and R. C. G. Swann, *Solid State Electron.* 8, 653 (1965).
28. R. C. Chittick, J. H. Alexander, and H. F. Sterling, *J. Electrochem. Soc.* 116, 77 (1969).
29. A. Triska, D. Dennison, and H. Fritzsche, *Bull. Am. Phys. Soc.* 20, 392 (1975).
30. H. Fritzsche, "Proceedings of the 7th International Conference on Amorphous and Liquid Semiconductors" (W. E. Spear, Ed.), p. 3. CICL, Edinburgh, 1977.
31. H. Fritsche, *Mater. Res. Soc. Proc. Symp.* 609, A17.1.1 (2000).
32. W. Spear and P. LeComber, *Solid State Commun.* 17, 1193 (1975).
33. I. Solomon, *Mater. Res. Soc. Proc. Symp.* 609, A17.3.1 (2000).
34. D. E. Carlson and C. R. Wronski, *Appl. Phys. Lett.* 28, 671 (1976).
35. H. Okamoto, Y. Nitta, T. Yamaguchi, and Y. Hamakawa, *Sol. Energy Mater.* 3, 313 (1980).
36. Y. Hamakawa, *Mater. Res. Soc. Proc. Symp.* 609, A17.2.1 (2000).
37. Y. Kuwano, T. Imai, M. Ohnishi, and S. Nakano, "Proceedings of the 14th IEEE Photovoltaic Specialists Conference, San Diego, 1980," p. 1408. IEEE, New York, 1980.
38. I. Shimizu, T. Komatsu, T. Saito, and E. Inoue, *J. Non-cryst. Solids* 35 & 36, 773 (1980).
39. S. Oda, Y. Saito, I. Shimizu, and E. Inoue, *Philos. Mag. B* 43, 1079 (1981).
40. P. G. LeComber, W. E. Spear, and A. Ghaith, *Electron. Lett.* 15, 179 (1979).
41. A. J. Snell, K. D. MacKenzie, W. E. Spear, P. G. LeComber, and A. J. Hughes, *J. Appl. Phys.* 24, 357 (1981).
42. M. J. Powell, B. C. Easton, and O. F. Hill, *Appl. Phys. Lett.* 38, 794 (1981).
43. R. L. Weisfield, *J. Non-cryst. Solids* 164–166, 771 (1993).
44. S. N. Kaplan, J. R. Morel, T. A. Mulera, V. Perez-Mendez, G. Schlurmacher, and R. A. Street, *IEEE Trans. Nucl. Sci.* NS-33, 351 (1986).
45. T. C. Chuang, L. E. Fennel, W. B. Jackson, J. Levine, M. J. Thompson, H. C. Tuan, and R. Weisfield, *J. Non-cryst. Solids* 97 & 98, 301 (1987).
46. L. E. Antonuk, C. F. Wild, J. Boundry, J. Jimenez, M. J. Longo, and R. A. Street, *IEEE Trans. Nucl. Sci.* NS-37, 165 (1990).
47. C. van Berkel, N. C. Bird, C. J. Curling, and I. D. French, *Mater. Res. Soc. Symp. Proc.* 297, 939 (1993).
48. W. H. Zachariasen, *J. Am. Chem. Soc.* 54, 3841 (1932).
49. D. E. Polk, *J. Non-cryst. Solids* 5, 365 (1971).
50. M. A. Paesler, D. E. Sayers, R. Tsu, and J. G. Hernandez, *Phys. Rev. B* 28, 4550 (1983).
51. D. Beeman, R. Tsu, and M. F. Thorpe, *Phys. Rev. B* 32, 874 (1985).
52. R. A. Street, *Phys. Rev. Lett.* 49, 1187 (1982).

53. N. F. Mott, *Philos. Mag.* 19, 835 (1969).
54. W. M. Arnold Bik, C. T. A. M. de Laat, and F. H. P. M. Habraken, *Nucl. Instrum. Methods Phys. Res. B* 64, 832 (1992).
55. M. H. Brodsky, *Thin Solid Films* 40, L23 (1977).
56. G. Lucovsky, R. J. Nemanich, and J. C. Knights, *Phys. Rev. B* 19, 2064 (1979).
57. H. Shanks, C. J. Fang, L. Ley, M. Cardona, F. J. Demond, and S. Kalbitzer, *Phys. Status Solidi B* 100, 43 (1980).
58. A. A. Langford, M. L. Fleet, B. P. Nelson, W. A. Lanford, and N. Maley, *Phys. Rev. B* 45, 13367 (1992).
59. W. Beyer and M. S. Abo Ghazala, *Mater. Res. Soc. Symp. Proc.* 507, 601 (1998).
60. A. H. Mahan, P. Raboisson, and R. Tsu, *Appl. Phys. Lett.* 50, 335 (1987).
61. D. A. G. Bruggeman, *Ann. Phys. (Leipzig)* 24, 636 (1935).
62. A. H. Mahan, D. L. Williamson, B. P. Nelson, and R. S. Crandall, *Sol. Cells* 27, 465 (1989)
63. H. Meiling, M. J. van den Boogaard, R. E. I. Schropp, J. Bezemer, and W. F. van der Weg, *Mater. Res. Soc. Symp. Proc.* 192, 645 (1990).
64. M. J. van den Boogaard, S. J. Jones, Y. Chen, D. L. Williamson, R. A. Hakvoort, A. van Veen, A. C. van der Steege, W. M. Arnold Bik, W. G. J. H. M. van Sark, and W. F. van der Weg, *Mater. Res. Soc. Symp. Proc.* 258, 407 (1992).
65. M. J. van den Boogaard, Ph.D. Thesis, Universiteit Utrecht, Utrecht, 1992.
66. S. Acco, D. L. Williamson, W. G. J. H. M. van Sark, W. C. Sinke, W. F. van der Weg, A. Polman, and S. Roorda, *Phys. Rev. B* 58, 12853 (1998).
67. A. H. Mahan, D. L. Williamson, B. P. Nelson, and R. S. Crandall, *Phys. Rev. B* 40, 12024 (1989).
68. P. C. Taylor, in "Semiconductors and Semimetals" (J. I. Pankove, Ed.), Vol. 21C, p. 99. Academic Press, Orlando, FL, 1984.
69. S. Acco, D. L. Williamson, P. A. Stolk, F. W. Saris, M. J. van den Boogaard, W. C. Sinke, W. F. van der Weg, and S. Roorda, *Phys. Rev. B* 53, 4415 (1996).
70. S. Acco, W. Beyer, E. E. Van Faassen, and W. F. van der Weg, *J. Appl. Phys.* 82, 2862 (1997).
71. R. S. Crandall, X. Liu, and E. Iwaniczko, *J. Non-cryst. Solids* 227–230, 23 (1998).
72. S. Acco, D. L. Williamson, S. Roorda, W. G. J. H. M. van Sark, A. Polman, and W. F. van der Weg, *J. Non-cryst. Solids* 227–230, 128 (1998).
73. P. J. Zanzucchi, C. R. Wronski, and D. E. Carlson, *J. Appl. Phys.* 48, 5227 (1977).
74. G. D. Cody, B. Abeles, C. Wronski, C. R. Stephens, and B. Brooks, *Sol. Cells* 2, 227 (1980).
75. J. Tauc, in "Optical Properties of Solids" (F. Abelès, Ed.), Chap. 5, p. 277. North-Holland, Amsterdam, the Netherlands, 1972.
76. R. H. Klazes, M. H. L. M. van den Broek, J. Bezemer, and S. Radelaar, *Philos. Mag. B* 45, 377 (1982).
77. T. Tiedje, J. M. Cebulka, D. L. Morel, and B. Abeles, *Phys. Rev. Lett.* 46, 1425 (1981).
78. K. Pierz, B. Hilgenberg, H. Mell, and G. Weiser, *J. Non-cryst. Solids* 97/98, 63 (1987).
79. K. Pierz, W. Fuhs, and H. Mell, *Philos. Mag. B* 63, 123 (1991).
80. N. Maley and J. S. Lannin, *Phys. Rev. B* 36, 1146 (1987).

81. R. C. Ross and J. Jaklik, *J. Appl. Phys.* 55, 3785 (1984).
82. H. Meiling, W. Lenting, J. Bezemer, and W. F. van der Weg, *Philos. Mag. B* 62, 19 (1990).
83. R. S. Crandall, *Sol. Cells* 24, 237 (1988).
84. A. J. M. Berntsen, W. F. van der Weg, P. A. Stolk, and F. W. Saris, *Phys. Rev. B* 48, 14656 (1993).
85. A. J. M. Berntsen, Ph.D. Thesis, Universiteit Utrecht, Utrecht, 1993.
86. W. E. Spear and P. G. LeComber, *Adv. Phys.* 26, 811 (1977).
87. D. L. Staebler and C. R. Wronski, *Appl. Phys. Lett.* 31, 292 (1977).
88. D. L. Staebler and C. R. Wronski, *J. Appl. Phys.* 51, 3262 (1980).
89. M. Stutzmann, W. B. Jackson, and C. C. Tsai, *Phys. Rev. B* 32, 23 (1985).
90. T. Kamei, N. Hata, A. Matsuda, T. Uchiyama, S. Amano, K. Tsukamoto, Y. Yoshioka, and T. Hirao, *Appl. Phys. Lett.* 68, 2380 (1996).
91. C. C. Tsai, J. C. Knights, R. A. Lujan, W. B. B. L. Stafford, and M. J. Thompson, *J. Non-cryst. Solids* 59 & 60, 731 (1983).
92. S. Yamasaki and J. Isoya, *J. Non-cryst. Solids* 164–166, 169 (1993).
93. P. Stradins and H. Fritzsche, *Philos. Mag. B* 69, 121 (1994).
94. C. Godet, P. Morin, and P. Roca i Cabarrocas, *J. Non-cryst. Solids* 198–200, 449 (1996).
95. J. Daey Ouwens and R. E. I. Schropp, *Phys. Rev. B* 54, 17759 (1996).
96. S. Zafar and E. A. Schiff, *Phys. Rev. B* 40, 5235 (1989).
97. C. Manfredotti, F. Fizzotti, M. Boero, P. Pastorino, P. Polesello, and E. Vittone, *Phys. Rev. B* 50, 18046 (1994).
98. H. M. Branz, *Solid State Commun.* 105, 387 (1998).
99. R. Biswas, Q. Li, B. C. Pan, and Y. Yoon, *Mater. Res. Soc. Symp. Proc.* 467, 135 (1997).
100. D. A. Anderson and W. E. Spear, *Philos. Mag.* 35, 1 (1977).
101. R. A. C. M. M. van Swaaij, Ph.D. Thesis, Universiteit Utrecht, Utrecht, 1994.
102. R. A. C. M. M. van Swaaij, A. J. M. Berntsen, W. G. J. H. M. van Sark, H. Herremans, J. Bezemer, and W. F. van der Weg, *J. Appl. Phys.* 76, 251 (1994).
103. J. Chevallier, H. Wieder, A. Onton, and C. R. Guarnieri, *Solid State Commun.* 24, 867 (1977).
104. S. Wagner, V. Chu, J. P. Conde, and J. Z. Liu, *J. Non-cryst. Solids* 114, 453 (1989).
105. C. M. Fortmann, in "Plasma Deposition of Amorphous Silicon-Based Materials" (G. Bruno, P. Capezzuto, and A. Madan, Eds.), Chap. 3, p. 131. Academic Press, Boston, 1995.
106. W. Paul, S. J. Jones, F. C. Marques, D. Pang, W. A. Turner, A. E. Wetsel, P. Wickboldt, and J. H. Chen, *Mater. Res. Soc. Symp. Proc.* 219, 211 (1991).
107. A. Matsuda and K. Tanaka, *J. Non-cryst. Solids* 97, 1367 (1987).
108. G. Bruno, P. Capezzuto, G. Cicala, and F. Cramarossa, *J. Mater. Res.* 4, 366 (1989).
109. G. Bruno, P. Capezzuto, M. Losurdo, P. Manodoro, and G. Cicala, *J. Non-cryst. Solids* 137&138, 799 (1991).
110. H. Herremans, W. Grevendonk, R. A. C. M. M. van Swaaij, W. G. J. H. M. van Sark, A. J. M. Berntsen, W. M. Arnold Bik, and J. Bezemer, *Philos. Mag. B* 66, 787 (1992).
111. M. P. Schmidt, I. Solomon, H. Tran-Quoc, and J. Bullot, *J. Non-cryst. Solids* 77–78, 849 (1985).

112. I. Solomon, M. P. Schmidt, and H. Tran-Quoc, *Phys. Rev. B* 38, 9895 (1988).
113. M. N. P. Carreño, I. Pereyra, M. C. A. Fantini, H. Takahashi, and R. Landers, *J. Appl. Phys.* 75, 538 (1994).
114. W. M. Arnoldbik and F. H. P. M. Habraken, *Rep. Prog. Phys.* 56, 859 (1993).
115. M. L. Oliveira and S. S. Camargo, Jr., *J. Appl. Phys.* 71, 1531 (1992).
116. F. Evangelisti, *J. Non-cryst. Solids* 164–166, 1009 (1993).
117. A. Grill, "Cold Plasma in Materials Fabrication—from Fundamentals to Applications." IEEE Press, Piscataway, NJ, 1994.
118. J. F. O'Hanlon, "A User's Guide to Vacuum Technology." Second Edition. Wiley, New York, 1989.
119. J. Bezemer, W. G. J. H. M. van Sark, M. B. von der Linden, and W. F. van der Weg, "Proceedings of the Twelfth E.C. Photovoltaic Solar Energy Conference, Amsterdam, the Netherlands, 1994" (R. Hill, W. Palz, and P. Helm, Eds.), p. 327. H.S. Stephens & Associates, Bedford, U.K., 1994.
120. J. Bezemer and W. G. J. H. M. van Sark, "Electronic, Optoelectronic and Magnetic Thin Films, Proceedings of the 8th International School on Condensed Matter Physics (ISCMP), Varna, Bulgaria, 1994," p. 219. Wiley, New York, 1995.
121. M. Böhm, A. E. Delahoy, F. B. Ellis, Jr., E. Eser, S. C. Gau, F. J. Kampas, and Z. Kiss, "Proceedings of the 18th IEEE Photovoltaic Specialists Conference, Las Vegas, 1985," p. 888. IEEE, New York, 1985.
122. C. R. Dickson, J. Pickens, and A. Wilczynski, *Sol. Cells* 19, 179 (1986).
123. K. Tanaka and A. Matsuda, *Mater. Sci. Rep.* 2, 139 (1987).
124. F. Jansen and D. Kuhman, *J. Vac. Sci. Technol.* 6, 13 (1988).
125. W. Y. Kim, H. Tasaki, M. Konagai, and K. Takahashi, *J. Appl. Phys.* 61, 3071 (1987).
126. A. Catalano and G. Wood, *J. Appl. Phys.* 63, 1220 (1988).
127. Y. Kuwano, M. Ohnishi, S. Tsuda, Y. Nakashima, and N. Nakamura, *Japan. J. Appl. Phys.* 21, 413 (1982).
128. R. A. Street, J. Zesch, and M. J. Thompson, *Appl. Phys. Lett.* 43, 672 (1983).
129. M. Ohnishi, H. Nishiwaki, E. Enomoto, Y. Nakashima, S. Tsuda, T. Takahama, H. Tarui, M. Tanaka, H. Dojo, and Y. Kuwano, *J. Non-cryst. Solids* 59/60, 1107 (1983).
130. S. Nakano, S. Tsuda, H. Tarui, T. Takahama, H. Haku, K. Watanabe, M. Nishikuni, Y. Hishikawa, and Y. Kuwano, *Mater. Res. Soc. Symp. Proc.* 70, 511 (1986).
131. S. Tsuda, T. Takahama, M. Isomura, H. Tarui, Y. Nakashima, Y. Hishikawa, N. Nakamura, T. Matsuoka, H. Nishiwaki, S. Nakano, M. Ohnishi, and Y. Kuwano, *Japan. J. Appl. Phys.* 26, 33 (1987).
132. W. Luft, Ed., "Photovoltaic Safety," AIP Conference Proceedings, Vol. 166. American Institute of Physics, New York, 1988.
133. P. J. Hargis, Jr., K. E. Greenberg, P. A. Miller, J. B. Gerardo, J. R. Torczynski, M. E. Riley, G. A. Hebner, J. R. Roberts, J. K. Olthoff, J. R. Whetstone, R. J. Van Brunt, M. A. Sobolewski, H. M. Anderson, M. P. Splichal, J. L. Mock, P. Bletzinger, A. Garscadden, R. A. Gottscho, G. Selwyn, M. Dalvie, J. E. Heidenreich, J. W. Butterbaugh, M. L. Brake, M. L. Passow, J. Pender, A. Lujan, M. E. Elta, D. B. Graves, H. H. Sawin, M. J. Kushner, J. T. Verdeyen, R. Horwath, and T. R. Turner, *Rev. Sci. Instrum.* 65, 140 (1994).
134. B. Chapman, "Glow Discharge Processes—Sputtering and Plasma Etching." Wiley, New York, 1980.

135. A. Gallagher, *Mater. Res. Soc. Symp. Proc.* 70, 3 (1986).
136. J. Perrin, Y. Takeda, N. Hirano, Y. Takeuchi, and A. Matsuda, *Surf. Sci.* 210, 114 (1989).
137. A. Matsuda, K. Nomoto, Y. Takeuchi, A. Suzuki, A. Yuuki, and J. Perrin, *Surf. Sci.* 227, 50 (1990).
138. A. Matsuda and K. Tanaka, *J. Appl. Phys.* 60, 2351 (1986).
139. T. Ichimura, T. Ihara, T. Hama, M. Ohsawa, H. Sakai, and Y. Uchida, *Japan. J. Appl. Phys.* 25, L276 (1986).
140. T. Hamasaki, M. Ueda, A. Chayahara, M. Hirose, and Y. Osaka, *Appl. Phys. Lett.* 44, 600 (1984).
141. T. Hamasaki, M. Ueda, A. Chayahara, M. Hirose, and Y. Osaka, *Appl. Phys. Lett.* 44, 1049 (1984).
142. P. Roca i Cabarrocas, J. B. Chevrier, J. Huc, A. Lloret, J. Y. Parey, and J. P. M. Schmitt, *J. Vac. Sci. Technol. A* 9, 2331 (1991).
143. M. Taniguchi, M. Hiorse, T. Hamasaki, and Y. Osaka, *Appl. Phys. Lett.* 37, 787 (1980).
144. M. Ohnishi, H. Nishiwaki, K. Uchihashi, K. Yoshida, M. Tanaka, K. Ninomiya, M. Nishikuni, N. Nakanura, S. Tsuda, S. Nakano, T. Yazaki, and Y. Kuwano, *Japan. J. Appl. Phys.* 27, 40 (1988).
145. F. N. Boulitrop, N. Proust, J. Magarino, E. Criton, J. F. Peray, and M. Dupre, *J. Appl. Phys.* 58, 3494 (1985).
146. H. Chatham and P. K. Bhat, *Mater. Res. Soc. Symp. Proc.* 149, 447 (1989).
147. H. Chatham, P. Bhat, A. Benson, and C. Matovich, *J. Non-cryst. Solids* 115, 201 (1989).
148. P. M. Meijer, J. D. P. Passchier, W. J. Goedheer, J. Bezemer, and W. G. J. H. M. van Sark, *Appl. Phys. Lett.* 64, 1780 (1994).
149. W. J. Goedheer, P. M. Meijer, J. Bezemer, J. D. P. Passchier, and W. G. J. H. M. van Sark, *IEEE Trans. Plasma Sci.* PS-23, 644 (1995).
150. W. G. J. H. M. van Sark, H. Meiling, J. Bezemer, M. B. von der Linden, R. E. I. Schropp, and W. F. van der Weg, *Sol. Energy Mater. Sol. Cells* 45, 57 (1997).
151. H. Meiling, W. G. J. H. M. van Sark, J. Bezemer, R. E. I. Schropp, and W. van der Weg, "Proceedings of the 25th IEEE Photovoltaic Specialists Conference, Washington, D.C., 1996," p. 1153. IEEE, New York, 1996.
152. J. P. M. Schmitt, J. Meot, P. Roubeau, and P. Parrens, "Proceedings of the Eighth E.C. Photovoltaic Solar Energy Conference, Florence, Italy, 1988" (I. Solomon, B. Equer, and P. Helm, Eds.), p. 964. Kluwer Academic, Dordrecht, 1989.
153. S. Tsuda, S. Sakai, and S. Nakano, *Appl. Surf. Sci.* 113/114, 734 (1997).
154. J. Yang, A. Banerjee, S. Sugiyama, and S. Guha, *Appl. Phys. Lett.* 70, 2975 (1997).
155. R. E. I. Schropp, C. H. M. van der Werf, M. Zeman, M. C. M. van de Sanden, C. I. M. A. Spee, E. Middelman, L. V. de Jonge-Meschaninova, A. A. M. van der Zijden, M. M. Besselink, R. J. Severens, J. Winkeler, and G. J. Jongerden, *Mater. Res. Soc. Symp. Proc.* 557, 713 (1999).
156. S. Fujikake, K. Tabuchi, T. Yoshida, Y. Ichikawa, and H. Sakai, *Mater. Res. Soc. Symp. Proc.* 377, 609 (1995).
157. T. Yoshida, S. Fujikake, S. Kato, M. Tanda, K. Tabuchi, A. Takano, Y. Ichikawa, and H. Sakai, *Sol. Energy Mater. Sol. Cells* 48, 383 (1997).

158. E. Middelman, E. van Andel, P. M. G. M. Peters, L. V. de Jonge-Meschaninova, R. J. Severens, G. J. Jongerden, R. E. I. Schropp, H. Meiling, M. Zeman, M. C. M. van de Sanden, A. Kuipers, and C. I. M. A. Spee, "Proceedings of the 2nd World Conference on Photovoltaic Energy Conversion, Vienna, Austria" (J. Schmid, H. A. Ossenbrink, P. Helm, H. Ehmann, and E. D. Dunlop, Eds.), p. 816. European Commision Joint Research Centre, Ispra, Italy, 1998.
159. A. Madan, P. Rava, R. E. I. Schropp, and B. von Roedern, *Appl. Surf. Sci.* 70/71, 716 (1993).
160. W. G. J. H. M. van Sark, J. Bezemer, M. Kars, M. Kuiper, E. A. G. Hamers, and W. F. van der Weg, "Proceedings of the Twelfth E.C. Photovoltaic Solar Energy Conference, Amsterdam, the Netherlands, 1994" (R. Hill, W. Palz, and P. Helm, Eds.), p. 350. H. S. Stephens & Associates, Bedford, U.K., 1994.
161. E. A. G. Hamers, W. G. J. H. M. van Sark, J. Bezemer, W. J. Goedheer, and W. F. van der Weg, *Int. J. Mass Spectrom. Ion Processes* 173, 91 (1998).
162. W. G. J H. M. van Sark, J. Bezemer, and W. J. Goedheer, unpublished results.
163. E. A. G. Hamers, Ph.D. Thesis, Universiteit Utrecht, Utrecht, 1998.
164. H. Shirai, D. Das, J. Hanna, and I. Shimizu, *Appl. Phys. Lett.* 59, 1096 (1991).
165. W. G. J. H. M. van Sark, J. Bezemer, P. G. van de Vliet, M. Kars, and W. F. van der Weg, "Proceedings of the Twelfth E.C. Photovoltaic Solar Energy Conference, Amsterdam, the Netherlands, 1994" (R. Hill, W. Palz, and P. Helm, Eds.), p. 335. H.S. Stephens & Associates, Bedford, U.K., 1994.
166. H. Kausche, M. Möller, and R. Plättner, "Proceedings of the Fifth E.C. Photovoltaic Solar Energy Conference, Athens, Greece, 1983." Reidel, Dordrecht, 1983.
167. A. Morimoto, M. Matsumoto, M. Kumeda, and T. Shimizu, *Japan. J. Appl. Phys.* 29, L1747 (1990).
168. A. Ricard, "Reactive Plasmas." Société Française du Vide, Paris, 1996.
169. H. S. Butler and G. S. Kino, *Phys. Fluids* 6, 1346 (1963).
170. K. Köhler, J. W. Coburn, D. E. Horne, E. Kay, and J. H. Keller, *J. Appl. Phys.* 57, 59 (1985).
171. M. A. Liebermann and S. E. Savas, *J. Vac. Sci. Technol. A* 16, 1632 (1990).
172. M. V. Alves, M. A. Liebermann, V. Vahedi, and C. K. Birdsall, *J. Appl. Phys.* 69, 3823 (1991).
173. J. Perrin, in "Plasma Deposition of Amorphous Silicon-Based Materials" (G. Bruno, P. Capezzuto, and A. Madan, Eds.), Chap. 4, p. 177. Academic Press, Boston, 1995.
174. P. M. Meijer and W. J. Goedheer, *IEEE Trans. Plasma Sci.* PS-19, 170 (1991).
175. P. M. Meijer, Ph.D. Thesis, Universiteit Utrecht, Utrecht, 1991.
176. K. Köhler, D. E. Horne, and J. W. Coburn, *J. Appl. Phys.* 58, 3350 (1985).
177. A. D. Kuypers and H. J. Hopman, *J. Appl. Phys.* 63, 1894 (1988).
178. W. J. Goedheer, private communication.
179. J. D. P. Passchier, Ph.D. Thesis, Universiteit Utrecht, Utrecht, 1994.
180. S. Rauf and M. J. Kushner, *J. Appl. Phys.* 83, 5087 (1998).
181. V. A. Godyak and R. B. Piejak, *J. Vac. Sci. Technol. A* 8, 3833 (1990).
182. C. M. Horwitz, *J. Vac. Sci. Technol. A* 1, 1795 (1983).
183. F. Finger, U. Kroll, V. Viret, A. Shah, W. Beyer, X.-M. Tang, J. Weber, A. A. Howling, and C. Hollenstein, *J. Appl. Phys.* 71, 5665 (1992).

184. C. Böhm and J. Perrin, *J. Phys. D* 24, 865 (1991).
185. K. M. H. Maessen, Ph.D. Thesis, Universiteit Utrecht, Utrecht 1988.
186. M. J. M. Pruppers, Ph.D. Thesis, Universiteit Utrecht, Utrecht 1988.
187. S. Ishihara, K. Masatoshi, H. Takashi, W. Kiyotaka, T. Arita, and K. Mori, *J. Appl. Phys.* 62, 485 (1987).
188. V. A. Godyak and R. B. Piejak, *Phys. Rev. Lett.* 65, 996 (1990).
189. G. J. Nienhuis, Ph.D. Thesis, Universiteit Utrecht, Utrecht, 1998.
190. M. J. Kushner, *J. Appl. Phys.* 63, 2532 (1988).
191. G. J. Nienhuis, W. J. Goedheer, E. A. G. Hamers, W. G. J. H. M. van Sark, and J. Bezemer, *J. Appl. Phys.* 82, 2060 (1997).
192. J. Perrin, O. Leroy, and M. C. Bordage, *Contrib. Plasma Phys.* 36, 3 (1996).
193. A. V. Phelps, *J. Phys. Chem. Ref. Data* 19, 653 (1990).
194. H. Tawara, Y. Itikawa, H. Nishimura, and M. Yoshino, *J. Phys. Chem. Ref. Data* 19, 617 (1990).
195. L. Layeillon, P. Duverneuil, J. P. Couderc, and B. Despax, *Plasma Sources Sci. Technol.* 3, 61 (1994).
196. L. Layeillon, P. Duverneuil, J. P. Couderc, and B. Despax, *Plasma Sources Sci. Technol.* 3, 72 (1994).
197. J. R. Doyle, D. A. Doughty, and A. Gallagher, *J. Appl. Phys.* 68, 4375 (1990).
198. P. Kae-Nune, J. Perrin, J. Guillon, and J. Jolly, *Japan. J. Appl. Phys.* 33, 4303 (1994).
199. E. R. Austin and F. W. Lampe, *J. Phys. Chem.* 81, 1134 (1977).
200. T. L. Pollock, H. S. Sandhu, A. Jodhan, and O. P. Strausz, *J. Am. Chem. Soc.* 95, 1017 (1973).
201. J. Perrin, J. P. M. Schmitt, G. De Rosny, B. Drévillon, J. Huc, and A. Lloret, *Chem. Phys.* 73, 383 (1982).
202. G. C. A. Perkins, E. R. Austin, and F. W. Lampe, *J. Am. Chem. Soc.* 101, 1109 (1979).
203. M. Hertl, N. Dorval, O. Leroy, J. Jolly, and M. Péalet, *Plasma Sources Sci. Technol.* 7, 130 (1998).
204. J. R. Doyle, D. A. Doughty, and A. Gallagher, *J. Appl. Phys.* 71, 4771 (1992).
205. A. G. Engelhardt and A. V. Phelps, *Phys. Rev.* 131, 2115 (1963).
206. E. Krishnakumar and S. K. Srivastava, *Contrib. Plasma Phys.* 35, 395 (1995).
207. H. Tawara and T. Kato, *At. Data Nucl. Data Tables* 36, 167 (1987).
208. P. Haaland, *J. Chem. Phys.* 93, 4066 (1990).
209. J. M. Wadehra and J. N. Bardsley, *Phys. Rev. Lett.* 41, 1795 (1978).
210. E. Krishnakumar, S. K. Srivastava, and I. Iga, *Int. J. Mass Spectrom. Ion Processes* 103, 107 (1991).
211. J. Perrin, *J. Phys. D* 26, 1662 (1993).
212. W. Kurachi and Y. Nakamura, *J. Phys. D* 22, 107 (1989).
213. H. Ehrhardt, L. Langhans, F. Linder, and H. Taylor, *Phys. Rev.* 173, 222 (1968).
214. A. P. Hickman, *J. Chem. Phys.* 70, 4872 (1979).
215. N. Itabashi, K. Kato, N. Nishiwaki, T. Goto, C. Yamada, and E. Hirota, *Japan. J. Appl. Phys.* 28, L325 (1989).
216. J. M. S. Henis, G. W. Steward, M. K. Tripodi, and P. P. Gaspar, *J. Chem. Phys.* 57, 389 (1972).
217. P. Kae-Nune, J. Perrin, J. Jolly, and J. Guillon, *Surf. Sci.* 360, L495 (1996).

218. T. Fuyuki, B. Allain, and J. Perrin, *J. Appl. Phys.* 68, 3322 (1990).
219. E. W McDaniel and E. A. Mason, "The Mobility and Diffusion of Ions in Gases." Wiley, New York, 1973.
220. J. D. P. Passchier and W. J. Goedheer, *J. Appl. Phys.* 74, 3744 (1993).
221. J. P. Boeuf and L. C. Pitchford, *Phys. Rev. E* 51, 1376 (1995).
222. C. K. Birdsall, *IEEE Trans. Plasma Sci.* 19, 65 (1991).
223. J. P. Boeuf and P. Belenguer, *J. Appl. Phys.* 71, 4751 (1992).
224. M. Yan and W. J. Goedheer, *Plasma Sources Sci. Technol.* 8, 349 (1999).
225. N. Sato and H. Tagashira, *IEEE Trans. Plasma. Sci.* PS-19, 10102 (1991).
226. N. V. Mantzaris, A. Boudouvis, and E. Gogolides, *J. Appl. Phys.* 77, 6169 (1995).
227. M. J. Kushner, *Solid State Technol.* 6, 135 (1996).
228. G. G. Lister, *Vacuum* 45, 525 (1994).
229. E. Gogolides and H. H. Sawin, *J. Appl. Phys.* 72, 3971 (1992).
230. P. M. Meijer, W. J. Goedheer, and J. D. P. Passchier, *Phys. Rev. A* 45, 1098 (1992).
231. I. P. Shkarofsky, T. W. Johnston, and M. P. Bachynski, "The Particle Kinetics of Plasmas." Addison-Wesley, Reading, MA 1966.
232. G. J. Nienhuis and W. J. Goedheer, *Plasma Sources Sci. Technol.* 8, 295 (1999).
233. A. D. Richards, B. E. Thompson, and H. H. Sawin, *Appl. Phys. Lett.* 50, 492 (1987).
234. D. P. Lymberopoulos and D. J. Economou, *J. Phys. D* 28, 727 (1995).
235. E. A. G. Hamers, W. G. J. H. M. van Sark, J. Bezemer, W. F. van der Weg, and W. J. Goedheer, *Mater. Res. Soc. Symp. Proc.* 420, 461 (1996).
236. M. Heintze and R. Zedlitz, *J. Non-cryst. Solids* 198–200, 1038 (1996).
237. E. W. McDaniel, "Collision Phenomena in Ionized Gases." Wiley, New York, 1964.
238. P. J. Chantry, *J. Appl. Phys.* 62, 1141 (1987).
239. M. S. Barnes, T. J. Cotler, and M. E. Elta, *J. Comput. Phys.* 77, 53 (1988).
240. W. L. Morgan, *Plasma Chem. Plasma Process.* 12, 477 (1992).
241. D. L. Sharfetter and H. K. Gummel, *IEEE Trans. Electron Devices* ED-16, 64 (1969).
242. W. Hackbusch, "Multigrid Methods and Applications." Springer-Verlag, Berlin, 1985.
243. J. D. P. Passchier, Ph.D. Thesis, Universiteit Utrecht, Utrecht, 1994.
244. A. A. Fridman, L. Boufendi, T. Hbid, B. V. Potapkin, and A. Bouchoule, *J. Appl. Phys.* 79, 1303 (1996).
245. J. Perrin, P. Roca i Cabarrocas, B. Allain, and J.-M. Friedt, *Japan. J. Appl. Phys.* 27, 2041 (1988).
246. J. L. Andújar, E. Bertran, A. Canillas, C. Roch, and J. L. Morenza, *J. Vac. Sci. Technol. A* 9, 2216 (1991).
247. W. G. J. H. M. van Sark, J. Bezemer, E. M. B. Heller, M. Kars, and W. F. van der Weg, *Mater. Res. Soc. Symp. Proc.* 377, 3 (1995).
248. G. Oversluizen and W. H. M. Lodders, *J. Appl. Phys.* 83, 8002 (1998).
249. M. Heintze and R. Zedlitz, *Prog. Photovolt. Res. Appl.* 1, 213 (1993).
250. A. A. Howling, J.-L. Dorier, C. Hollenstein, U. Kroll, and F. Finger, *J. Vac. Sci. Technol. A* 10, 1080 (1992).
251. W. G. J. H. M. van Sark, J. Bezemer, and W. F. van der Weg, *Surf. Coat. Technol.* 74–75, 63 (1995).
252. L. Sansonnens, A. A. Howling, and C. Hollenstein, *Plasma Sources Sci. Technol.* 7, 114 (1998).

253. M. Capitelli, C. Gorse, R. Winkler, and J. Wilhelm, *Plasma Chem. Plasma Process.* 8, 399 (1988).
254. H. Curtins, N. Wyrsch, N. Favre, and A. V. Shah, *Plasma Chem. Plasma Process.* 7, 267 (1987).
255. L. A. Pinnaduwga, W. Ding, and D. L. McCorkle, *Appl. Phys. Lett.* 71, 3634 (1997).
256. L. A. Pinnaduwga and P. G. Datskos, *J. Appl. Phys.* 81, 7715 (1997).
257. O. Leroy, G. Gousset, L. L. Alves, J. Perrin, and J. Jolly, *Plasma Sources Sci. Technol.* 7, 348 (1998).
258. J. P. M. Schmitt, *Thin Solid Films* 174, 193 (1989).
259. L. Sansonnens, D. Franz, C. Hollenstein, A. A. Howling, J. Schmitt, E. Turlot, T. Emeraud, U. Kroll, J. Meier, and A. Shah, "Proceedings of the 13th European Photovoltaic Solar Energy Conference, Nice, France, 1995" (W. Freiesleben, W. Palz, H. A. Ossenbrink, and P. Helm, Eds.), p. 276. H.S. Stephens & Associates, Bedford, U.K., 1995.
260. D. Vender and R. W. Boswell, *IEEE Trans. Plasma Sci.* 18, 725 (1990).
261. C. K. Birdsall, *IEEE Trans. Plasma. Sci.* 19, 65 (1991).
262. C. K. Birdsall and A. B. Langdon, "Plasma Physics via Computer Simulation." Adam Hilger, Bristol, 1991.
263. V. Vahedi, "XPDP1 Manual." University of California, Berkeley, CA, 1994.
264. J. D. P. Passchier, Technical Report No. GF95.08, Utrecht University, Utrecht, 1995.
265. W. G. J. H. M. van Sark and J. Bezemer, Technical Report, Utrecht University, Utrecht, 1995.
266. Y. Ohmori, M. Shimozuma, and H. Tagashira, *J. Phys. D* 19, 1029 (1986).
267. E. A. Mason and E. W. McDaniel, "Transport Properties of Ions in Gases." Wiley, New York, 1988.
268. D. B. Graves, *J. Appl. Phys.* 62, 88 (1987).
269. J.-P. Boeuf, *Phys. Rev. A* 36, 2782 (1987).
270. E. Gogolides and H. H. Sawin, *J. Appl. Phys.* 72, 3971 (1992).
271. A. Manenschijn and W. J. Goedheer, *J. Appl. Phys.* 69, 2923 (1991).
272. V. Vahedi, C. K. Birdsall, M. A. Lieberman, G. Dipeso, and T. D. Rognlien, *Phys. Fluids B* 5, 2719 (1993).
273. K. Nanbu and Y. Kitatani, *Vacuum* 47, 1023 (1996).
274. A. K. Jain and D. G. Thompson, *J. Phys. B* 20, L389 (1987).
275. H. Tawara, Y. Itikawa, H. Nishimura, and M. Yoshino, *J. Phys. Chem. Ref. Data* 19, 621 (1990).
276. M. R. Wertheimer and M. Moisan, *J. Vac. Sci. Technol. A* 3, 2643 (1985).
277. H. Curtins, N. Wyrsch, M. Favre, and A. V. Shah, *Plasma Chem. Plasma Process.* 7, 267 (1987).
278. V. A. Godyak, *Sov. Phys. Tech. Phys.* 16, 1073 (1973).
279. M. Heintze and R. Zedlitz, *J. Phys. D* 26, 1781 (1993).
280. M. Heintze and R. Zedlitz, *J. Non-cryst. Solids* 164–166, 55 (1993).
281. J. Dutta, W. Bacsa, and C. Hollenstein, *J. Appl. Phys.* 72, 3220 (1992).
282. P. Mérel, M. Chaker, M. Moisan, A. Ricard, and M. Tabbal, *J. Appl. Phys.* 72, 3220 (1992).
283. S. Vepřek and M. G. J. Vepřek-Heijman, *Appl. Phys. Lett.* 56, 1766 (1990).

284. J. Dutta, U. Kroll, P. Chabloz, A. Shah, A. A. Howling, J.-L. Dorier, and C. Hollenstein, *J Appl. Phys.* 72, 3220 (1992).
285. T. Takagi, *J. Vac. Sci. Technol. A* 2, 282 (1983).
286. G. Turban, B. Drévillon, D. S. Mataras, and D. E. Rapakoulias, in "Plasma Deposition of Amorphous Silicon-Based Materials" (G. Bruno, P. Capezzuto, and A. Madan, Eds.), Chap. 2, p. 63. Academic Press, Boston, 1995.
287. A. Matsuda and N. Hata, in "Glow-Discharge Hydrogenated Amorphous Silicon" (K. Tanaka, Ed.), Chap. 2, p. 9. KTK Scientific Publishers, Tokyo, 1989.
288. A. Matsuda, T. Kaga, H. Tanaka, L. Malhotra, and K. Tanaka, *Japan. J. Appl. Phys.* 22, L115 (1983).
289. D. Mataras, S. Cavadias, and D. Rapakoulias, *J. Appl. Phys.* 66, 119 (1989).
290. D. Mataras, S. Cavadias, and D. Rapakoulias, *J. Vac. Sci. Technol. A* 11, 664 (1993).
291. N. Itabashi, K. Kato, N. Nishiwaki, T. Goto, C. Yamada, and E. Hirota, *Japan. J. Appl. Phys.* 27, L1565 (1988).
292. N. Itabashi, N. Nishiwaki, M. Magane, S. Naito, T. Goto, A. Matsuda, C. Yamada, and E. Hirota, *Jpn. J. Appl. Phys.* 29, L505 (1990).
293. N. Hershkowitz, M. H. Cho, C. H. Nam, and T. Intrator, *Plasma Chem. Plasma Process.* 8, 35 (1988).
294. P. Awakowicz, *Mater. Sci. Forum* 287–288, 3 (1998).
295. R. van der Heijden, Master's Thesis, Utrecht University, Utrecht, 1996.
296. E. R. Mosburg, R. C. Kerns, and J. R. Abelson, *J. Appl. Phys.* 54, 4916 (1983).
297. G. Bruno, P. Capezzuto, G. Cicala, P. Manodoro, and V. Tassielli, *IEEE Trans. Plasma Sci.* PS-18, 934 (1990).
298. S. Vepřek and M. Heintze, *Plasma Chem. Plasma Process.* 10, 3 (1990).
299. J. Cárabe, J. J. Gandía, and M. T. Gutiérrez, *J. Appl. Phys.* 73, 4618 (1993).
300. N. Spiliopoulos, D. Mataras, and D. E. Rapakoulias, *J. Electrochem. Soc.* 144, 634 (1997).
301. E. A. G. Hamers, W. G. J. H. M. van Sark, J. Bezemer, H. Meiling, and W. F. van der Weg, *J. Non-cryst. Solids* 226, 205 (1998).
302. G. Nolet, *J. Electrochem. Soc.* 122, 1030 (1975).
303. G. Turban, Y. Catherine, and B. Grolleau, *Thin Solid Films* 60, 147 (1979).
304. J. J. Wagner and S. Vepřek, *Plasma Chem. Plasma Process.* 2, 95 (1982).
305. G. Turban, Y. Catherine, and B. Grolleau, *Plasma Chem. Plasma Process.* 2, 61 (1982).
306. P. A. Longeway, H. A. Weakliem, and R. D. Estes, *J. Appl. Phys.* 57, 5499 (1985).
307. R. Robertson, D. Hils, H. Chatham, and A. Gallagher, *Appl. Phys. Lett.* 43, 544 (1983).
308. M. Heintze and S. Vepřek, *Appl. Phys. Lett.* 54, 1320 (1989).
309. G. Turban, Y. Catherine, and B. Grolleau, *Thin Solid Films* 77, 287 (1981).
310. G. Turban, *Pure Appl. Chem.* 56, 215 (1984).
311. P. Kae-Nune, J. Perrin, J. Guillon, and J. Jolly, *Plasma Sources Sci. Technol.* 4, 250 (1995).
312. R. Robertson and A. Gallagher, *J. Appl. Phys.* 59, 3402 (1986).
313. F. A. Baiocchi, R. C. Wetzel, and R. S. Freund, *Phys. Rev. Lett.* 53, 771 (1984).
314. R. S. Freund, R. C. Wetzel, R. J. Shul, and T. R. Hayes, *Phys. Rev. A* 41, 3575 (1990).

315. Hiden Analytical Ltd., 240 Europa Boulevard, Gemini Business Park, Warrington WA5 5UN, England.

316. P. Ho, W. G. Breiland, and R. J. Buss, *J. Chem. Phys.* 91, 2627 (1989).

317. J. Perrin, M. Shiratani, P. Kae-Nune, H. Videlot, J. Jolly, and J. Guillon, *J. Vac. Sci. Technol. A* 16, 278 (1998).

318. J. Perrin, C. Böhm, R. Eternadi, and A. Lloret, *Plasma Sources Sci. Technol.* 3, 252 (1994).

319. G. Turban, Y. Catherine, and B. Grolleau, *Thin Solid Films* 67, 309 (1980).

320. H. Chatham, D. Hils, R. Robertson, and A. Gallagher, *J. Chem. Phys.* 81, 1770 (1984).

321. A. A. Howling, L. Sansonnens, J.-L. Dorier, and C. Hollenstein, *J. Appl. Phys.* 75, 1340 (1994).

322. L. Boufendi and T. Bouchoule, A. Hbid, *J. Vac. Sci. Technol. A* 14, 572 (1996).

323. B. E. Thompson, K. D. Allen, A. D. Richards, and H. H. Sawin, *J. Appl. Phys.* 59, 1890 (1986).

324. J. Liu, G. L. Huppert, and H. H. Sawin, *J. Appl. Phys.* 68, 3916 (1990).

325. C. Wild and P. Koidl, *J. Appl. Phys.* 69, 2909 (1991).

326. R. J. M. M. Snijkers, M. J. M. Sambeek, G. M. W. Kroesen, and F. J. de Hoog, *Appl. Phys. Lett.* 63, 308 (1993).

327. J. K. Olthoff, R. J. van Brunt, S. B. Radovanov, J. A. Rees, and R. Surowiec, *J. Appl. Phys.* 75, 115 (1994).

328. M. Fivaz, S. Brunner, W. Swarzenbach, A. A. Howling, and C. Hollenstein, *Plasma Sources Sci. Technol.* 4, 373 (1995).

329. D. A. Dahl, J. E. Delmore, and A. D. Appelhans, *Rev. Sci. Instrum.* 61, 607 (1990).

330. M. L. Mandich, W. D. Reents, and K. D. Kolenbrander, *J. Chem. Phys.* 92, 437 (1990).

331. W. D. Reents and M. L. Mandich, *J. Chem. Phys.* 96, 4429 (1992).

332. E. A. G. Hamers, J. Bezemer, H. Meiling, W. G. J. H. M. van Sark, and W. F. van der Weg, *Mater. Res. Soc. Symp. Proc.* 467, 603 (1997).

333. R. J. M. M. Snijkers, Ph.D. Thesis, Technische Universiteit Eindhoven, Eindhoven, The Netherlands, 1993.

334. E. R. Fisher and P. B. Armentrout, *J. Chem. Phys.* 93, 4858 (1990).

335. W. N. Allen, T. M. H. Cheng, and F. W. Lampe, *J. Chem. Phys.* 66, 3371 (1977).

336. R. M. A. Azzam and N. M. Bashara, "Ellipsometry and Polarized Light." North Holland, Amsterdam, 1977.

337. B. Drévillon, J. Perrin, R. Marbot, A. Violet, and J. L. Dalby, *Rev. Sci. Instrum.* 53, 969 (1982).

338. N. Blayo and B. Drévillon, *Appl. Phys. Lett.* 59, 950 (1991).

339. R. W. Collins, *Rev. Sci. Instrum.* 61, 2029 (1990).

340. H. Fujiwara, Y. Toyoshima, M. Kondo, and A. Matsuda, *Phys. Rev. B* 60, 13598 (1999).

341. R. Brenot, B. Drévillon, P. Bulkin, P. Roca i Cabarrocas, and R. Vanderhaghen, *Appl. Surf. Sci.* 154–155, 283 (2000).

342. R. W. Collins, J. Koh, A. S. Ferlauto, P. I. Rovira, Y. Lee, R. J. Koval, and C. R. Wronski, *Thin Solid Films* 364, 129 (2000).

343. B. Drévillon, *Prog. Cryst. Growth Charact. Mater.* 27, 1 (1993).
344. R. Ossikovski and B. Drévillon, *Phys. Rev. B* 54, 10530 (1996).
345. H. Fujiwara, J. Koh, P. I. Rovira, and R. W. Collins, *Phys. Rev. B* 61, 10832 (2000).
346. D. E. Aspnes, J. B. Theeten, and F. Hottier, *Phys. Rev. B* 20, 3292 (1979).
347. A. Canillas, E. Bertran, J. L. Andújar, and B. Drévillon, *J. Appl. Phys.* 68, 2752 (1990).
348. Y. Lu, S. Kim, M. Gunes, Y. Lee, C. R. Wronski, and R. W. Collins, *Mater. Res. Soc. Symp. Proc.* 336, 595 (1994).
349. D. V. Tsu, B. S. Chao, S. R. Ovshinsky, S. Guha, and J. Yang, *Appl. Phys. Lett.* 71, 1317 (1997).
350. J. Koh, Y. Lee, H. Fujiwara, C. R. Wronski, and R. W. Collins, *Appl. Phys. Lett.* 73, 1526 (1998).
351. S. Guha, J. Yang, D. L. Williamson, Y. Lubianiker, J. D. Cohen, and A. H. Mahan, *Appl. Phys. Lett.* 74, 1860 (1999).
352. A. H. Mahan, J. Yang, S. Guha, and D. L. Williamson, *Phys. Rev. B* 61, 1677 (2000).
353. H. Fujiwara, J. Koh, C. R. Wronski, and R. W. Collins, *Appl. Phys. Lett.* 70, 2150 (1997).
354. H. Fujiwara, J. Koh, C. R. Wronski, and R. W. Collins, *Appl. Phys. Lett.* 72, 2993 (1998).
355. H. Fujiwara, J. Koh, C. R. Wronski, and R. W. Collins, *Appl. Phys. Lett.* 74, 3687 (1999).
356. N. Blayo and B. Drévillon, *J. Non-cryst. Solids* 137 & 138, 771 (1991).
357. A. Matsuda, *J. Vac. Sci. Technol. A* 16, 365 (1998).
358. B. Drévillon, J. Perrin, J. M. Siefert, J. Huc, A. Lloret, G. de Rosny, and J. P. M. Schmitt, *Appl. Phys. Lett.* 42, 801 (1983).
359. R. A. Street, J. C. Knights, and D. K. Biegelsen, *Phys. Rev. B* 19, 3027 (1978).
360. J. C. Knights and R. A. Lujan, *Appl. Phys. Lett.* 35, 244 (1979).
361. S. Nishikawa, H. Kakinuma, T. Watanabe, and K. Nihei, *Japan. J. Appl. Phys.* 24, 639 (1985).
362. G. D. Cody, T. Tiedje, B. Abeles, B. Brooks, and Y. Goldstein, *Phys. Rev. Lett.* 47, 1480 (1981).
363. K. M. H. Maessen, M. J. M. Pruppers, J. Bezemer, F. H. P. M. Habraken, and W. F. van der Weg, *Mater. Res. Soc. Symp. Proc.* 95, 201 (1987).
364. K. Tanaka, K. Nakagawa, A. Matsuda, M. Matsumura, H. Yamamoto, S. Yamasaki, H. Okushi, and S. Izima, *Japan. J. Appl. Phys.* 20, 267 (1981).
365. P. Sichanugrist, M. Konagai, and K. Takahashi, *Japan. J. Appl. Phys.* 25, 440 (1986).
366. M. Hirose, *J. Phys. (Paris), Colloq.* 42, C4 (1981).
367. P. Chaudhuri, S. Ray, and A. K. Barua, *Thin Solid Films* 113, 261 (1984).
368. J. Shirafuji, S. Nagata, and M. Kuwagaki, *J. Appl. Phys.* 58, 3661 (1985).
369. S. Okamoto, Y. Hishikawa, and S. Tsuda, *Japan. J. Appl. Phys.* 35, 26 (1996).
370. H. Meiling, R. E. I. Schropp, W. G. J. H. M. van Sark, J. Bezemer, and W. F. van der Weg, "Proceedings of the Tenth E.C. Photovoltaic Solar Energy Conference, Lisbon, Portugal, 1991" (A. Luque, G. Sala, W. Palz, G. Dos Santos, and P. Helm, Eds.), p. 359. Kluwer Academic, Dordrecht, 1991.
371. J. P. M. Schmitt, *J. Non-cryst. Solids* 59 & 60, 649 (1983).

372. C. C. Tsai, G. B. Anderson, and R. Thompson, *J. Non-cryst. Solids* 137–138, 637 (1991).
373. P. Roca i Cabarrocas, S. Hamma, S. N. Sharma, G. Viera, E. Bertran, and J. Costa, *J. Non-cryst. Solids* 227–230, 871 (1998).
374. C. Longeaud, J. P. Kleider, P. Roca i Cabarrocas, S. Hamma, R. Meaudre, and M. Meaudre, *J. Non-cryst. Solids* 227–230, 96 (1998).
375. J. C. Knights, *Japan. J. Appl. Phys.* 18, 101 (1979).
376. R. B. Wehrspohn, S. C. Deane, I. D. French, I. Gale, J. Hewett, M. J. Powell, and J. Robertson, *J. Appl. Phys.* 87, 144 (2000).
377. A. A. Howling, J.-L. Dorier, and C. Hollenstein, *Appl. Phys. Lett.* 62, 1341 (1993).
378. A. Bouchoule, Ed., "Dusty Plasmas: Physics, Chemistry, and Technological Impact in Plasma Processing." Wiley, New York, 1999.
379. P. Roca i Cabarrocas, *J. Non-cryst. Solids* 266–269, 31 (2000).
380. M. Otobe, T. Kanai, T. Ifuku, H. Yajima, and S. Oda, *J. Non-cryst. Solids* 198–200, 875 (1996).
381. D. M. Tanenbaum, A. L. Laracuente, and A. Gallagher, *Appl. Phys. Lett.* 68, 705 (1996).
382. P. Roca i Cabarrocas, *Mater. Res. Soc. Symp. Proc.* 507, 855 (1998).
383. E. Stoffels, W. W. Stoffels, G. M. W. Kroesen, and F. J. de Hoog, *Electron Technol.* 31, 255 (1998).
384. C. Courteille, J.-L. Dorier, J. Dutta, C. Hollenstein, A. A. Howling, and T. Stoto, *J. Appl. Phys.* 78, 61 (1995).
385. S. Schmitt-Rink, D. A. B. Miller, and D. S. Chemla, *Phys. Rev. B* 35, 8113 (1987).
386. A. J. Read, R. J. Needs, K. J. Nash, L. T. Canham, P. D. J. Calcott, and A. Qteish, *Phys. Rev. Lett.* 69, 1232 (1992).
387. P. Roca i Cabarrocas, P. Gay, and A. Hadjadj, *J. Vac. Sci. Technol. A* 14, 655 (1996).
388. A. Fontcuberta i Morral, R. Brenot, E. A. G. Hamers, R. Vanderhaghen, and P. Roca i Cabarrocas, *J. Non-cryst. Solids* 266–269, 48 (2000).
389. R. Butté, S. Vignoli, M. Meaudre, R. Meaudre, O. Marty, L. Saviot, and P. Roca i Cabarrocas, *J. Non-cryst. Solids* 266–269, 263 (2000).
390. Corning 7059 Material Information, Corning Glass Works, NY.
391. J. Wallinga, D. Knoesen, E. A. G. Hamers, W. G. J. H. M. van Sark, W. F. van der Weg, and R. E. I. Schropp, *Mater. Res. Soc. Symp. Proc.* 420, 57 (1996).
392. M. B. Schubert and G. H. Bauer, "Proceedings of the Twenty-First IEEE Photovoltaic Specialists Conference, Kissimee, U.S.A., 1990," p. 1595. IEEE, New York, 1990.
393. Y. Hishikawa, M. Ohnishi, and Y. Kuwano, *Mater. Res. Soc. Symp. Proc.* 192, 3 (1990).
394. A. Shin and S. Lee, *J. Non-cryst. Solids* 260, 245 (1999).
395. N. Maley and J. S. Lannin, *Phys. Rev. B* 31, 5577 (1985).
396. A. J. M. Berntsen, W. G. J. H. M. van Sark, and W. F. van der Weg, *J. Appl. Phys.* 78, 1964 (1995).
397. R. W. Collins and J. M. Cavese, *J. Appl. Phys.* 61, 1869 (1987).
398. Y. Toyoshima, K. Arai, A. Matsuda, and K. Tanaka, *Appl. Phys. Lett.* 56, 1540 (1990).
399. G. Ganguly and A. Matsuda, *J. Non-cryst. Solids* 164–166, 31 (1993).
400. S. Veprek, F.-A. Sarrott, S. Rambert, and E. Taglauer, *J. Vac. Sci. Technol. A* 7, 2614 (1989).

401. B. J. Garrison, M. T. Miller, and D. W. Bremmen, *Chem. Phys. Lett.* 146, 513 (1988).
402. M. Kitabatake, P. Fons, and J. E. Greene, *J. Vac. Sci. Technol. A* 8, 3726 (1990).
403. P. Roca i Cabarrocas, *Mater. Res. Soc. Symp. Proc.* 149, 33 (1989).
404. G. Ganguly, T. Ikeda, I. Sakata, and A. Matsuda, *Mat. Res. Soc. Proc. Symp.* 420, 347 (1996).
405. A. I. Kosarev, A. S. Smirnov, A. S. Abramov, A. J. Vinogradov, A. Y. Ustavschikov, and M. V. Shutov, *J. Vac. Sci. Technol. A* 15, 298 (1997).
406. A. S. Abramov, A. Y. Vinogradov, A. I. Kosarev, M. V. Shutov, A. S. Smirnov, and K. E. Orlov, *Tech. Phys.* 43, 180 (1998).
407. J. P. Harbison, A. J. Williams, and D. V. Lang, *J. Appl. Phys.* 55, 946 (1984).
408. K. H. Müller, *J. Appl. Phys.* 62, 1796 (1987).
409. J. Robertson, *J. Non-cryst. Solids* 164–166, 1115 (1993).
410. J. R. Abelson, *Appl. Phys. A* 56, 493 (1993).
411. R. Biswas, *J. Appl. Phys.* 73, 3295 (1993).
412. W. Möller, *Appl. Phys. A* 56, 527 (1993).
413. M. Hoedemaker, J. van der Kuur, E. J. Melker, and B. J. Thijsse, *Nucl. Instrum. Methods Phys. Res. B* 127/128, 888 (1997).
414. J. F. Ziegler, J. P. Biersack, and U. Littmarck, "The Stopping and Range of Ions in Solids." Pergamon, New York, 1985.
415. B. Drévillon, J. Huc, and N. Boussarssar, *J. Non-cryst. Solids* 59/60, 735 (1983).
416. B. Drévillon and M. Toutlemonde, *Appl. Phys. Lett.* 59, 950 (1985).
417. J. Perrin, Y. Takeda, N. Hirano, H. Matsuura, and A. Matsuda, *Japan. J. Appl. Phys.* 28, 5 (1989).
418. P. Roca i Cabarrocas, P. Morin, V. Chu, J. P. Conde, J. Z. Liu, H. R. Park, and S. Wagner, *J. Appl. Phys.* 69, 2942 (1991).
419. E. A. G. Hamers, J. Bezemer, and W. F. van der Weg, *Appl. Phys. Lett.* 75, 609 (1999).
420. D. A. Doughty, J. R. Doyle, G. H. Lin, and A. Gallagher, *J. Appl. Phys.* 67, 6220 (1990).
421. J. K. Hirvonen, in "Materials and Processes for Surface and Interface Engineering" (Y. Pauleau, Ed.), Chap. 9, p. 307. Kluwer Academic, Dordrecht, 1995.
422. J. E. Potts, E. M. Peterson, and J. A. McMillan, *J. Appl. Phys.* 52, 6665 (1981).
423. H. Meiling, J. Bezemer, R. E. I. Schropp, and W. F. van der Weg, *Mater. Res. Soc. Proc. Symp.* 467, 459 (1997).
424. S. Habermehl and G. Lucovsky, *J. Vac. Sci. Technol. A* 14, 3024 (1996).
425. S. R. Kasi, H. Kang, C. S. Sass, and J. W. Rabalais, *Surf. Sci. Rep.* 10, 1 (1989).
426. L. Hanley, H. Lim, D. G. Schultz, S. B. Wainhaus, P. de Sainte Claire, and W. L. Hase, *Nucl. Instrum. Methods Phys. Res. B* 125, 218 (1997).
427. H. Sugai, Y. Mitsuoka, and H. Toyoda, *J. Vac. Sci. Technol. A* 16, 290 (1998).
428. K. H. Müller, *Phys. Rev. B* 35, 7906 (1987).
429. J. W. Rabalais, A. H. Al-Bayati, K. J. Boyd, D. Marton, J. Kulik, Z. Zhang, and W. K. Chu, *Phys. Rev. B* 53, 10781 (1996).
430. K. H. Müller, *J. Appl. Phys.* 59, 2803 (1986).
431. J. E. Yehoda, B. Yang, K. Vedam, and R. Messier, *J. Vac. Sci. Technol. A* 6, 1631 (1988).
432. R. J. Severens, M. C. M. van de Sanden, H. J. M. Verhoeven, J. Bastiaansen, and D. C. Schram, *Mater. Res. Soc. Symp. Proc.* 420, 341 (1996).

433. M. C. M. van de Sanden, R. J. Severens, W. M. M. Kessels, F. van de Pas, L. van Ijzendoorn, and D. C. Schram, *Mater. Res. Soc. Proc. Symp.* 467, 621 (1997).
434. E. H. C. Ullersma, Ph.D. Thesis, Universiteit Utrecht, Utrecht, 1998.
435. R. Robertson and A. Gallagher, *J. Chem. Phys.* 85, 3623 (1986).
436. A. Gallagher, *J. Appl. Phys.* 63, 2406 (1988).
437. J. L. Guizot, K. Nomoto, and A. Matsuda, *Surf. Sci.* 244, 22 (1991).
438. A. J. Flewitt, J. Robertson, and W. I. Milne, *J. Appl. Phys.* 85, 8032 (1999).
439. J. Robertson, *J. Appl. Phys.* 87, 2608 (2000).
440. G. Ganguly and A. Matsuda, *Phys. Rev. B* 47, 3661 (1993).
441. K. Winer, *Phys. Rev. B* 41, 7952 (1990).
442. R. A. Street, *Phys. Rev. B* 43, 2454 (1991).
443. R. A. Street, *Phys. Rev. B* 44, 10610 (1991).
444. R. A. Street and K. Winer, *Phys. Rev. B* 40, 6236 (1989).
445. M. J. Powell and S. C. Deane, *Phys. Rev. B* 48, 10815 (1993).
446. M. J. Powell and S. C. Deane, *Phys. Rev. B* 53, 10121 (1996).
447. K. Gleason, K. S. Wang, M. K. Chen, and J. A. Reimer, *J. Appl. Phys.* 61, 2866 (1987).
448. M. J. McCaughy and M. J. Kushner, *J. Appl. Phys.* 65, 186 (1989).
449. S. Shirafuji, S. Nakajima, Y. F. Wang, T. Genji, and K. Tachibana, *Japan. J. Appl. Phys.* 32, 1546 (1993).
450. J. Robertson, *Mater. Res. Soc. Proc. Symp.* 609, A1.4.1 (2000).
451. C. C. Tsai, J. C. Knights, G. Chang, and B. Wacker, *J. Appl. Phys.* 59, 2998 (1986).
452. A. J. Flewitt, J. Robertson, and W. I. Milne, *J. Non-cryst. Solids* 266–269, 74 (2000).
453. A. von Keudell and J. R. Abelson, *Japan. J. Appl. Phys.* 38, 4002 (1999).
454. S. Ramalingam, D. Maroudas, and E. S. Aydill, *J. Appl. Phys.* 86, 2872 (1999).
455. S. Yamasaki, T. Umeda, J. Isoya, and K. Tanaka, *Appl. Phys. Lett.* 70, 1137 (1997).
456. D. A. Porter and K. E. Easterling, "Phase Transformations in Metals and Alloys." Chapman & Hall, London, 1992.
457. P. Roca i Cabarrocas, Y. Bouizem, and M. L. Theye, *Philos. Mag. B* 65, 1025 (1992).
458. C. M. Greenlief, S. M. Gates, and P. A. Holbert, *J. Vac. Sci. Technol. A* 7, 1845 (1989).
459. W. B. Jackson, *J. Non-cryst. Solids* 164–166, 263 (1993).
460. W. Beyer, *J. Non-Cryst. Solids* 198–200, 40 (1996).
461. W. Beyer and U. Zastrow, *Mater. Res. Soc. Symp. Proc.* 420, 497 (1996).
462. C. G. Van de Walle, *Phys. Rev. B* 49, 4579 (1994).
463. J. Robertson, C. W. Chen, M. J. Powell, and S. C. Deane, *J. Non-cryst. Solids* 227–230, 138 (1998).
464. K. J. Chang and J. D. Chadi, *Phys. Rev. Lett.* 62, 937 (1989).
465. K. J. Chang and J. D. Chadi, *Phys. Rev. B* 40, 11644 (1990).
466. Y. Miyoshi, Y. Yoshida, S. Miyazaki, and M. Hirose, *J. Non-cryst. Solids* 198–200, 1029 (1996).
467. J. Robertson and M. J. Powell, *Thin Solid Films* 337, 32 (1999).
468. W. M. M. Kessels, R. J. Severens, M. C. M. van de Sanden, and D. C. Schram, *J. Non-cryst. Solids* 227–230, 133 (1998).
469. G. H. Lin, J. R. Doyle, M. Z. He, and A. Gallagher, *J. Appl. Phys.* 64, 341 (1988).

470. P. Roca i Cabarrocas, *Appl. Phys. Lett.* 65, 1674 (1995).
471. H. Shirai, J. Hanna, and I. Shimizu, *Japan. J. Appl. Phys.* 30, L679 (1991).
472. M. C. M. van de Sanden, W. M. M. Kessels, R. J. Severens, and D. C. Schram, *Plasma Phys. Controlled Fusion* 41, A365 (1999).
473. M. C. M. van de Sanden, W. M. M. Kessels, A. H. Smets, B. A. Korevaar, R. J. Severens, and D. C. Schram, *Mater. Res. Soc. Symp. Proc.* 557, 13 (1999).
474. M. Stutzmann, *Philos. Mag. B* 60, 531 (1989).
475. A. Shah, J. Dutta, N. Wyrsch, K. Prasad, H. Curtins, F. Finger, A. Howling, and C. Hollenstein, *Mater. Res. Soc. Symp. Proc.* 258, 15 (1992).
476. C. M. Ferreira and J. Loureiro, *J. Phys. D* 17, 1175 (1984).
477. J. Kuske, U. Stephan, O. Steinke, and S. Röhlecke, *Mater. Res. Soc. Symp.* Proc. 377, 27 (1995).
478. U. Kroll, J. Meier, M. Goetz, A. Howling, J.-L. Dorier, J. Dutta, A. Shah, and C. Hollenstein, *J. Non-cryst. Solids* 164–166, 59 (1993).
479. H. Meiling, J. F. M. Westendorp, J. Hautala, Z. Saleh, and C. T. Malone, *Mater. Res. Soc. Symp. Proc.* 345, 65 (1995).
480. S. Röhlecke, R. Tews, A. Kottwitz, and K. Schade, *Surf. Coat. Technol.* 74–75, 259 (1995).
481. C. Beneking, *J. Appl. Phys.* 68, 4461 (1990).
482. C. Beneking, *J. Appl. Phys.* 68, 5435 (1990).
483. C. Beneking, F. Finger, and H. Wagner, "Proceedings of the Eleventh E.C. Photovoltaic Solar Energy Conference, Montreux, Switzerland 1992" (L. Guimarães, W. Palz, C. de Reyff, H. Kiess, and P. Helm, Eds.), p. 586. Harwood, Chur, Switzerland, 1993.
484. M. J. Colgan, M. Meyyappan, and D. E. Murnick, *Plasma Sources Sci. Technol.* 3, 181 (1994).
485. U. Kroll, Y. Ziegler, J. Meier, H. Keppner, and A. Shah, *Mater. Res. Soc. Symp. Proc.* 336, 115 (1994).
486. E. A. G. Hamers, W. G. J. H. M. van Sark, J. Bezemer, W. F. van der Weg, G. J. Nienhuis, and W. J. Goedheer, "Proceedings of the Thirteenth European Sectional Conference on the Atomic and Molecular Physics of Ionized Gases," ESCAMPIG 96, Poprad, Slovakia, *Europhys. Conf. Abs.* 20E, 27 (1996).
487. W. G. J. H. M. van Sark, H. Meiling, E. A. G. Hamers, J. Bezemer, and W. F. van der Weg, *J. Vac. Sci. Technol. A* 15, 654 (1997).
488. H. Meiling, W. G. J. H. M. van Sark, J. Bezemer, and W. van der Weg, *J. Appl. Phys.* 80, 3546 (1996).
489. G. E. N. Landweer and J. Bezemer, in "Amorphous Silicon and Related Materials, Advances in Disordered Semiconductors 1" (H. Fritzsche, Ed.). World Scientific, Singapore, 1989.
490. M. Zhu, M. B. von der Linden, J. Bezemer, R. E. I. Schropp, and W. F. van der Weg, *J. Non-cryst. Solids* 137–138, 355 (1991).
491. M. Zhu, M. B. von der Linden, and W. F. van der Weg, *Mater. Res. Soc. Symp. Proc.* 336, 449 (1994).
492. M. B. von der Linden, Ph.D. Thesis, Universiteit Utrecht, Utrecht, 1994.
493. W. G. J. H. M. van Sark, J. Bezemer, and W. F. van der Weg, *J. Mater. Res.* 13, 45 (1998).

494. R. E. I. Schropp, H. Meiling, W. G. J. H. M. van Sark, J. Stammeijer, J. Bezemer, and W. F. van der Weg, "Proceedings of the Tenth E.C. Photovoltaic Solar Energy Conference, Lisbon, Portugal, 1991" (A. Luque, G. Sala, W. Palz, G. Dos Santos, and P. Helm, Eds.), p. 1087. Kluwer Academic, Dordrecht, 1991.

495. S. J. Jones, D. L. Williamson, T. Liu, X. Deng, and M. Izu, *Mater. Res. Soc. Proc. Symp.* 609, A7.4.1 (2000).

496. C. C. Tsai, R. Thompson, C. Doland, F. A. Ponce, G. B. Anderson, and B. Wacker, *Mater. Res. Soc. Symp. Proc.* 118, 49 (1988).

497. Y. S. Tsuo, R. Weil, S. Asher, A. Nelson, Y. Xu, and R. Tsu, "Proceedings of the 19th IEEE Photovoltaic Specialists Conference, New Orleans, 1987," p. 705. IEEE, New York, 1987.

498. H. Shirai, D. Das, J. Hanna, and I. Shimizu, *Appl. Phys. Lett.* 59, 1096 (1991).

499. A. Asano, T. Ichimura, and H. Sakai, *J. Appl. Phys.* 65, 2439 (1989).

500. A. Asano, *Appl. Phys. Lett.* 56, 533 (1990).

501. G. Parsons, *IEEE Electron Device Lett.* 13, 80 (1992).

502. N. Layadi, P. Roca i Cabarrocas, B. Drévillon, and I. Solomon, *Phys. Rev. B* 52, 5136 (1995).

503. P. Roca i Cabarrocas, N. Layadi, B. Drévillon, and I. Solomon, *J. Non-cryst. Solids* 198–200, 871 (1996).

504. P. Roca i Cabarrocas, S. Hamma, A. Hadjadi, J. Bertomeu, and J. Andreu, *Appl. Phys. Lett.* 69, 529 (1996).

505. N. Shibata, K. Fukada, H. Ohtoshi, J. Hanna, S. Oda, and I. Shimizu, *Mater. Res. Soc. Symp. Proc.* 95, 225 (1987).

506. Y. H. Yang, M. Katiyar, G. F. Feng, N. Maley, and J. R. Abelson, *Appl. Phys. Lett.* 65, 1769 (1994).

507. S. Hamma and P. Roca i Cabarrocas, *J. Appl. Phys.* 81, 7282 (1997).

508. P. Roca i Cabarrocas, R. Brenot, P. Bulkin, R. Vanderhaghen, B. Drévillon, and I. French, *J. Appl. Phys.* 86, 7079 (1999).

509. K. Saito, M. Kondo, M. Fukawa, T. Nishimiya, A. Matsuda, W. Futako, and I. Shimizu, *Appl. Phys. Lett.* 71, 3403 (1997).

510. C. Anandan, C. Mukherjee, T. Seth, P. N. Dixit, and R. Bhattacharyya, *Appl. Phys. Lett.* 66, 85 (1995).

511. C. Mukherjee, C. Anandan, T. Seth, P. N. Dixit, and R. Bhattacharyya, *Appl. Phys. Lett.* 68, 194 (1996).

512. A. C. W. Biebericher, J. Bezemer, W. F. van der Weg, and W. J. Goedheer, *Appl. Phys. Lett.* 76, 2002 (2000).

513. H. Kirimura, H. Maeda, H. Murakami, T. Nakahigashi, S. Ohtani, T. Tabata, T. Hayashi, M. Kobayashi, Y. Mitsuda, N. Nakamura, H. Kuwahara, and A. Doi, *Japan. J. Appl. Phys.* 33, 4389 (1994).

514. S. Morrison, J. Xi, and A. Madan, *Mater. Res. Soc. Symp. Proc.* 507, 559 (1998).

515. A. Madan, S. Morrison, and H. Kuwahara, *Sol. Energy Mater. Sol. Cells* 59, 51 (1999).

516. G. Scarsbrook, I. P. Llewellyn, S. M. Ojha, and R. A. Heinecke, *Vacuum* 38, 627 (1988).

517. L. J. Overzet and J. T. Verdeyen, *Appl. Phys. Lett.* 48, 695 (1986).

518. J. P. Booth, G. Cunge, N. Sadeghi, and R. W. Boswell, *J. Appl. Phys.* 82, 552 (1997).

519. A. C. W. Biebericher, J. Bezemer, W. F. van der Weg, and W. J. Goedheer, *Mater. Res. Soc. Proc. Symp.* 609, A4.1.1 (2000).

520. A. C. W. Biebericher, J. Bezemer, W. F. van der Weg, and W. J. Goedheer, unpublished results.

521. H. Wiesmann, A. K. Ghosh, T. McMahon, and M. Strongin, *J. Appl. Phys.* 50, 3752 (1979).

522. H. Matsumura, *Japan. J. Appl. Phys.* 25, L949 (1986).

523. H. Matsumura, *Appl. Phys. Lett.* 51, 804 (1987).

524. H. Matsumura, *J. Appl. Phys.* 65, 4396 (1989).

525. J. R. Doyle, R. Robertson, G. H. Lin, M. Z. He, and A. Gallagher, *J. Appl. Phys.* 64, 3215 (1988).

526. F. Jansen, I. Chen, and M. A. Machonkin, *J. Appl. Phys.* 66, 5749 (1989).

527. A. H. Mahan and M. Vaněček, "Amorphous Silicon Materials and Solar Cells," AIP Conference Proceedings, Vol. 234 (B. Stafford, Ed.), p. 195. American Institute of Physics, New York, 1991.

528. E. C. Molenbroek, A. H. Mahan, E. Johnson, and A. C. Gallagher, *J. Appl. Phys.* 79, 7278 (1996).

529. M. Heintze, R. Zedlitz, H. N. Wanka, and M. B. Schubert, *J. Appl. Phys.* 79, 2699 (1996).

530. E. C. Molenbroek, A. H. Mahan, and A. C. Gallagher, *J. Appl. Phys.* 82, 1909 (1997).

531. K. F. Feenstra, R. E. I. Schropp, and W. F. van der Weg, *J. Appl. Phys.* 85, 6843 (1999).

532. P. Papadopoulos, A. Scholz, S. Bauer, B. Schröder, and H. Öchsner, *J. Non-cryst. Solids* 164–166, 87 (1993).

533. B. P. Nelson, E. Iwaniczko, R. E. I. Schropp, H. Mahan, E. C. Molenbroek, S. Salamon, and R. S. Crandall, "Proceedings of the Twelfth E.C. Photovoltaic Solar Energy Conference, Amsterdam, the Netherlands, 1994" (R. Hill, W. Palz, and P. Helm, Eds.), p. 679. H. S. Stephens & Associates, Bedford, U.K., 1994.

534. A. H. Mahan, E. Iwaniczko, B. P. Nelson, R. C. Reedy, Jr., R. S. Crandall, S. Guha, and J. Yang, "Proceedings of the 25th IEEE Photovoltaic Specialists Conference, Washington DC, 1996," p. 1065. IEEE, New York, 1996.

535. R. E. I. Schropp, K. F. Feenstra, E. C. Molenbroek, H. Meiling, and J. K. Rath, *Philos. Mag. B* 76, 309 (1997).

536. R. S. Crandall, A. H. Mahan, B. Nelson, M. Vaněček, and I. Balberg, "AIP Conference Proceedings" (R. Noufi, Ed.), Vol. 268, p. 81. American Institute of Physics, Denver, 1992.

537. Q. Wang, E. Iwaniczko, Y. Xu, W. Gao, B. P. Nelson, A. H. Mahan, R. S. Crandall, and H. M. Branz, *Mater. Res. Soc. Proc. Symp.* 609, A4.3.1 (2000).

538. H. Meiling and R. E. I. Schropp, *Appl. Phys. Lett.* 70, 2681 (1997).

539. V. Chu, J. Jarego, H. Silva, T. Silva, M. Reissner, P. Brogueira, and J. P. Conde, *Appl. Phys. Lett.* 70, 2714 (1997).

540. H. Meiling, A. M. Brockhoff, J. K. Rath, and R. E. I. Schropp, *J. Non-cryst. Solids* 227–230, 1202 (1998).

541. J. Cifre, J. Bertomeu, J. Puigdollers, M. C. Polo, J. Andreu, and A. Lloret, *Appl. Phys. A* 59, 645 (1994).

542. A. R. Middya, S. Hazra, S. Ray, C. Longeaud, and J. P. Kleider, "Proceedings of the Thirteenth European Photovoltaic Solar Energy Conference, Nice, France, 1995" (W. Freiesleben, W. Palz, H. A. Ossenbrink, and P. Helm, Eds.), p. 679. H. S. Stephens & Associates, Bedford, U.K., 1995.
543. P. Brogueira, J. P. Conde, S. Arekat, and V. Chu, *J. Appl. Phys.* 79, 8748 (1996).
544. J. K. Rath, H. Meiling, and R. E. I. Schropp, *Sol. Energy Mater. Sol. Cells* 48, 269 (1997).
545. H. N. Wanka, R. Zedlitz, M. Heintze, and M. B. Schubert, "Proceedings of the Thirteenth European Photovoltaic Solar Energy Conference, Nice, France, 1995" (W. Freiesleben, W. Palz, H. A. Ossenbrink, and P. Helm, Eds.), p. 1753. H.S. Stephens & Associates, Bedford, U.K., 1995.
546. V. Brogueira, P. Chu, A. C. Ferro, and J. P. Conde, *J. Vac. Sci. Technol. A* 15, 2968 (1997).
547. P. Alpuim, V. Chu, and J. P. Conde, *Mater. Res. Soc. Proc. Symp.* 609, A22.6.1 (2000).
548. K. F. Feenstra, Ph.D. Thesis, Universiteit Utrecht, Utrecht, 1998.
549. C. Horbach, W. Beyer, and H. Wagner, *J. Non-cryst. Solids* 137 & 138, 661 (1991).
550. S. Bauer, B. Schröder, and H. Oechsner, *J. Non-cryst. Solids* 227–230, 34 (1998).
551. A. H. Mahan, Y. Chen, D. L. Williamson, and G. D. Mooney, *J. Non-cryst. Solids* 137–138, 65 (1991).
552. Y. Wu, J. T. Stephen, D. X. Han, J. M. Rutland, R. S. Crandall, and A. H. Mahan, *Phys. Rev. Lett.* 77, 2049 (1996).
553. R. P. Muller, J. K. Holt, D. G. Goodwin, and W. A. Goddard, III, *Mater. Res. Soc. Proc. Symp.* 609, A6.1.1 (2000).
554. J. K. Holt, M. Swiatek, D. G. Goodwin, and H. A. Atwater, *Mater. Res. Soc. Proc. Symp.* 609, A6.2.1 (2000).
555. B. Schröder and S. Bauer, *J. Non-cryst. Solids* 266–269, 1115 (2000).
556. A. T. M. Wilbers, G. M. W. Kroesen, C. J. Timmermans, and D. C. Schram, *Meas. Sci. Technol.* 1, 1326 (1990).
557. J. W. A. M. Gielen, P. R. M. Kleuskens, M. C. M. van de Sanden, L. J. van IJzendoorn, D. C. Schram, E. H. A. Dekempeneer, and J. Meneve, *J. Appl. Phys.* 80, 5986 (1996).
558. J. J. Beulens, A. J. M. Buuron, and D. C. Schram, *Surf. Coat. Technol.* 47, 401 (1991).
559. M. J. de Graaf, R. J. Severens, R. P. Dahiya, M. C. M. van de Sanden, and D. C. Schram, *Phys. Rev. E* 48, 2098 (1993).
560. R. F. G. Meulenbroeks, M. F. M. Steenbakkers, Z. Qing, M. C. M. van de Sanden, and D. C. Schram, *Phys. Rev. E* 49, 2272 (1994).
561. W. M. M. Kessels, A. H. M. Smets, B. A. Korevaar, G. J. Adriaenssens, M. C. M. van de Sanden, and D. C. Schram, *Mater. Res. Soc. Symp. Proc.* 557, 25 (1999).
562. B. A. Korevaar, G. J. Adriaenssens, A. H. M. Smets, W. M. M. Kessels, H.-Z. Song, M. C. M. van de Sanden, and D. C. Schram, *J. Non-cryst. Solids* 266–269, 380 (2000).
563. M. C. M. van de Sanden, R. J. Severens, W. M. M. Kessels, R. F. G. Meulenbroeks, and D. C. Schram, *J. Appl. Phys.* 84, 2426 (1998).
564. M. C. M. van de Sanden, R. J. Severens, W. M. M. Kessels, R. F. G. Meulenbroeks, and D. C. Schram, *J. Appl. Phys.* 85, 1243 (1999).

565. J. J. Beulens, M. J. de Graaf, and D. C. Schram, *Plasma Sources Sci. Technol.* 2, 180 (1993).
566. M. C. M. van de Sanden, J. M. de Regt, and D. C. Schram, *Plasma Sources Sci. Technol.* 3, 501 (1994).
567. W. M. M. Kessels, C. M. Leewis, M. C. M. van de Sanden, and D. C. Schram, *J. Appl. Phys.* 86, 4029 (1999).
568. R. F. G. Meulenbroeks, R. A. H. Engeln, M. N. A. Beurskens, R. M. J. Paffen, M. C. M. van de Sanden, J. A. M. van der Mullen, and D. C. Schram, *Plasma Sources Sci. Technol.* 4, 74 (1995).
569. W. M. M. Kessels, M. C. M. van de Sanden, and D. C. Schram, *J. Vac. Sci. Technol. A* 18, 2153 (2000).
570. W. M. M. Kessels, M. C. M. van de Sanden, R. J. Severens, L. J. van Ijzendoorn, and D. C. Schram, *Mater. Res. Soc. Symp. Proc.* 507, 529 (1998).
571. W. M. M. Kessels, Ph.D. Thesis, Technische Universiteit Eindhoven, Eindhoven, the Netherlands, 2000.
572. W. M. M. Kessels, R. J. Severens, A. H. M. Smets, B. A. Korevaar, G. J. Adriaenssens, D. C. Schram, and M. C. M. van de Sanden, *J. Appl. Phys.* 89, 2404 (2001).
573. W. M. M. Kessels, J. P. M. Hoefnagels, M. G. H. Boogderts, D. C. Schram, and M. C. M. van de Sanden, *J. Appl. Phys.* 89, 2065 (2001).
574. H. S. Nalwa, Ed., "Handbook of Advanced Electronic and Photonic Materials and Devices." Academic Press, Boston, 2000.
575. M. A. Green, "Solar Cells, Operating Principles, Technology and System Applications." University of New South Wales, Kensington, NSW, Australia, 1992.
576. M. Hack and M. Shur, *J. Appl. Phys.* 58, 997 (1985).
577. A. Catalano, in "Amorphous and Microcrystalline Semiconductor Devices—Optoelectronic Devices" (J. Kanicki, Ed.), Chap. 2, p. 9. Artech House, Norwood, MA, 1991.
578. A. Polman, W. G. J. H. M. van Sark, W. C. Sinke, and F. W. Saris, *Sol. Cells* 17, 241 (1986).
579. H. Hartnagel, A. Dawar, A. Jain, and C. Jagadish, "Semiconducting Transparent Thin Films." Institute of Physics Publishing, Bristol, 1995.
580. E. Yablonovitch and G. D. Cody, *IEEE Trans. Electron. Dev.* ED-29, 300 (1982).
581. H. W. Deckman, C. R. Wronski, H. Witzke, and E. Yablonovitch, *Appl. Phys. Lett.* 42, 968 (1983).
582. J. Daey Ouwens, R. E. I. Schropp, J. Wallinga, W. F. van der Weg, M. Ritala, M. Leskelä, andM. Hyrvärinen, "Proceedings of the Twelfth E.C. Photovoltaic Solar Energy Conference, Amsterdam, the Netherlands, 1994" (R. Hill, W. Palz, and P. Helm, Eds.), p. 1296. H.S. Stephens & Associates, Bedford, U.K., 1994.
583. M. Kubon, E. Böhmer, F. Siebke, B. Rech, C. Beneking, and H. Wagner, *Sol. Energy Mater. Sol. Cells* 41/42, 485 (1996).
584. Y. Tawada, H. Okamoto, and Y. Hamakawa, *Appl. Phys. Lett.* 39, 237 (1981).
585. K. Miyachi, N. Ishiguro, T. Miyashita, N. Yanagawa, H. Tanaka, M. Koyama, Y. Ashida, and N. Fukuda, "Proceedings of the Eleventh E.C. Photovoltaic Solar Energy Conference, Montreux, Switzerland, 1992" (L. Guimarães, W. Palz, C. de Reyff, H. Kiess, and P. Helm, Eds.), p. 88. Harwood, Chur, Switzerland, 1993.

586. R. R. Arya, A. Catalano, and R. S. Oswald, *Appl. Phys. Lett.* 49, 1089 (1986).
587. C. Beneking, B. Rech, T. Eickhoff, Y. G. Michael, N. Schultz, and H. Wagner, "Proceedings of the Twelfth E.C. Photovoltaic Solar Energy Conference, Amsterdam, the Netherlands, 1994," p. 683. Kluwer Academic, Dordrecht, 1994.
588. G. E. N. Landweer, B. S. Girwar, C. H. M. van der Werf, J. W. Metselaar, and R. E. I. Schropp, "Proceedings of the Twelfth E.C. Photovoltaic Solar Energy Conference, Amsterdam, the Netherlands, 1994," p. 1300. Kluwer Academic, Dordrecht, 1994.
589. A. Banerjee and S. Guha, *J. Appl. Phys.* 69, 1030 (1991).
590. Y. Kuwano, M. Ohnishi, H. Nishiwaki, S. Tsuda, T. Fukatsu, K. Enomoto, Y. Nakashima, and H. Tarui, "Proceedings of the 16th IEEE Photovoltaic Specialists Conference, San Diego, 1982," p. 1338. IEEE, New York, 1982.
591. S. Guha, J. Yang, A. Pawlikiewicz, T. Glatfelter, R. Ross, and S. R. Ovshinsky, *Appl. Phys. Lett.* 54, 2330 (1989).
592. J. Zimmer, H. Stiebig, and H. Wagner, *J. Appl. Phys.* 84, 611 (1998).
593. R. E. Rochelau, M. Tun, and S. S. Hegedus, "Proceedings of the 26th IEEE Photovoltaic Specialists Conference, Anaheim, 1996," p. 703. IEEE, New York, 1997.
594. J. K. Rath, R. E. I. Schropp, and W. Beyer, *J. Non-cryst. Solids* 227–230, 1282 (1998).
595. J. Bruns, M. Choudhury, and H. G. Wagemann, "Proceedings of the Thirteenth European Photovoltaic Solar Energy Conference, Nice, France, 1995" (W. Freiesleben, W. Palz, H. A. Ossenbrink, and P. Helm, Eds.), p. 230. H.S. Stephens & Associates, Bedford, U.K., 1995.
596. M. Zeman, J. A. Willemen, S. Solntsev, and G. W. Metselaar, *Sol. Energy Mater. Sol. Cells* 34, 557 (1994).
597. M. Zeman, R. A. C. M. M. van Swaaij, E. Schroten, L. L. A. Vosteen, and J. W. Metselaar, *Mater. Res. Soc. Symp. Proc.* 507, 409 (1998).
598. M. Zeman, J. A. Willemen, L. L. A. Vosteen, G. Tao, and G. W. Metselaar, *Sol. Energy Mater. Sol. Cells* 46, 81 (1997).
599. X. Xu, J. Yang, and S. Guha, *Appl. Phys. Lett.* 62, 1399 (1993).
600. Y. Lee, L. Jiao, Z. Lu, R. W. Collins, and C. R. Wronski, *Sol. Energy Mater. Sol. Cells* 49, 149 (1997).
601. B. W. Faughnan and R. Crandall, *Appl. Phys. Lett.* 44, 537 (1984).
602. Z. E. Smith, S. Wagner, and B. W. Faughnan, *Appl. Phys. Lett.* 46, 1078 (1985).
603. A. Catalano, R. Bennett, R. Arya, K. Rajan, and J. Newton, "Proceedings of the 18th IEEE Photovoltaic Specialists Conference, Las Vegas, 1985," p. 1378. IEEE, New York, 1985.
604. M. Stutzmann, W. B. Jackson, and C. C. Tsai, *Phys. Rev. B* 34, 63 (1986).
605. R. E. I. Schropp, A. Sluiter, M. B. von der Linden, and J. Daey Ouwens, *J. Non-cryst. Solids* 164–166, 709 (1993).
606. D. E. Carlson, K. Rajan, R. R. Arya, F. Willing, and L. Yang, *J. Mater. Res.* 13, 2754 (1998).
607. Y. Hamakawa, W. Ma, and H. Okamoto, in "Plasma Deposition of Amorphous Silicon-Based Materials" (G. Bruno, P. Capezzuto, and A. Madan, Eds.), Chap. 6, p. 283. Academic Press, Boston, 1995.
608. L. L. Kazmerski, *Renew. Sustain. Energy Rev.* 1, 71 (1997).
609. A. V. Shah, R. Platz, and H. Keppner, *Sol. Energy Mater. Sol. Cells* 38, 501 (1995).

610. T. Strand, L. Mrig, R. Hansen, and K. Emery, *Sol. Energy Mater. Sol. Cells* 41/42, 617 (1996).

611. R. Ruther and J. Livingstone, *Sol. Energy Mater. Sol. Cells* 36, 29 (1996).

612. J. Merten and J. Andreu, *Sol. Energy Mater. Sol. Cells* 52, 11 (1998).

613. J. Yang, A. Banerjee, T. Glatfelter, K. Hoffman, X. Xu, and S. Guha, Proceedings of the 24th IEEE Photovoltaic Specialists Conference, Waikoloa HI, 1994," p. 380. IEEE, New York, 1995.

614. S. Guha, J. Yang, A. Banerjee, T. Glatfelter, and S. Sugiyama, *Sol. Energy Mater. Sol. Cells* 48, 365 (1997).

615. S. Kiyama, S. Nakano, Y. Domoto, H. Hirano, H. Tarui, K. Wakisaka, M. Tanaka, S. Tsuda, and S. Nakano, *Sol. Energy Mater. Sol. Cells* 48, 373 (1997).

616. Y. Hamakawa, *Appl. Surf. Sci.* 142, 215 (1999).

617. S. M. Sze, "Semiconductor Devices, Physics and Technology." Wiley, New York, 1985.

618. M. J. Powell, *IEEE Trans. Elec. Dev.* ED-36, 2753 (1989).

619. K. Suzuki, in "Amorphous and Microcrystalline Semiconductor Devices—Optoelectronic Devices" (J. Kanicki, Ed.), Chap. 3, p. 77. Artech House, Norwood, MA, 1991.

620. A. Iqbal, W. B. Jackson, C. C. Tsai, J. W. Allen, and C. W. Bates, *J. Appl. Phys.* 61, 2947 (1987).

621. C. van Berkel, in "Amorphous and Microcrystalline Semiconductor Devices—Materials and Device Physics" (J. Kanicki, Ed.), Chap. 8, p. 397. Artech House, Norwood, MA, 1992.

622. W. B. Jackson and M. Moyer, *Mater. Res. Soc. Symp. Proc.* 118, 231 (1988).

623. J. Kakalios, R. A. Street, and W. B. Jackson, *Phys. Rev. B* 59, 1037 (1987).

624. W. B. Jackson, J. M. Marshall, and M. D. Moyer, *Phys. Rev. B* 39, 1164 (1989).

625. A. M. Brockhoff, E. H. C. Ullersma, H. Meiling, F. H. P. M. Habraken, and W. F. van der Weg, *Appl. Phys. Lett.* 73, 3244 (1998).

626. C. van Berkel and M. J. Powell, *J. Appl. Phys.* 60, 1521 (1986).

627. M. Katayama, *Thin Solid Films* 341, 140 (1999).

628. H. C. Tuan, *J. Non-cryst. Solids* 115, 132 (1989).

629. H. Matsumura, H. Hayama, Y. Nara, and K. Ishibashi, *Japan. J. Appl. Phys.* 20, 311 (1980).

630. R. L. Weisfield, *IEEE Trans. Electron. Dev.* ED-36, 2935 (1989).

631. I. Shimizu, S. Shirai, and E. Inoue, *J. Appl. Phys.* 52, 2776 (1981).

632. I. Shimizu, in "Semiconductors and Semimetals" (J. I. Pankove, Ed.), Vol. 21D, p. 55. Academic Press, Orlando, FL, 1984.

633. R. Schaffert, "Electrophotography." Focal, London, 1975.

634. J. Kodama, S. Araki, M. Kimura, and T. Inagaki, *Japan. J. Appl. Phys.* 29, L867 (1990).

635. R. A. C. M. M. van Swaaij, W. P. M. Willems, J. Bezemer, H. J. P. Lokker, and W. F. van der Weg, *Mater. Res. Soc. Symp. Proc.* 297, 559 (1993).

636. R. A. C. M. M. van Swaaij, S. J. Elmer, W. P. M. Willems, J. Bezemer, J. M. Marshall, and A. R. Hepburn, *J. Non-cryst. Solids* 164–166, 533 (1993).

637. T. Takeda and S. Sano, *Mater. Res. Soc. Symp. Proc.* 118, 399 (1988).

638. E. Fortunato, M. Vieira, I. Ferreira, C. N. Carvalho, G. Lavareda, and R. Martins, *Mater. Res. Soc. Symp. Proc.* 297, 981 (1993).

639. E. M. C. Fortunato, D. Brida, I. M. M. Ferreira, H. M. B. Águas, P. Nunes, A. Cabrita, F. Guiliani, Y. Nunes, M. J. P. Maneira, and R. Martins, *Mater. Res. Soc. Proc. Symp.* 609, A12.7.1 (2000).

640. H. Stiebig and N. Bohm, *J. Non-cryst. Solids* 164–166, 789 (1993).

641. R. Brüggemann, T. Neidlinger, and M. B. Schubert, *J. Appl. Phys.* 81, 7666 (1997).

642. G. de Cesare, F. Irrera, F. Lemmi, and F. Palma, *Appl. Phys. Lett.* 66, 1178 (1995).

643. M. Topič, F. Smole, J. Furlan, and W. Kusian, *J. Non-cryst. Solids* 227–230, 1326 (1998).

644. J. Krč, M. Topič, F. Smole, D. Knipp, and H. Stiebig, *Mater. Res. Soc. Proc. Symp.* 609, A12.5.1 (2000).

645. D. Caputo and G. de Cesare, *Sensors and Actuators A* 78, 108 (1999).

646. D. Caputo, G. de Cesare, A. Nascetti, F. Palma, and M. Petri, *Appl. Phys. Lett.* 72, 1229 (1998).

647. R. A. Street, X. D. Wu, R. Weisfield, S. Ready, R. Apte, M. Nguyen, and P. Nylen, *Nucl. Instrum. Methods Phys. Res. A* 380, 450 (1996).

648. R. A. Street, R. B. Apte, T. Granberg, M. P., S. E. Ready, K. S. Shah, and R. L. Weisfield, *J. Non-cryst. Solids* 227–230, 1306 (1998).

649. J.-P. Moy, *Nucl. Instrum. Methods Phys. Res. A* 442, 26 (2000).

650. J. C. Chou, Y.-F. Wang, and J. S. Lin, *Sensors and Actuators B* 62, 92 (2000).

651. J. C. Chou, H. M. Tsai, C. N. Shiao, and J. S. Lin, *Sensors and Actuators B* 62, 97 (2000).

652. S. Jamash, S. D. Collins, and R. L. Smith, *IEEE Trans. Electron. Dev.* ED-45, 1239 (1998).

653. P. F. Ruths, S. Askok, S. J. Fonash, and J. M. Ruths, *IEEE Trans. Electron. Dev.* ED-28, 1003 (1981).

654. E. Fortunato, A. Malik, A. Sêco, I. Ferreira, and R. Martins, *J. Non-cryst. Solids* 227–230, 1349 (1998).

655. F. V. Hunt, "Electroacoustics, the Analysis of Transduction and Its Historical Background." American Institute of Physics, New York, 1982.

656. M. Smits, private communication.

657. R. E. I. Schropp, M. Smits, H. Meiling, W. G. J. H. M. van Sark, M. M. Boone, and W. F. van der Weg, *Mater. Res. Soc. Symp. Proc.* 219, 519 (1991).

658. P. J. Mears, L. Reekie, I. M. Jauncey, and D. N. Payne, *Electron. Lett.* 23, 1026 (1987).

659. E. Desurvire, R. J. Simpson, and P. C. Becker, *Opt. Lett.* 12, 888 (1987).

660. G. N. van den Hoven, A. Polman, C. van Dam, J. W. M. van Uffelen, and M. K. Smit, *Appl. Phys. Lett.* 68, 1886 (1996).

661. A. Polman, *J. Appl. Phys.* 82, 1 (1997).

662. F. Priolo, G. Franzò, S. Coffa, and A. Carnera, *Mat. Chem. Phys.* 54, 273 (1998).

663. F. Priolo, G. Franzò, S. Coffa, and A. Carnera, *Phys. Rev. B* 57, 4443 (1998).

664. G. N. van den Hoven, J. H. Shin, A. Polman, S. Lombardo, and S. U. Campisano, *J. Appl. Phys.* 78, 2642 (1995).

665. J. H. Shin, R. Serna, G. N. van den Hoven, A. Polman, W. G. J. H. M. van Sark, and A. M. Vredenberg, *Appl. Phys. Lett.* 68, 697 (1996).

666. A. Polman, J. H. Shin, R. Serna, G. N. van den Hoven, W. G. J. H. M. van Sark, A. M. Vredenberg, S. Lombardo, and S. U. Campisano, *Mater. Res. Soc. Symp. Proc.* 422, 239 (1996).
667. S. Coffa, G. Franzò, A. Polman, and R. Serna, *Phys. Rev. B* 49, 16313 (1994).
668. P. G. Kik, M. J. A. de Dood, K. Kikoin, and A. Polman, *Appl. Phys. Lett.* 70, 1721 (1997).
669. F. Priolo, G. Franzò, S. Coffa, A. Polman, S. Libertino, R. Barklie, and D. Carey, *J. Appl. Phys.* 78, 3874 (1995).
670. G. Franzò, F. Priolo, S. Coffa, A. Polman, and A. Carnera, *Appl. Phys. Lett.* 64, 2235 (1994).
671. B. Zheng, J. Michel, F. Y. G. Ren, L. C. Kimerling, D. C. Jacobson, and J. M. Poate, *Appl. Phys. Lett.* 64, 2842 (1994).
672. J. Palm, F. Gan, B. Zheng, J. Michel, and L. C. Kimerling, *Phys. Rev. B* 54, 17603 (1996).
673. S. Lombardo, S. U. Campisano, G. N. van den Hoven, and A. Polman, *J. Appl. Phys.* 77, 6504 (1995).
674. L. H. Slooff, Master's thesis, FOM—Institute for Atomic and Molecular Physics, Amsterdam, 1996.
675. J. Stimmer, A. Reittinger, J. F. Nützel, G. Abstreiter, H. Holzbrecher, and C. Buchal, *Appl. Phys. Lett.* 68, 3290 (1996).
676. W. Fuhs, I. Ulber, G. Weiser, M. S. Bresler, O. B. Gusev, A. N. Kuznetsov, V. K. Kudoyarova, E. I. Terukov, and I. N. Yassiervich, *Phys. Rev. B* 56, 9545 (1997).
677. E. I. Terukov, O. I. Kon'kov, V. K. Kudoyarova, O. B. Gusev, and G. Weiser, *Semiconductors* 24, 884 (1998).
678. E. I. Terukov, M. M. Kazanin, O. I. Kon'kov, V. K. Kudoyarova, K. V. Kougiya, Y. A. Nikulin, and Kazanskii, *Semiconductors* 34, 829 (2000).
679. V. B. Voronkov, V. G. Golubev, A. V. Medvedev, A. B. Pevtsov, N. A. Feoktistov, N. I. Gorshkov, and D. N. Suglobov, *Phys. Solid State* 40, 1301 (1998).
680. J. H. Shin and M. Kim, *J. Vac. Sci. Technol. A* 17, 3230 (1999).
681. A. Masuda, J. Sakai, H. Akiyama, O. Eryu, K. Nakashima, and H. Matsumura, *J. Non-cryst. Solids* 226–229, 136 (2000).
682. T. Förster, *Discuss. Faraday Soc.* 27, 7 (1959).
683. H. Kühne, G. Weiser, E. I. Terukov, A. N. Kuznetsov, and V. K. Kudoyarova, *J. Appl. Phys.* 86, 1896 (1999).
684. M. B. Schubert, A. Hierzenberger, H. J. Lehner, and J. H. Werner, *Sensors Actuators A* 74, 193 (1999).
685. E. Betzig and J. K. Trautmann, *Science* 257, 189 (1992).
686. E. Betzig, P. L. Finn, and J. S. Weiner, *Appl. Phys. Lett.* 60, 2484 (1992).
687. S. Madsen, M. Mullenborn, K. Birkelund, and F. Grey, *Appl. Phys. Lett.* 69, 544 (1996).
688. S. Madsen, S. I. Bozhevolnyi, K. Birkelund, M. Mullenborn, J. M. Hvam, and F. Grey, *J. Appl. Phys.* 82, 49 (1997).
689. M. K. Herndon, R. T. Collins, R. E. Hollingsworth, P. R. Larson, and M. B. Johnson, *Appl. Phys. Lett.* 74, 141 (1999).
690. W. Gao, S. H. Lee, J. Bullock, Y. Xu, D. K. Benson, S. Morrison, and H. M. Branz, *Sol. Energy Mater. Sol. Cells* 59, 243 (1999).

Growth, Structure, and Properties of Plasma-Deposited Amorphous Hydrogenated Carbon–Nitrogen Films

D. F. FRANCESCHINI

Instituto de Física, Universidade Federal Fluminense, Avenida Litorânea s/n, Niterói, RJ, 24210-340, Brazil

2.1. Introduction

Since the suggestion by Liu and Cohen [1] of a hypothetical compound, carbon nitride or β-C_3N_4 (with the β-Si_3N_4 structure), whose mechanical properties would be similar to that of crystalline diamond, a strong research effort has been dedicated to the study of carbon–nitrogen-based solids [2, 3]. This effort constituted one of the first initiatives for a new way of doing research in Materials Science: the modeling of a new material with specific properties, for subsequent synthesis attempts. Up to this moment, no successful synthesis of a single-phase crystalline solid with the β-carbon nitride structure has been reported. Nevertheless, this research effort led to the development of an entirely new class of materials, which have also found practical applications. This is case of the use of amorphous carbon nitride as a wear and corrosion protective coating for magnetic recording media [4]

The work on carbon nitride solids is strongly related to research on diamond-like carbon (DLC) materials [5, 6]. DLC materials are thin film amorphous metastable carbon-based solids, pure or alloyed with hydrogen, which have properties similar to that of crystalline diamond (high hardness, low friction coefficient, high resistance to wear and chemical attack). This resemblance to diamond is due to the DLC structure, which is characterized by a high fraction of highly cross-linked sp^3-hybridized carbon atoms. To obtain this diamond-like structure

THIN FILMS AND NANOSTRUCTURES
Vol. 30

ISBN 0-12-533030-8
ISSN 1079-4050/02 $35.00

at ambient conditions, DLC film deposition techniques always involve the bombarding of the film growing surface by fast particles with kinetic energy of the order of 100 eV. Since in β-C_3N_4 the carbon atoms should be sp^3 hybridized, one of the ways to search for carbon nitride was the study of nitrogen incorporation into DLC films.

One of the branches of this research has been the study of nitrogen incorporation in plasma-deposited hard amorphous hydrogenated carbon (a-C:H) films, the subject of this chapter. The research on plasma-deposited a-C(N):H films started even before Liu and Cohen's suggestion, with the study of electronic doping of a-C:H films reported by Jones and Stewart [7]. After that, followed the pioneering works of Han and Feldman [8], Amir and Kalish [9], and Kaufman, Metin, and Saperstein [10], which first reported important effects of nitrogen incorporation on the structure and the properties of a-C:H films. Further research on the field revealed some aspects of the nitrogen incorporation process in a-C:H films to place obstacles for obtaining an amorphous hydrogenated carbon nitride solid structurally analogous to β-carbon nitride. The first one is the limited nitrogen uptake observed in a-C(N):H films. No more than 30-at.% N incorporation could be achieved up to now. In fact, nitrogen contents lower than about 15-at.% N are the more frequent occurrence. The second one is that nitrogen incorporation results in a strong decrease in the sp^3-carbon atom fraction [11], which is the main responsibility for a-C:H film rigidity. Finally, one has the preferential bonding of hydrogen atoms to nitrogen atoms, which makes difficult the formation of a carbon–nitrogen extended network.

Despite such limitations, plasma-deposited a-C(N):H films were found to be used in a number of applications. The stress reduction induced by nitrogen incorporation [12] and consequent adhesion improvement, allowed the development of a-C(N):H antireflective coatings for Ge-based infrared detectors [13]. It was also found that N can electronically dope a-C:H films, and can strongly decrease the defect density, which gives prospects on its use as a semiconductor material [14]. Nitrogen incorporation was also found to decrease the threshold electric field in electron-field emission process [15], making possible the use of a-C(N):H films as an overcoat on emission tips in flat-panel display devices [16].

Research on plasma-deposited a-C(N):H films has been frequently included in the general discussion of carbon nitride solids [2, 3]. However, the presence of hydrogen in its composition, and the complexity of the deposition process, which introduces the nitrogen species in the already intricate hydrocarbon plasma-deposition mechanism, make a-C(N):H films deserve special consideration. This is the aim of the present work: to review and to discuss the main results on the growth, structure, and properties of plasma-deposited a-C(N):H films. As this subject is closely related to a-C:H films, a summary of the main aspects relative to the plasma deposition of a-C:H films, their structure, and the relationship between the main process parameters governing film structure and properties is presented in Section 2.2. Section 2.2 is not intended to be a review, but only a guide to the main concepts and to the up-to-date key literature in the field.

Because they have been found to be strongly related, the nitrogen incorporation process and the growth kinetics are treated together in Section 2.3. The discussion of these themes includes the presentation of the main experimental results concerning nitrogen incorporation, the related effects of plasma chemistry and surface processes during film growth, and an attempt to model the growth kinetics.

In Section 2.4, the main aspects of the nitrogen-induced structural changes are presented, by the discussion of the most important characterization techniques. This presentation is complemented by an overview of a-C(N):H structure. Finally, in Sections 2.5 and 2.6, respectively, results concerning the mechanical properties, and the electrical and optical properties of a-C(N):H films are presented. As long as possible, they will be correlated with the observed structure changes.

2.2. Amorphous Hydrogenated Carbon Films

2.2.1. a-C:H Film Structure

The hard amorphous hydrogenated carbon films (a-C:H) belong to the class of materials called diamond-like carbon (DLC). These materials are designated as DLC due to their superlative properties, similar to that of crystalline diamond, such as high hardness, low friction coefficient, high chemical and wear resistance, and high electrical resistivity. These properties are closely related to the DLC film structure, which is formed mainly by sp^3- and sp^2-carbon atoms, with high sp^3 fraction (up to 80%), with a high degree of cross-linking of the amorphous network. Amorphous hydrogenated carbon films have a lower sp^3 fraction (up to 50%), in addition to the presence of a remarkable fraction of hydrogen atoms bonded to the network, in the 10- to 50-at.% range [5, 6].

The structure of the a-C:H films may be viewed, in a first approach, as a random covalent network [17, 18]. In this model structure, the fourfold sp^3 carbon, the threefold sp^2 carbon, and the singly coordinated hydrogen atoms, randomly bond with each other, forming a fully constrained network. To fulfil all of the constraints, the mean coordination number of the network should be equal to approximately 2.45 [17, 18]. Networks with higher mean coordination number are said to be overconstrained. The high sp^3-carbon atom fraction of hard a-C:H films leads to mean coordination numbers that are very often near 3 [19]. In this case, the formation of chemical bonds occurs with considerable distortion in chemical bond length and angle, giving rise to a strained amorphous network. This feature has been identified as the source of the high internal compressive stress [17], which is observed in all kinds of DLC. In DLC films, the internal stress can reach several GPa [5, 6], placing an obstacle to the production of thick films necessary for some applications. On the other hand, networks with mean coordination numbers smaller than 2.45 are said to be underconstrained, and are mechanically soft, like hydrocarbon polymers. The lower sp^2 fraction and the presence of a relatively high concentration of hydrogen atoms in their composition, make them less rigid and stressed than other kinds of DLC, such as highly tetrahedral, pure or hydrogenated amorphous carbons (ta-C and ta-C:H). These films can have hardness and density approaching that of crystalline diamond. In the C-H-sp^3-fraction ternary phase diagram, a-C:H films are situated near the rigidity border, at the side of hydrocarbon polymers [5].

Despite the diamond-like mechanical properties and electrical resistivity owned by a-C:H films, their optical properties are not comparable with that of crystalline diamond. Although a-C:H films were shown to be transparent in the infrared wavelength range (unless for the absorption bands due to, e.g., C—H groups), the transparency in the UV-vis wavelength range is remarkably smaller than that of diamond. The comparison of the bandgap as determined by optical absorption measurements—the optical gap—clearly shows the difference. The gap width of crystalline diamond is about 5.4 eV, whereas the gap width of hydrogenated diamond-like films are situated in the 1- to 4-eV [20] range. This relatively small gap observed in hydrogenated diamond-like (as also in other DLC materials) has been ascribed to the dominant role of the π electrons, arising from bond formation between sp^2-carbon atoms. Robertson and O'Reilly [21] and Robertson [22] proposed that the sp^2-carbon atoms in DLC films tend to cluster in the form of condensed sixfold aromatic rings, which also tend to cluster in the form of graphitic islands. Within this model, the bandgap width of DLC films is strongly correlated with the size of the π-bonded clusters, following an $M^{-1/2}$ dependence, where M is number of atoms in the cluster.

The structure of a-C:H films may be thus pictured as sp^2-carbon atoms in condensed aromatic clusters, dispersed in an sp^3-rich matrix, which confers to the network its characteristic rigidity. This situation can also be regarded as a random covalent network in which the sp^2 clusters of a defined size take part in the structure as an individual composed "atom" with its corresponding coordination number [17]. Such kinds of models have been successfully used to describe the dependence of a-C:H film mechanical properties on composition, hybridization, and sp^2 clustering [23].

2.2.2. a-C:H Film Deposition

It is well known that the stable crystalline form of carbon at ambient conditions is graphite, which is fully sp^2 hybidized. The synthesis of the fully sp^3-hybridized crystalline diamond is performed at high temperatures and pressures. So, the production of metastable carbon solids with a high fraction of sp^3-hybridized carbon at low temperatures and pressures is expected to require the use of out-of-equilibrium processes. In the case of DLC synthesis, this is achieved by submitting the growing film to the bombardment by energetic particles, as first reported by Eisenberg and Chabot [24]. This has been generally achieved by direct ion-beam deposition of C^+ ions accelerated with the required kinetic energy ($\sim$100 eV), by assisting the growth from a carbon vapor source with a noble gas ion beam, or by extracting and accelerating ions from a hydrocarbon plasma.

Roughly speaking, the condensation of the energy and momentum from the fast particle to the growing layer, locally generates conditions similar to the high temperatures and pressures needed for diamond synthesis. A more rigorous explanation is given by the subplantation model [25].

A version of the subplantation model [26] states different roles for the bombarding particles, as a function of their kinetic energies. Thinking on pure carbon deposition, for an intermediate energy range, at about 100 eV, the incoming fast particle penetrates into the subsurface, and stops at a lattice interstice, generating a local high-density state (sp^3 hybridization). For a lower energy range, the particle cannot penetrate into the surface, and is thus reflected or sticks to it, in the stable sp^2 configuration. For the higher energy range, the incoming ion generates a collision cascade, resulting in atom desorption and phonon activation, without generating a dense phase. This mechanism causes the structural parameters (sp^3 fraction, atomic density) and the mechanical properties (hardness, stress) to assume maximum values within a narrow ion-beam energy range around 100 eV.

Deposition of a-C:H films by plasma fundamentally retains the structure dependence on the bombarding ion energy. However, its mechanism is to some extent more complex, due to the presence in the deposition process species other than the atomic C^+ ions. In plasma deposition of a-C:H films, the diamond-like structure and the properties are determined by the hydrocarbon-derived molecular ions. However, film growth proceeds mainly by the attachment of neutral reactive radicals that result from hydrocarbon fragmentation [27]. So, in addition to the strong dependence of film structure on the bombarding ion energy, a strong dependence on the hydrocarbon used was also reported [28].

The reactions taking place at the growing surface of plasma-deposited a-C:H were reviewed by Jacob [29], and from this discussion emerged a framework to understand the a-C:H film deposition mechanism. This framework is to some extent equivalent to the subplantation model, since it emphasizes the role of energetic molecular ions. In addition, it takes into account the role of neutral radicals.

The previously mentioned picture of a-C:H deposition is based on some kinds of species coming from the plasma, and interacting with the film surface: Carbon-carrying neutral radicals and atomic hydrogen (slow fragments), plus carbon-carrying ions and hydrogen-derived ions. Each one of these species may play different roles in film growth, superimposing or competing with the other species. *Carbon-carrying ions* have low penetration depth and can stick to the

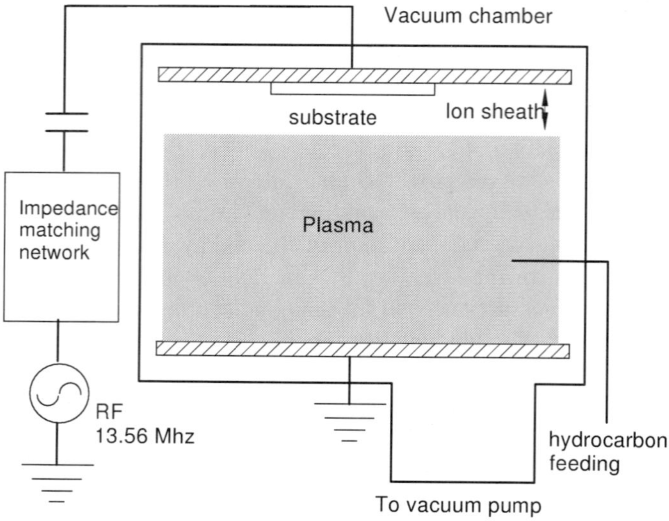

FIG. 1. Schematic of a-C:H plasma deposition by radio-frequency glow-discharge (RFPECVD).

surface, can displace hydrogen atoms (thus generating dangling bonds to which carbon-carrying radicals may stick), can recombine with dangling bonds, can activate lattice vibrations from direct energy transfer to film carbon atoms (consequently promoting carbon atom cross-linking). *Hydrogen ions* have a higher penetration depth than carbon (their range defines the thickness of the growth layer), can displace hydrogen atoms generating dangling bonds, and after stopping can hydrogenate sp^2-C atoms or can saturate dangling bonds. *Atomic hydrogen* can produce dangling bonds (by abstracting hydrogen atoms or by hydrogenating sp^2-carbon atoms), can erode carbon atoms at higher deposition temperatures, and its main role, can saturate dangling bonds. *Carbon-carrying radicals* react with surface dangling bonds contributing to the film growth, and thus saturate dangling bonds. Thus, the growth of a-C(H) films from hydrocarbon plasmas is a complex process and depends, in addition to the ion-beam energy, strongly on plasma chemistry details.

The radio-frequency glow-discharge method [30–34] has been the most used method in the study of a-C:H films. In this chapter, it is referred to as RFPECVD (radio frequency plasma enhanced chemical vapor deposition). Film deposition by RFPECVD is usually performed in a parallel-plate reactor, as shown in Figure 1. The plasma discharge is established between an RF-powered electrode and the other one, which is maintained at ground potential. The hydrocarbon gas or vapor is fed at a controlled flow to the reactor, which is previously evacuated to background pressures below 10^{-5} Torr. The RF power is fed to the substrate electrode

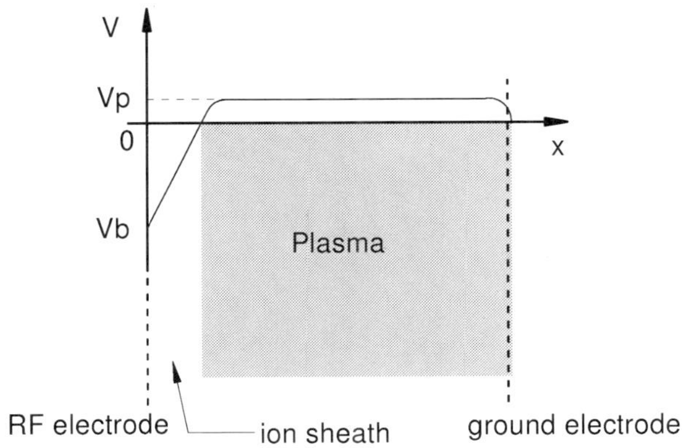

FIG. 2. Variation of the averaged in time electric potential across the interelectrode space.

through an impedance matching network, to assure maximum power transmission to the plasma. The coupling is made through a blocking capacitor, which is part of the matching network.

The presence of the blocking capacitor in the circuit gives rise to a great asymmetry on the electric potential within the interelectrode space. RF glow-discharge plasmas are divided into two regions, the plasma region (which emits light), and the ion sheath or cathode dark space. As referred to ground, the averaged-in-time electrical potential in the plasma region is small and positive, being zero at the grounded electrode. Across the ion sheath, there is a strong drop in the potential, which becomes negative, assuming the value V_B at the RF-powered electrode, as shown in Figure 2. The latter is the so-called self-bias potential V_B, which allows the extraction and the acceleration of the ions from the plasma toward the substrate electrode, giving rise to the ion bombardment necessary for diamond-like film growth. This effect is due to the much greater mobility of the plasma electrons, as compared to the ionic species. During an RF cycle, the electrons can move forth and back between the two electrodes, whereas the ions tend to be more concentrated near the RF electrode. This leads to an average-in-time positive net space charge, which polarizes the RF electrode negatively. The difference between the electron and ion motions also leads the traveling electrons to hit the RF electrode, neutralizing the positive charge built up at the RF electrode by the bombardment of the highly insulating a-C:H film. Otherwise, this positive charging would extinguish the discharge. These two effects—the self-bias potential and the charge neutralization—are essential to diamond-like a-C:H film growth, and are the reasons for the so widespread use of the RFPECVD method.

The value of the self-bias potential was reported to strongly depend on the RF power and deposition pressure, following a $(W/p)^{1/2}$ dependence [30]. The

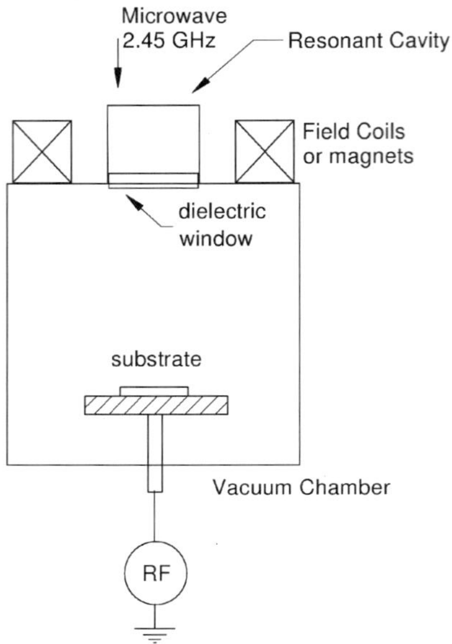

FIG. 3. Schematic of plasma deposition by electron cyclotron resonance with an RF-biased substrate (ECRRF).

reactor geometry also influences the value of V_B. As the area ratio between the RF electrode to the grounded electrode decreases, V_B is found to increase [30–34].

Although there are great advantages for the RFPECVD method, the direct current (dc) glow discharge has also been used for a-C:H film deposition [35]. By conveniently choosing a sufficiently high dc bias, a-C:H films as thick as 3 μm could be grown. This is possible because the electrical resistivity of a-C:H films greatly decreases with the applied electrical field, allowing the dc bias to reach the substrate surface. A variation of the RF flow-discharge method was reported, in which the substrate is placed in a negatively dc-biased electrode parallel and opposed to the RF-powered one [10].

There has been an increased interest in the use of denser plasmas, such as electron cyclotron resonance (ECR) plasmas for a-C:H film deposition [34, 36]. In ECR plasmas, the plasma discharge is sustained by microwave radiation (2.45 GHz), which is fed into the deposition chamber through a dielectric window placed at the end of a resonant cavity, as shown in Figure 3. In addition, a steady magnetic field is applied in the region near the window, to generate cyclotron resonance conditions for the electrons in the plasma. This greatly enhances the collisional process, thus increasing plasma ionization and fragmentation. This allows performing deposition at pressures lower than is possible in RFPECVD. The ion energies coming from ECR plasmas are far below 100 eV, the typical energy for the production of diamond-like films. To give the required kinetic energy to the plasma ions, the substrate is placed over a dc- or RF-biased electrode. In this way, the ions are extracted from the plasma and are accelerated toward the substrate. In addition to the increased plasma density, this method also allows the variation of bombarding ions energy, without introducing great changes in the energy fed to the plasma, as happens in RFPECVD. In this chapter, ECR plasmas with RF substrate biasing are referred to as ECRRF.

The use of the dense plasmas of a helical resonator plasma source [37] was reported, also together with RF biasing of the substrate electrode. A variation of the RF-glow-discharge method, in which a steady magnetic field is perpendicularly imposed to the RF-biased electrode surface, was used to deposit films with a gaseous mixture largely diluted in helium, to increase the plasma electron temperature [14]. Production of a-C:H films was also performed by using highly ionized plasma sources such as the plasma beam source [38, 39] or hydrocarbon and nitrogened ion-beam sources [40], to deposit or to assist the deposition process.

Since the main parameter influencing diamond-like carbon film structure is the energy of bombarding ions, it is expected that the same happens with a-C:H films. In fact, it was found that in RFPECVD deposition of a-C:H films, the variation of substrate self-bias results in strong changes of film growth, composition, structure, and properties.

First, as the variation of self-bias in RFPECVD is performed by changing the RF power delivered to the plasma, in addition to changing the ion energy, it also changes the plasma fragmentation. So, when the self-bias increases, an increase on the production of ions as well as of reactive neutral radicals is also expected. As a consequence, the deposition rate is found to increase for increasing V_B [31–33, 33]. For higher values of V_B, a decrease in the deposition rate has been reported [33] and was attributed to the onset of film sputtering by ions with higher energy.

The hydrogen content of a-C:H films was found to continuously decrease with increasing self-bias, for films deposited by RFPECVD and ECRRF dual plasma, as a consequence of the increasing bombarding energy [31–33, 41]. As expected, the C-atom hybridization was found to strongly correlate with the self-bias, although in a more complex way than hydrogen-free DLC films.

Tamor, Vassel, and Carduner [19] measured by ^{13}C-nuclear magnetic resonance spectroscopy (^{13}C-NMR) the evolution of the C-atoms hybridization against the self-bias, in the up to -1000-V range, for a-C:H films deposited by RFPECVD in CH_4 atmospheres. As the self-bias increased, the overall carbon sp^3 fraction was found to continuously decrease. This in principle is contradictory with the subplantation model, from which an sp^3 fraction should pass through a maximum. However, the authors were able to extract from the NMR spectra, the contribution of each kind of carbon atoms present in the sample. They reported, upon increasing self-bias, a continuous increase on the sp^2 fraction of hydrogenated carbon atoms, together with the corresponding decrease of the sp^3 fraction of these atoms. On the other hand, the sp^3 fraction of nonhydrogenated carbon atoms was found to increase until reaching a maximum value of 20% at -200-V self-bias, and finally to decrease for higher self-bias. At the same time, the sp^2 fraction of nonhydrogenated carbon atoms was found to continuously increase. It must be thus noted that in plasma-deposited a-C:H films, the generation of quaternary carbon atom bonds occurs in parallel with the generation of sp^2 bonds that follows the dehydrogenation process.

The mechanical properties were generally found to follow the behavior described earlier. Amorphous hydrogenated carbon films deposited with very low self-bias, say below -100 V, are reported to be highly hydrogenated, mechanically soft, and nonstressed, being known as polymer-like a-C:H [31–33]. As the self-bias increases, the films become hard and stressed. Films deposited from methane show a clear maximum in mechanical hardness, in the -200- to -400-V self-bias range [32, 41]. As expected, the film density also follows the passing through a maximum behavior [41].

Films deposited from hydrocarbons with a higher number of carbon atoms, e.g., C_6H_6, show shifts in the position of the hardness and density maximums to higher self-bias [41]. Such behavior was ascribed to the fact the most abundant molecular ions that bombard the film growing surface are different for different hydrocarbons. In methane, the most abundant ions are expected to be the $CH_n{}^+$ ions, for acetylene, the most abundant ions are expected to be the $C_2H_n{}^+$ ions, for benzene, the most abundant ions are expected to be the $C_6H_6{}^+$ ions, and so on. Upon colliding with the surface, the molecular ions break up, being the carried energy shared between their constituent atoms. In this way, the energy delivered to the film surface is smaller as high is the number of carbon atoms, giving rise to a displacement in the optimum self-bias position to higher values [26].

The optical gap of a-C:H films was found to continuously decrease with increasing self-bias [42]. The gap shrinking was found to be strongly correlated to the variation of Raman spectra that is related to the increase of the graphitic clusters present in the a-C:H films. Accordingly, the electrical resistivity of C:H films was found to strongly decrease with substrate bias [43].

The deposition pressure also affects the structure and the properties of a-C:H, since it may significantly alter the ion energy. When accelerated across the ion sheath, the ions may collide with gas molecules and may lose energy, which no longer is equal to eV_B. Bubenzer et al. [31] suggested the ion energy to vary proportionally to $(V_B/p^{1/2})$. The main effect of the pressure increase in RFPECVD a-C:H deposition is to make the films more polymer-like, i.e., with increased hydrogen content and lower density, hardness and stress [32, 33]. To obtain hard a-C:H films, lower pressures are needed, assuring in this way a more effective bombardment.

2.3. Nitrogen Incorporation Into a-C:H Films

2.3.1. DETERMINATION OF CHEMICAL COMPOSITION

The study of a-C(N):H films requires methods of chemical analysis capable of determining carbon, hydrogen, nitrogen, and major contaminants, and at the same time give information on the composition profile through the films. Although there have been reports on the use of conventional, bulk chemical analysis methods, like combustion analysis, surface and thin film analysis methods are preferred. This includes methods based on electron spectroscopy, like X-ray photoelectron spectroscopy (XPS) and Auger electron spectroscopy (AES) [44–46]; and nuclear techniques based on MeV-energy ion-beam scattering [44, 47]. The methods based on electron spectroscopy suffer from some limitations. First, they probe the very surface chemical composition, which may differ from film bulk composition. In addition, they cannot detect hydrogen. Their advantage is the additional ability to probe changes on the electronic structure, due to changes in the chemical bond environment.

The use of nuclear techniques allows the determination of C, N, H, O, and heavier contaminants relative fractions with great accuracy, and of the elements depth profile with moderate resolution (typically 10 nm). Rutherford backscattering spectroscopy (RBS) of light ions (like alpha particles) is used for the determination of carbon and heavier elements. Hydrogen contents are measured by forward scattering of protons by incident alpha particles (ERDA) elastic recoil detection analysis [44, 47].

Since the nuclear and electronic scattering cross sections for alpha particles are well known, the relative concentrations of the elements and their depth profiles can be easily obtained. The relative element concentrations are determined by the relative scattering intensities. The depth profile is obtained from the energy spread of the scattered particles, which lose energy before and after the nuclear collision, by inelastic scattering with electrons. The knowledge of the elements areal density and of the film thickness allows the determination of film density.

Figure 4 shows an RBS spectrum obtained from an a-C(N):H sample, together with spectrum simulation by a computer code, which allows the determination of the areal density of each element. To make this kind of analysis, previous knowledge of the hydrogen content is necessary [47]. Figure 5 shows ERDA spectra of a-C(N):H films, as deposited and after high-energy nitrogen ion implantation. In this figure, the hydrogen depletion profile is caused by the ion implantation process. Nitrogen contents of a-C(N):H films are very often rather low, making the nitrogen RBS signal small as compared to the carbon signal, and even smaller as compared to the substrate (usually Si single crystal wafers), as shown in Figure 4. To improve the accuracy in the determination of nitrogen, sometimes a nuclear reaction is used to obtain an independent measure of nitrogen content [46, 47].

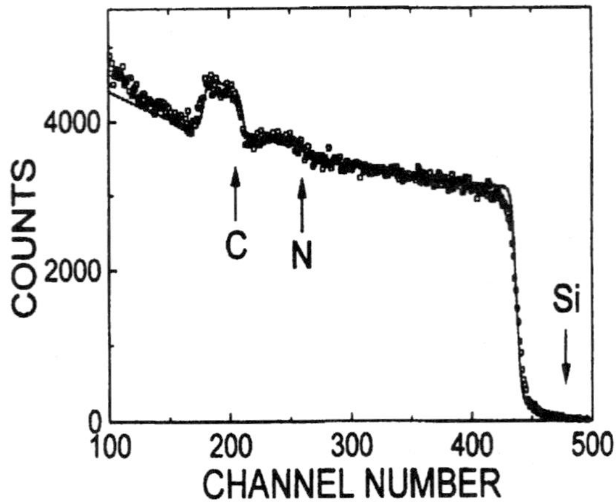

FIG. 4. RBS spectrum of a 12.5-at.% N a-C(N):H film. The continuum line is a computer simulation of the spectrum. (Reproduced from [59].)

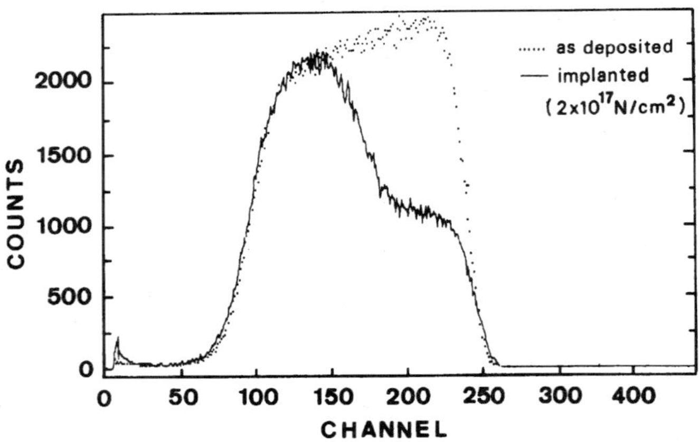

FIG. 5. ERDA spectra for as-deposited and N-implanted a-C(N):H films. (Reproduced from [47].)

This is the case for the $^{15}N(d, p)^{14}N$ reaction, which is exited by a deuteron beam with a 610-keV energy. Figure 6 shows the products of the nuclear reactions taking place upon deuteron irradiation, for the same samples shown in Figure 5. As occurs with hydrogen, nitrogen is also depleted by ion implantation treatment.

2.3.2. Nitrogen Incorporation and Growth Kinetics

RFPECVD (see Section 2.2.1) in atmospheres composed of mixtures of a hydrocarbon gas or vapor and a nitrogen-containing gas has been the most used

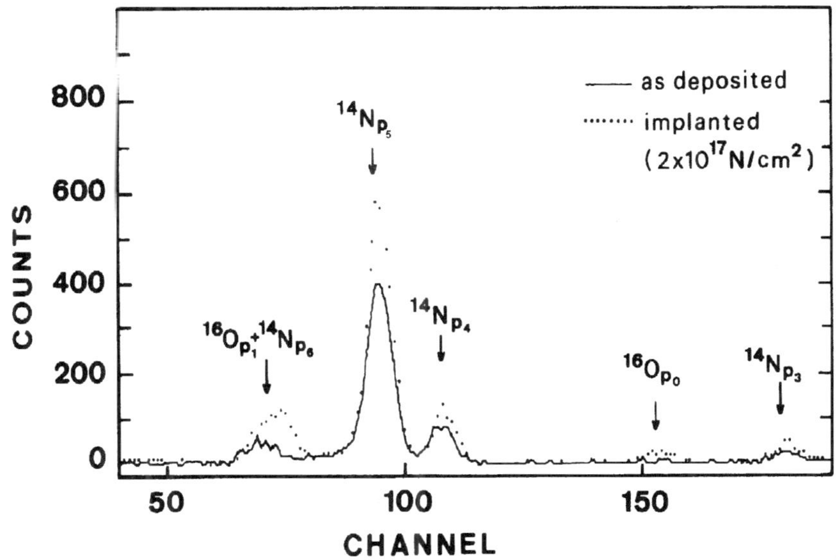

FIG. 6. Nuclear reaction spectra of as-deposited and implanted a-C(N):H films. (Reproduced from [47].)

deposition technique in the investigation of a-C(N):H films [8–10, 12, 14, 48–58]. Several hydrocarbon gases and vapors (methane, CH_4, acetylene, C_2H_2, butadiene, C_4H_{10}, cyclopentane, C_5H_{10}, benzene, C_6H_6) and nitrogen-containing gases and vapors (N_2, NH_3, CH_3NH_2) have been used in the precursor mixtures, within a wide range of deposition parameters. Due to the strong changes introduced on the deposition process and on film composition, structure and properties, the main focus of most works has been the study of the influence of nitrogen precursor partial pressure. Relatively few works have considered the effects of substrate bias or total deposition pressure.

In Table I are shown representative data concerning plasma deposition of a-C(N):H films, produced by different methods and under different gaseous mixtures and deposition parameters. In this table are displayed data on chemical composition (maximum N content, and range of variation of H content), deposition details (deposition pressure, self-bias, and atmosphere composition), and the method used for chemical composition determination.

In addition to the intended nitrogen incorporation, the addition of a nitrogen-containing gas to the deposition atmosphere also affects the film growth kinetics. It is always reported that a-C(N):H film deposition rates steeply decrease with increasing nitrogen-containing gas partial pressure [56], as shown in Figure 7. It must be noted that the observed decrease in the deposition rate is not a conse-

TABLE I
CHEMICAL COMPOSITION AND DEPOSITION PARAMETERS OF PLASMA-DEPOSITED A-C(N):H FILMS

Reference	Deposition method	Hydrocarbon	Nitrogen precursor/partial pressure	Substrate bias (V) or {ion energy (eV)}	Pressure (Pa)	Maximum N content (at.%)	H content (at.%)	Chemical composition determination method
Grill and Patel [48]	DcPECVD	C_2H_2	N_2	500	20	14	19–37	RBS, ERDA
Amir and Kalish [9]	Dual chamber DcPECVD	CH_6H_6	N_2	150	2	9.5 (C and N)	—	NRA, AES
Chan et al. [49]	ECRRFPECVD	CH_4/Ar	N_2 (0–63%)	120	—	11	—	AES
Silva et al. [14]	Magnetic field enhanced RFPECVD	CH_4/He	N_2 (0–80%)	—	13	15	12–21	RBS, ERDA, NR
Rodil et al. [50]	Plasma wave ECR	C_2H_2	N_2	{80}	10^{-2}	30	32–26	RBS, ERDA
Kaufman, Metin, and Saperstein [10]	RFPECVDdc bias	C_5H_{10}/Ar	N_2 (0–95%)	380	3.5	11		RBS, XPS
Lacerda et al. [52]	RFPECVD	C_2H_2	CH_3NH_2 (0–40%)	350	3	12	12–16	RBS, ERDA, NR
Schwan et al. [53]	RFPECVD	C_2H_2	N_2 (0–70%)	—	7	20	30–38	Combustion analysis

Jacobsohn et al. [54]	RFPECVD	C_2H_2	N_2 (0–90%)	300	3	22	11–13	RBS, ERDA
Lee, Eun, and Rhee [55]	RFPECVD	C_6H_6	N_2 (0–80%)	500	13	10	22–27	Combustion analysis
			NH_3 (0–80%)	500	13	8	27	Combustion analysis
Franceschini, Achete, and Freire [12]	RFPECVD	CH_4	N_2 (0–50%)	370	8	11	12–17	RBS, ERDA, NR
Han and Feldman [8]	RFPECVD	CH_4	N_2	0 (Substrate in anode)	30	18	43–48	Combustion analysis
Freire and Franceschini [56]	RFPECVD	CH_4	NH_3 (0–25%)	370	8	13	16–20	RBS, ERDA, NR
Wood, Weydeven, and Tsuji [57]	RFPECVD	Not quoted	N_2 N/C ratio (~0–5)	1340	7	16	10–16	RBS, secondary ion mass spectroscopy (SIMS)

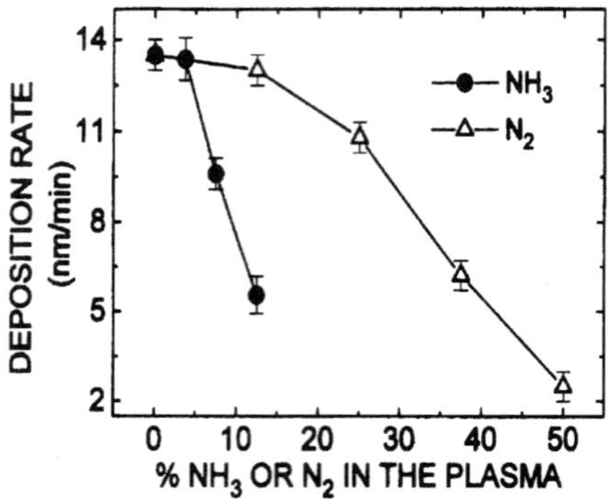

FIG. 7. Deposition rate of a-C(N):H films deposited by RFPECVD, as a function of NH_3 or N_2 partial pressure. (Reproduced from [56].)

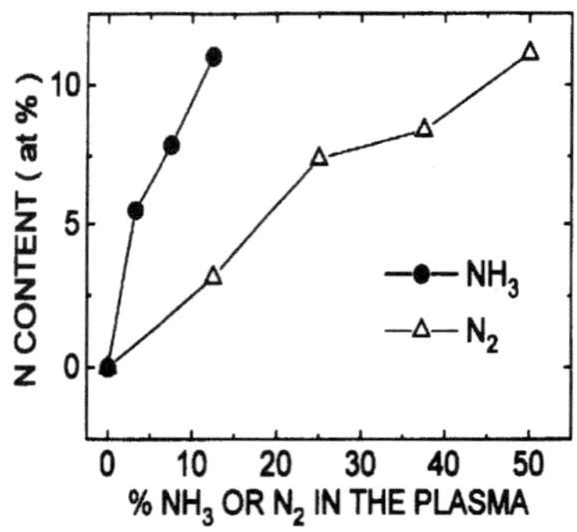

FIG. 8. Nitrogen content of a-C(N):H films deposited by RFPECVD, as a function of NH_3 or N_2 partial pressure. (Reproduced from [56].)

quence of a pure dilution effect, as one could conclude from the fact that solid nitrogen does not exist at room temperature. As an example, the deposition rate decrease observed in a-C:H film deposition by RFPECVD in hydrocarbon-argon atmospheres occurs in a slower way than in deposition from methane-nitrogen or methane-ammonia atmosphere [59]. This deposition rate drop is the responsible for the relatively low nitrogen content of a-C(N):H films, which was not reported to be higher than about 20 at.%, as shown in Table I.

Figure 8 shows the evolution of nitrogen content for the same samples shown in Figure 7, also as a function of the N-precursor partial pressure. Comparison of both N-content variations make clear that the nitrogen incorporation yield is quite higher for the NH_3-derived film, if we consider the nominal nitrogen content in the deposition atmosphere. However, the clearly high reactivity of NH_3 plasma species does not lead to a higher maximum nitrogen incorporation, which is limited by the deposition rate.

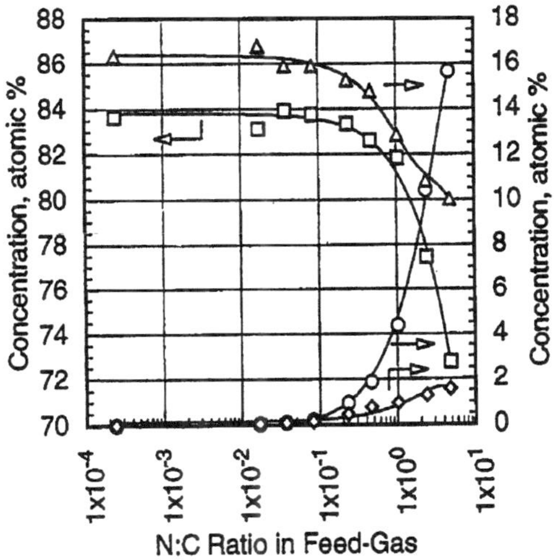

FIG. 9. Variation of C, N, H, and O contents as functions of an N/C ratio in the plasma. (Reproduced from [57].)

Concerning the hydrogen content, some authors report a continuous decrease upon nitrogen incorporation [11, 48, 50, 53, 57], as depicted in Figure 9, which shows a typical evolution of the chemical composition of plasma-deposited a-C(N):H films. Other authors [12, 54] report only slight variations of the hydrogen content, as shown in Table I. A more drastic effect on the hydrogen bonding in a-C(N):H films, the preferential bonding to nitrogen atoms, is discussed in Section 2.4.

Examination of Table I shows that in most cases the maximum N content is lower than 16 at.%. Nitrogen content in a higher range, like 20 at.% or more, were only observed for depositions performed with a C_2H_2-N_2 precursor mixture, suggesting a possible dependence of the nitrogen incorporation process on the hydrocarbon used. Comparison of the growth behaviors observed with use of different precursor mixtures confirms the suggestion. Figure 10 shows the deposition rates of RF plasma-deposited a-C(N):H films as a function of the nitrogen content in the film, normalized to the deposition rate obtained by the corresponding nitrogen-free film, taken from [54]. All the plots show a clear decrease of deposition rate upon nitrogen incorporation. The plots for CH_4-N_2, CH_4-NH_3, and C_2H_2-CH_3NH_2, seem to follow a very similar dependence, with the deposition rate having a trend to vanish at about 13 at.%. On the other hand, the deposition rate corresponding to C_2H_2-N_2 mixtures changes in a slower fashion as compared to the other ones. For about 20-at.% N incorporated into the film, the deposition rate is still half of the initial one. This behavior clearly shows that the higher ni-

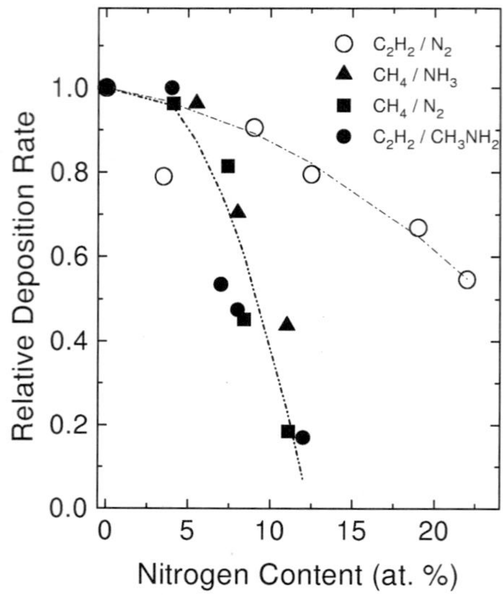

FIG. 10. Relative deposition rates of a-C(N):H deposited from several precursors as functions of the film nitrogen content. (Reproduced from [54].)

trogen uptake of a-C(N):H films deposited by C_2H_2-N_2 plasmas is due to kinetic reasons.

Another point that one has to observe from analysis of Figure 10, is that despite the different precursor atmospheres, and consequently different N precursor partial pressures in the deposition, there is a coincidence of the deposition rate behavior upon nitrogen content (for mixtures other than C_2H-N_2). This points to a strong dependence of growth kinetics with nitrogen content.

In which, concerns about the quite different behavior shown in film deposition using acetylene as the hydrocarbon is not exclusive of a-C(N):H film deposition. In a-C:H film deposition, the use of acetylene has also shown to lead to higher deposition rates as compared to methane [32, 33], and somewhat different film composition, structure, and properties as compared to several hydrocarbons [28]. One of the factors that leads to this situation is its relatively low hydrogen content, as compared to other hydrocarbons. Since the main role of the hydrogen species in a-C:H film deposition at room temperature is to saturate dangling bonds on the film surface, thus blocking surface sites initially available for the attachment of hydrocarbon reactive radicals [29]. Following this, we must expect that an increased hydrogen flux to the surface may lead to a decreased deposition rate. In fact, as Figure 11 shows, the nominal hydrogen content in the deposition atmosphere owned by the C_2H_2-N_2 mixture is much lower than that of other mixtures displayed in Figure 10.

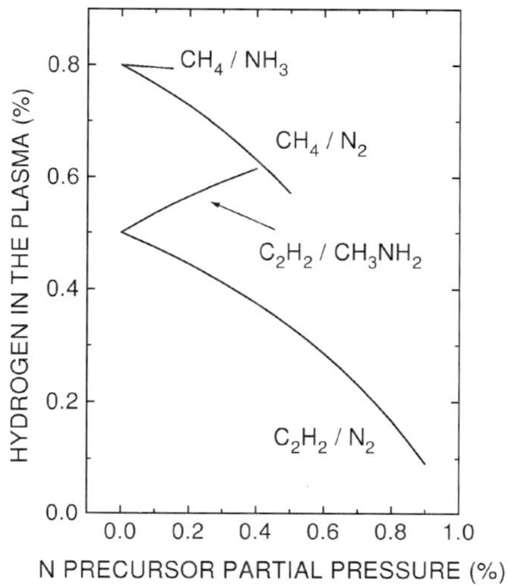

FIG. 11. Nominal hydrogen fraction in the deposition atmosphere against the nitrogen precursor partial pressure. (Reproduced from [54].)

The previously discussed dependence of the deposition rate on the hydrogen flux was also reported for a-C(N):H films deposited by an association of graphite-target N_2^+ ion-beam sputtering, and N_2^+/H_2^+ ion-beam assistance [60]. Although this deposition process may be not regarded as a typical plasma-deposition process, it retains some of its ingredients. In this work, increasing H_2 partial pressure in the deposition chamber resulted in a strong decrease in the deposition rate.

Table I includes results concerning a-C(N):H film deposition by ECRRF dual plasmas [49]. Despite the higher ionization and fragmentation of ECRRF plasmas, the results obtained are similar to those of glow-discharge plasmas. This is also the case for deposition performed by highly ionized methods such as plasma beam deposition [39].

Results from ECRRF plasma deposition of a-C(N):H films [61, 62] are useful to show the role of the bombarding ion energy on film growth. In the ECRRF plasmas, differently from RFPECVD, the energy of the bombarding ions can be varied—by the variation of the substrate RF self-bias—without great changes in the plasma. Inoue et al. [62] studied the deposition of a-C(N):H films by dual ECRRF plasmas in CH_4-N_2 atmospheres. In Figure 12 are plotted variations of the deposition rate as functions of self-bias, for several values of N_2 partial pressure, being the ECR plasma power fixed. Beginning with the nitrogen-free film, an increase in the deposition rate is observed for increasing self-bias, reflecting ion-beam assisting of the growth process. For increasing nitrogen partial pressure,

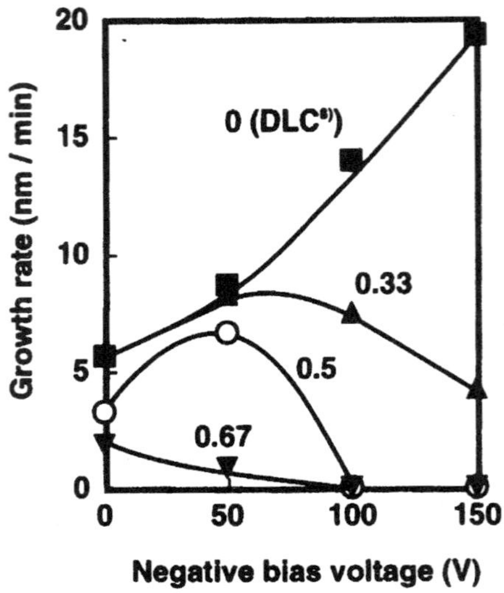

FIG. 12. Growth rate of a-C(N):H films deposited by ECRRF as a function of RF self-bias, for several N_2 partial pressures. (Reproduced from [62].)

the deposition rate initially increases, passes through a maximum, and after falls down. As the N_2 fraction in the discharge increases, the deposition rate vanishing occurs for lower self-bias. For the higher N_2 partial pressure (0.67), only a decreasing behavior of deposition rate is observed. From the analysis of these plots, one can conclude that either nitrogen fraction (as in RFPECVD) or self-bias increase induces a deposition rate decrease. The observed deposition rate decrease with increasing self-bias was ascribed to film etching by N_2^+ fast ions.

2.3.3. Plasma and Surface Processes Affecting Film Growth

To understand the deposition mechanism of a-C(N):H films by plasma processes, one needs information on the plasma species that contribute to the growth process, and on the interaction of these species with the film growing surface.

Knowledge on the plasma species can be obtained by the use of plasma diagnostics techniques, such as optical emission spectroscopy (OES) and mass spectroscopy (MS). Both techniques are able to probe atomic and molecular, neutral or ionized species present in plasmas. OES is based on measuring the light emission spectrum that arises from the relaxation of plasma species in excited energy states. MS, on the other hand, is generally based on the measurement of mass spectra of ground state species.

A typical OES spectrum taken from CH_4-N_2 plasma used in a-C(N):H film deposition is shown in Figure 13 [63]. This figure also shows band assignment of the

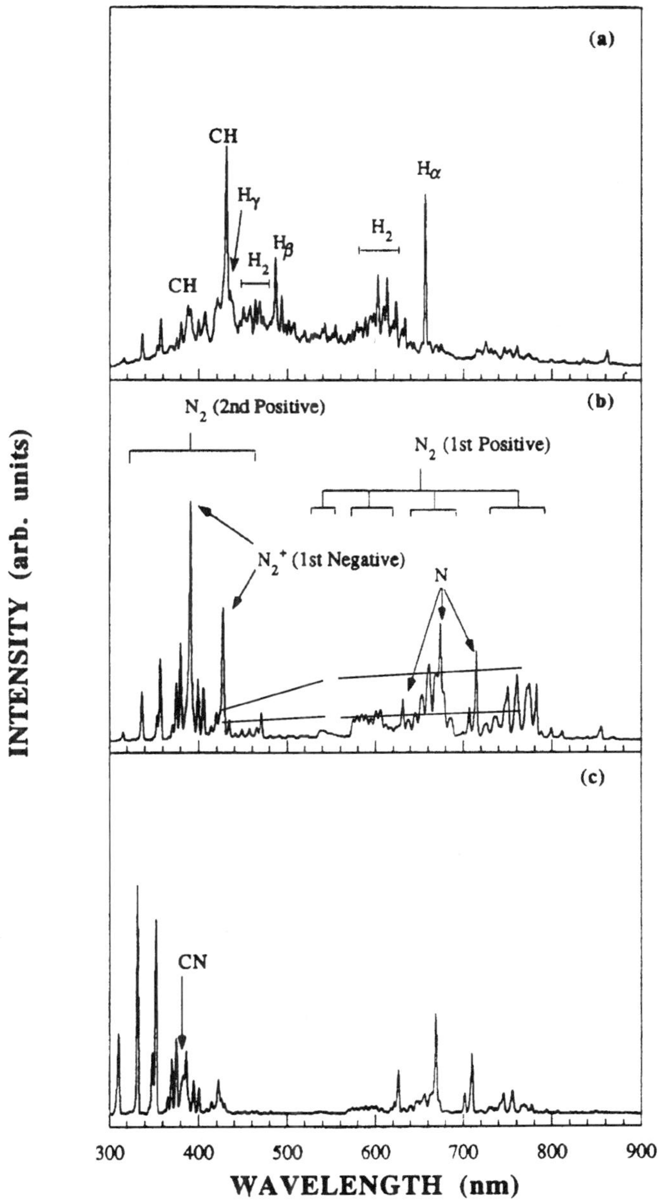

FIG. 13. Optical emission spectra taken from (a) CH_4, (b) N_2, and (c) CH_4/N_2 plasmas. (Reproduced from [63].)

most relevant emitting spectroscopic systems for CH_4, N_2, and CH_4/N_2 plasmas. Figure 14, also taken from Ref. [63], shows the variation of the emission intensity of selected plasma species, as a function the of N_2 fraction in the discharge atmosphere, RF power, and deposition pressure. In Ref. [63], the analysis was focused on specific features of N_2, $N_2{}^+$, atomic N, CH, and CN species emission spectra, some of them designated in Figures 13 and 14 by the related spectroscopic nomenclature. Figure 14 shows that upon the increase of N_2 partial pressure, all the bands associated with nitrogen gas (N_2, $N_2{}^+$, and atomic N) were found to increase. In addition, a decrease of the C$-$H and an increase of the CN emission intensities were observed. This behavior has been identified as a possible source for the deposition rate decrease of a-C(N):H films, due to the decrease of C_mH_n radical flux toward the growing surface. On the other hand, as the RF power increased, bands related to all species showed increased intensity, including the $N_2{}^+$ systems. This feature was associated to lack of film deposition observed above a critical power, by $N_2{}^+$ etching of the film.

The OES of CH_4-N_2 plasmas was also studied by Bhatasharyya, Granier, and Turban [64] (ECRRF) and Zhang et al. [65] (RFPECVD). The previously discussed decreasing behavior of CH emission intensity, and the increasing behavior of the CN emission intensity, for increasing N_2 partial pressure were also reported in both studies. In both works, the plasma was also probed by mass spectroscopy. In accordance with OES results, both works reported a decrease on the CH_3 fragment and an increase in the CN fragment mass signals. Bhatasharyya, Granier, and Turban [64] suggested that the increase of the CN abundance in the plasma would be responsible for the nitrogen incorporation, by C≡N bonds formation in the film, as suggested by the transmission infrared spectra taken from the film (see Section 2.4.2). Bhatasharyya, Granier, and Turban also ascribed the deposition rate decrease to the diminishing of the CH radical abundance.

The correlation between the plasma diagnostics results with film growth process must be carefully discussed. Beginning with the reduction of CH radical abundance and the increase of CN radical abundance. Since nitrogen is being admixed in the deposition atmosphere as the expenses of hydrocarbon flow, the decrease of CH radical abundance should be expected of course, even in the absence of the formation of CN radicals. However, the deposition rate decrease may not be ascribed to this fact, at least as the only reason. First, the drop observed in the deposition rate of a-C(N):H films is too sharp to be attributed to a dilution effect. In addition, it was reported that the CH radicals are not the main source of a-C:H film growth [66], instead the C_2H radical was pointed as the most relevant one, even in a-C:H deposition from methane. From this point of view, the work of Bhatasharyya, Granier, and Turban [64] gives important additional information. Their neutral radicals MS results showed, in addition to the decrease of CH abundance, a twofold increase in the abundance of the C_2H radical upon an N_2 partial

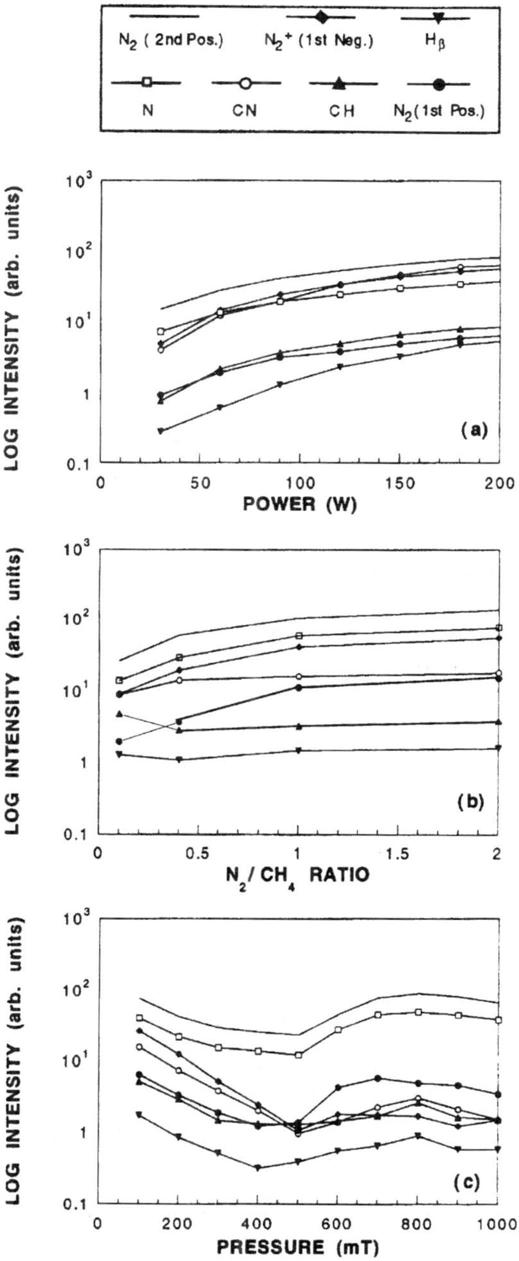

FIG. 14. Emission intensity of the main spectroscopic systems for power, N_2/CH_4 ratio and deposition pressure variations. (Reproduced from [63].)

pressure increase to 40%. These facts show the need of a more detailed study on the role of carbon-carrying radicals in a-C(N):H deposition.

This is also the case of the CN radicals in the nitrogen incorporation process. Bhatasharyya, Granier, and Turban proposed, by comparing the film deposition rate with CN radical abundance, that CN radicals were responsible for the nitrogen incorporation, by reaction with the film surface [64]. In addition, a successful attempt to synthesize amorphous a-C(N):H and a-C(N) directly from CN radicals was reported [67]. On the other hand, failure was reported in the attempt to deposit a-C(N):H films from pure methylamine [52] CH_3NH_2, a precursor that may readily produce CN radicals. Under pressure and self-bias conditions that resulted in a-C(N):H growth from CH_4-N_2 plasma, no deposit was obtained using pure methylamine. Deposition performed under higher RF power and total pressure resulted in about a 30-at.% nitrogen content, but with a very low deposition rate.

It must be taken into account that parallel to the incidence of CN radicals, the growing surface is intensely bombarded by N_2^+ fast ions. As pointed out by Zhang et al. [65], if CN radicals had a dominant role in the deposition process, no great differences would be found between film growth behavior on the RF powered and the grounded electrodes of a parallel plate RFPECVD deposition system. Since they are exposed to the same plasma, similar incidence of CN radicals is expected. In fact, Zhang et al. reported great differences in the growth behavior between the two electrodes [65]. At the RF-powered electrode, which is exposed to intense ion bombardment, the deposition rate strongly dropped with N_2 partial pressure. At the grounded electrode, where no bombardment is expected, an almost constant behavior of the deposition rate was found as the N_2 partial pressure increased.

To complement the preceding discussion, a more detailed discussion on the role of N_2^+ ion bombarding in the growth process is developed in the following paragraphs.

The relevance of N_2^+ ions bombarding was evidenced by the studies carried out by Hammer and Gissler [68] and Hammer et al. [69]. They reported the effects arising from a 150-eV N_2^+ ion bombardment of a pure carbon film. Hammer and Gissler [68] reported that N_2^+ ion bombardment results in the removal of carbon atoms from the films at rates as high as 0.5 C atoms per N_2^+ ion. Such a process was identified as a "chemical sputtering" since the sputtering yield was found to be substantially higher than that obtained [69] by noble ion bombardment with equivalent energies. This erosion process was found to result in the evolution of CN-containing molecules, such as CN and C_2N_2, as shown in Figure 15. This figure shows the mass spectrum of the evolved fragments in the N_2^+ ion bombardment process. In addition to the importance of the effect—high etch rates with relatively low energy beams—also the strong evolution of CN molecules, and even of the heavier C_2N_2 molecules must be noted.

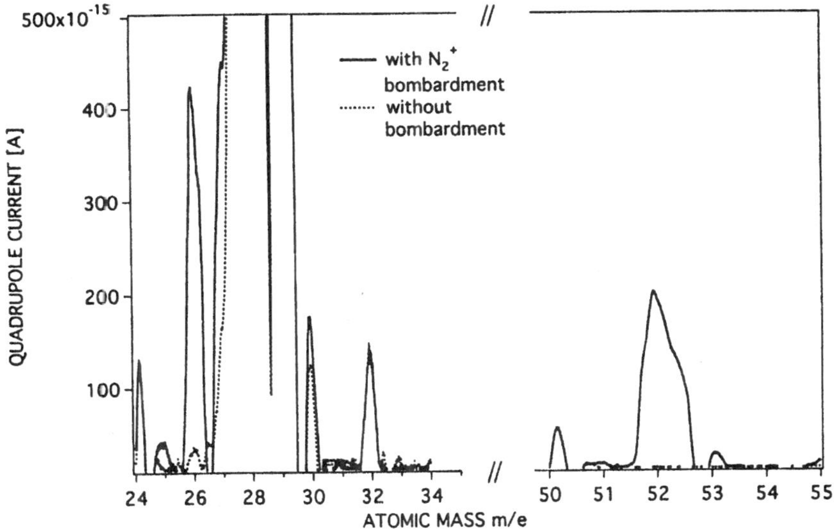

FIG. 15. Mass spectrum of molecules evolved from N_2^+ ion bombardment of a pure carbon film. (Reproduced from [68].)

One may observe that in plasma deposition of a-C(N):H films, as the nitrogen precursor gas partial pressure is increased, the bombardment of the film surface by N_2^+ ions will certainly occur. As this situation was shown to generate carbon erosion, and evolution of CN molecules, there is no reason to believe that CN fragments from the plasma would survive the bombardment, resting attached to the film.

A similar process was observed in the study of the etching process of a-C:H films by N_2 ECRRF plasmas [70]. It was found that RF-biased, previously grown a-C:H films were eroded by an N_2 plasma, with RF self-bias in the range of values of the results discussed in the previous session. As shown in Figure 16, the fragments evolved upon N_2^+ ion bombardment of a-C:H films are very similar to those presented by pure carbon films. This work also reports that ion bombardment also results in nitrogen incorporation. In another article, Hong and Turban reported that this process might indeed be more effective for nitrogen incorporation in amorphous carbon films. By alternating CH_4 and N_2 plasmas in a dual ECRRF plasma deposition system, they produced a-C(N):H films with a nitrogen content twice that could be obtained with CH_4-N_2 plasmas [71].

As a final comment on surface processes, the evaporation of N_2 molecules must not be neglected. Since one cannot have solid nitrogen at ambient temperature, as the nitrogen content increases, there will be an increased probability of N—N

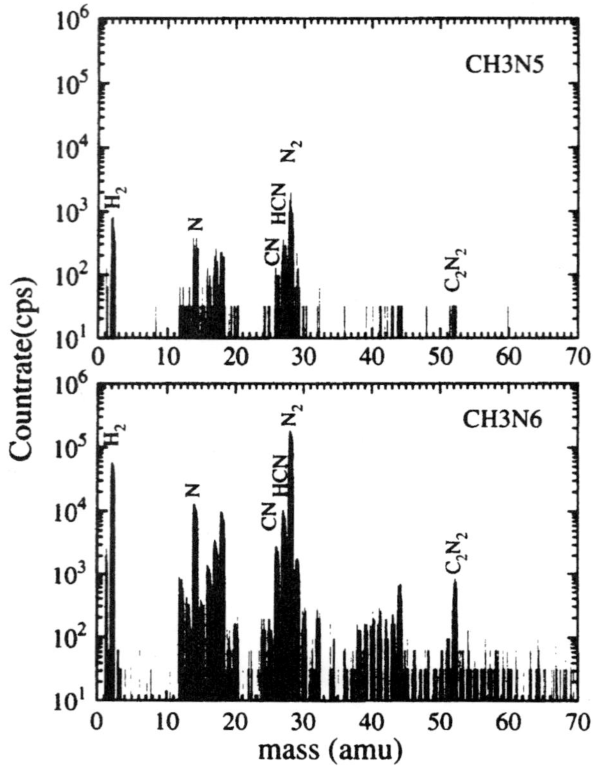

FIG. 16. Mass spectra of molecules evolved from ECRRF N_2 plasma treatment of an a-C:H film, for different treatment conditions. (Reproduced from [70].)

bonds formation. As pointed out by Marton, Boyd, and Rabalais [2], N—N bond formation should result in N_2 gas evaporation.

From the foregoing discussion, one may suggest that nitrogen ions play the dominant role in the a-C(N):H deposition process. Stronger evidence of this suggestion may only be obtained by more detailed research on this theme, such as the measurement of the absolute sticking factors of the species involved.

2.3.4. Modeling of a-C(N):H Film Growth

A simple model for a-C(N):H film growth kinetics was proposed [72], and was based on some aspects of the discussion developed in the previous subsection. The model is based on the incidence of only two species over a deposit, one representing C-carrying radicals and ions, while the other represents energetic N_2^+ ions. It is also based on the fact that C-containing radicals are the main channel for

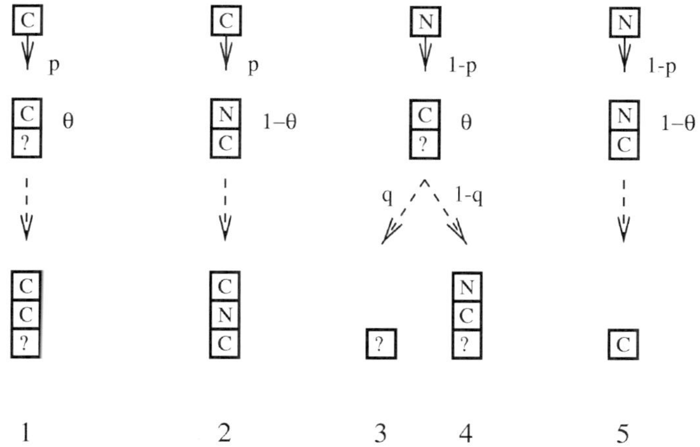

FIG. 17. Schematics of the process involved in the interaction of C and N species incident upon a carbon–nitrogen deposit. (Reproduced from [72].)

film growth in plasma-deposited a-C:H films [27]. In addition, it is supposed that nitrogen incorporation occurs mainly by the aggregation of the nitrogen atoms that are products of the breaking up of N_2^+ fast ions after colliding and penetrating into the film subsurface. These two species were, respectively, C and N, which are impinged to the model deposit with complementary probabilities p and $(1 - p)$, being the variation of p analogous to the variation of N_2 partial pressures in hydrocarbon—N_2 plasmas. Upon falling over the surface of the deposit, these species may take part in five different processes at the surface, as shown in Figure 17, and described as follows:

(1) the aggregation of the C species over a surface C species, with probability equal to $p\theta$ (θ being the C surface coverage);
(2) the aggregation of a C species over a surface N species, with probability equal to $p(1 - \theta)$;
(3) the erosion of a surface C species by an impinging N species, with probability $(1 - p)q\theta$, q being an interaction parameter related to the carbon sputtering yield by fast N_2^+ ions;
(4) the aggregation of an N species over a surface C species, with probability $(1 - p)(1 - q)\theta$; and
(5) the evaporation of an N_2 molecule, as a consequence of an N species falling over a surface N species with probability $(1 - p)(1 - \theta)$.

Todorov et al. [73] made equivalent assumptions in the modeling of a-CN_x film growth within another context. They focused their attention on the calculation of the collision cascade details following subsurface penetration of C^+ and N^+ ions

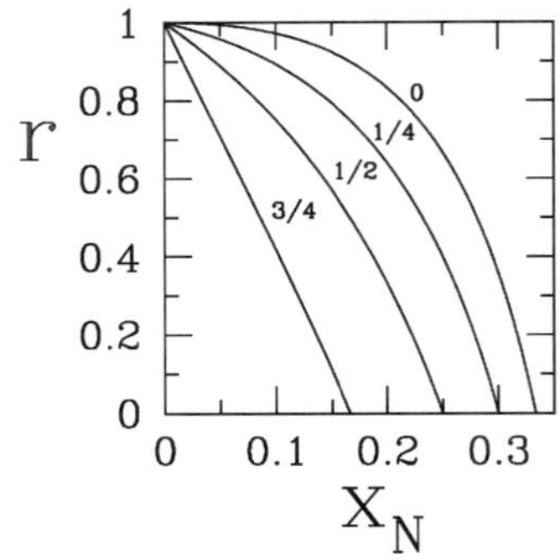

FIG. 18. Calculated relative deposition rate as a function of N content in film, for several values of the interaction parameter *a*. (Reproduced from [74].)

in direct ion-beam deposition of carbon–nitrogen films. Because of the presence of both carbon and nitrogen ionic species, they also included the erosion of nitrogen by fast C^+ ions, the process complementary to process (3).

Simple analytical calculations based on the previously discussed assumptions lead, among other results, to the determination of the film relative deposition rate as a function of the nitrogen content in the film, for several values of the interaction parameter q, as shown in Figure 18.

Several features of Figure 18 are related to the experimental results discussed in Section 2.3.2. For a range of the q interaction parameter ($q < 3/4$), the curves are very similar in shape with experimental a-C(N):H deposition rate vs N content plots. As q is varied, a wide range of maximum nitrogen (or zero deposition rate) contents is spanned. In addition, it is interesting to note that in this model nitrogen evaporation seems to be the major factor limiting nitrogen incorporation. The figure shows that even for $q = 0$ (no carbon removal by nitrogen ions) the deposition rate shows a strong fall upon nitrogen incorporation, vanishing at about 33-at.% N.

Comparison of experimental deposition rate results from a-C(N):H deposition with C_2H_2-N_2 RFPECVD [54] showed a very good agreement (see Fig. 19) for a q parameter equal to 0.25. This q value is closely related to the 0.5 C atom per N_2^+ ion previously reported [68]. Within this model, no agreement was obtained for the deposition rate plots owning to other gas mixtures, which had deposition rate vanishing at about 13–15 at.% N (Fig. 10). As shown in Figure 18, for higher q values the deposition rate curves show a trend to change curvature, being qualitatively different from a-C(N):H experimental deposition rate plots.

Further improvement of the previously described model [74] allowed the inclusion of a sticking probability α superimposed to the aggregation of both carbon and nitrogen (C and N species in the model). This was done to represent the block-

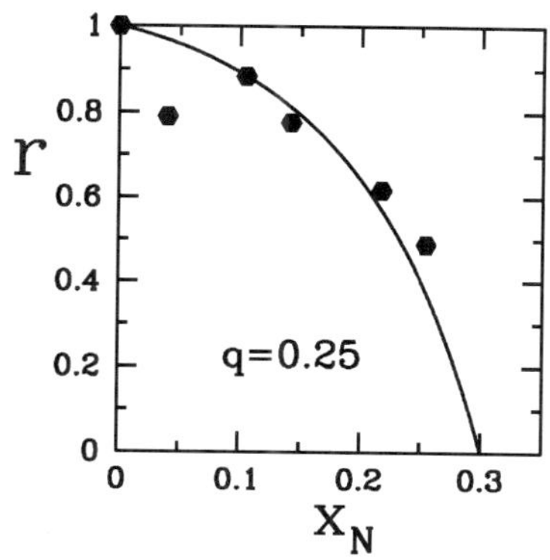

FIG. 19. Comparison of the two species model deposition rate as a function of N content with experimental results obtained from C_2H_2/N_2 plasmas. (Reproduced from [74].)

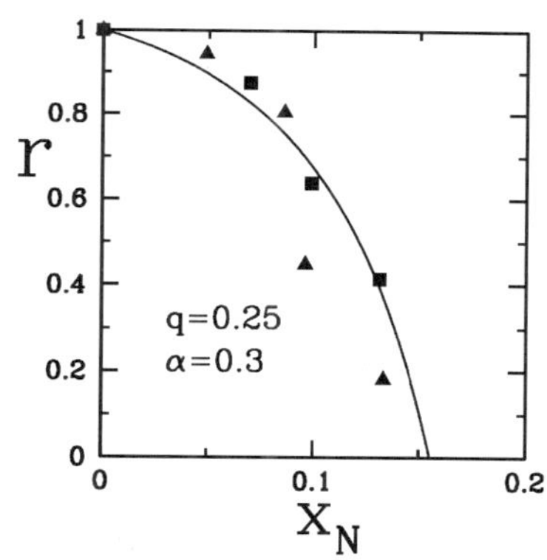

FIG. 20. Comparison of the two species model deposition rate as a function of N content, including the blocking of surface sites, with experimental results obtained from CH_4/N_2 plasmas. (Reproduced from [74].)

ing of surface sites to the sticking of hydrocarbon neutral radicals by an increased hydrogen flux. In this way, it was possible to fit the deposition rate vs nitrogen content plots of a-C(N):H films deposited by RFPECVD in methane–nitrogen atmospheres, by supposing a sticking factor α equal to 0.3, whereas in C_2H_2-N_2 modeling α equals one. Figure 20 shows the comparison of CH_4-N_2 experimental and model results of the deposition rate.

The previously discussed model suggests how realistic is the assumption of N_2^+ ion bombardment as the main driving force of growth kinetics and nitrogen incorporation in a-C(N):H film growth.

2.4. Characterization of a-C(N):H Film Structure

2.4.1. Raman Spectroscopy

Raman scattering spectroscopy is used to probe the vibrational excitations of a sample, by measuring the wavelength change of a scattered monochromatic light beam. This is usually performed by impinging a monochromatic laser beam to the sample surface, and by recording the scattered beam spectrum.

Raman spectroscopy has been widely used in carbon materials characterization. The main features owing to the Raman spectra of carbon materials are the so-called D and G bands, located, respectively, at about 1350 and 1580 cm^{-1} wave numbers. The G band is the only feature in the first-order Raman spectrum of single crystalline graphite. It corresponds to the excitation of the in-plane E_{2g} and E_{1u} vibration modes of graphite, which are represented in Figure 21. The D band (D from disorder) is the additional feature that is present in the Raman spectrum of microcrystalle and disordered graphite. It arises from the breaking of translational symmetry within the graphene planes. In microcrystalline graphite, the ratio of the

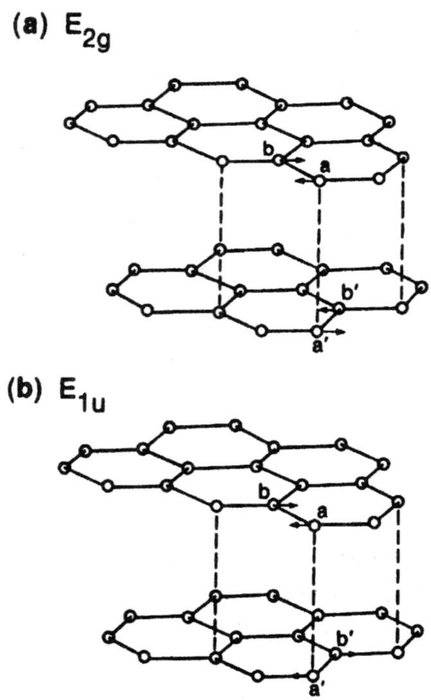

FIG. 21. E_{2g} and E_{1u} vibration modes of crystalline graphite. (Reproduced from [10].)

D to the G band intensities I_D/I_G was found to be inversely proportional to the crystal size across the basal plane.

In amorphous carbon films, the G and D bands are highly broadened, with the band intensity ratio I_D/I_G, maximum positions ω_G and ω_D, and line breadths Γ_G and Γ_D varying in a relatively broad range. The interpretation of the changes in the spectra parameters of DLC films is not straightforward as in crystalline graphite. Instead of direct relations with structural characteristics, empirical procedures have been used to extract useful information from Raman spectra of DLC materials. A usual approach is that coming from the investigation conducted by Dillon, Woolam, and Katkanant [75]. They performed Raman analysis in thermally treated (up to 1000 °C) a-C:H films deposited by RFPECVD and ion-beam deposition. To analyze the great changes introduced in the broad spectra, they were deconvoluted in their constituent D and G bands. The changes observed in the spectra parameters of the heat-treated samples were the following:

(a) The increase in the I_D/I_G integrated band intensity ratio up to about 600 °C, followed by a decrease;
(b) The increase in the position of the G and D lines until stabilization for temperatures higher than 600 °C and;
(c) The narrowing of the D and G bands until stabilization for temperatures higher than 600 °C.

Such results were supposed to correspond to an increase in the size and/or number of the sp^2-carbon condensed ring domains present in the films upon increasing temperature, up to graphitization at 600 °C. After that, the expected behavior for graphitized carbon, the decrease of I_D/I_G upon the increase of crystal size is observed.

This picture was found to be consistent with the comparison of Raman spectra and optical gap of a-C:H films deposited by RFPECVD, with increasing self-bias [41]. It was found that both, the band intensity ratio I_D/I_G and the peak position ω_G increased upon increasing self-bias potential. At the same time, a decrease on the optical gap was observed. Within the cluster model for the electronic structure of amorphous carbon films, a decrease in the optical gap is expected for the increase of the sp^2-carbon clusters size. From this, one can admit that in a-C:H films, the modifications mentioned earlier in the Raman spectra really correspond to an increase in the graphitic clusters size.

In reality, several factors were mentioned as being responsible for this behavior, such as variations in bond angle distortion, in the internal stress or in the hydrogen content [40, 76], but all of them are also strongly correlated with the variation of optical gap width in amorphous carbon films. Theoretical work on Raman spectroscopy on DLC materials gave additional support for Dillon's interpretation [77].

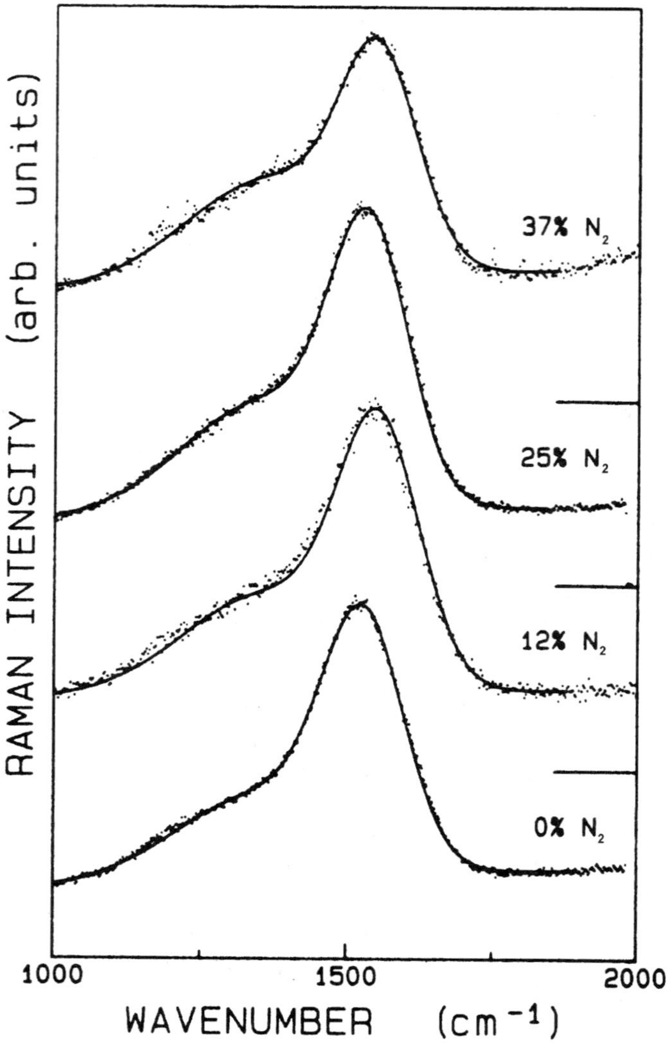

FIG. 22. Raman spectra taken from a-C(N):H films deposited by RFPECVD in CH_4-N_2 atmospheres, for several values of the N_2 partial pressure. (Reproduced from [78].)

Raman spectroscopy has been applied to probe the structural evolution of sp^2-hybidized carbon aggregates in plasma-deposited a-C(N):H films [12, 14, 52, 54, 56]. Very often, due to the great breadth of D and G bands, the spectra deconvolution was done by fitting two Gaussian lines, as reported by Mariotto, Freire, and Achete [78]. They studied the Raman spectra of a-C(N):H films deposited by RFPECVD in CH_4-N_2 atmospheres. The obtained spectra are shown in Figure 22

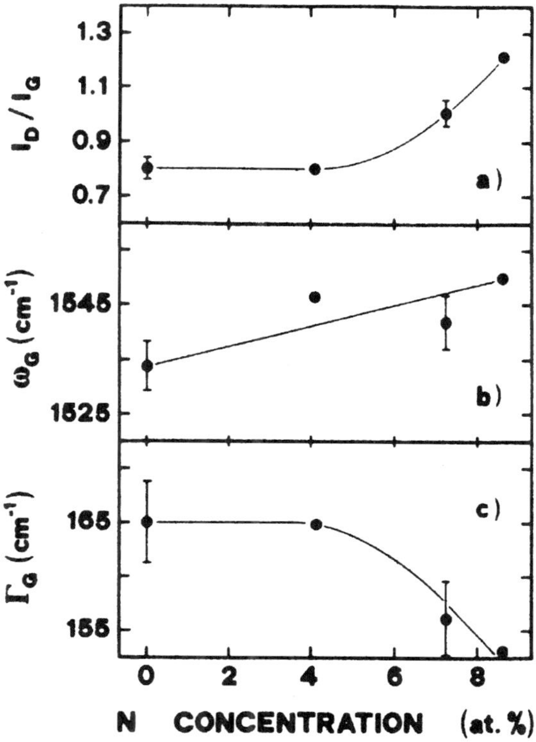

FIG. 23. Variation of the integrated intensity ratio I_D/I_G, the maximum position of the G band, ω_G, and its width Γ_G against film nitrogen content. (Reproduced from [78].)

and the corresponding results from the fitting procedure are shown in Figure 23. The analysis of the spectra showed an increase on the I_D/I_G band ratio from about 0.8 to about 1.3. At the same time, the position of the maximum of the G band was found to shift from 1530 to 1547 cm^{-1}, and to be narrowed from 165 to 154 cm^{-1}. All of these changes occurred within a 12.5-at.% N incorporation, and were ascribed to an increase in the size or number of graphitic domains.

Other studies showed the same trends discussed earlier. Films obtained by RF-PECVD, deposited at about the same conditions in other precursor atmospheres [79], showed the same behavior, with the I_D/I_G and ω_G following about the same dependence on nitrogen content as the one shown in Figure 23.

On the other hand, films deposited by RFPECVD in CH_2-N_2 atmospheres showed a different behavior [54]. In this study, in addition to the reported enhanced maximum nitrogen incorporation of 22 at.% (see Section 2.3.1), an almost constant behavior of the Raman parameters was observed for nitrogen content up to about 12 at.%. Above this limit, I_D/I_G and ω_G were shown to increase, and Γ_G was shown to decrease. This is evidence that the details related to the growth process, may give rise to differentiated sp^2-clusters size evolution.

The Raman parameters reported for a-C(N):H films deposited by magnetic field enhanced RFPECVD in CH_4-N_2-He atmospheres also showed an intermediate N content range (about 7 at.%), with almost constant Raman parameters. In this case, the behavior was found to be associated with a nontypical variation of the C and N atom hybridization state, as discussed in Section 2.4.3.

Results on Raman spectroscopy thus show that nitrogen incorporation, at least for a large enough N content, results in the increase of the graphitic clusters. This is contrary to the formation of an amorphous solid related to the β-C_3N_4 phase, which presumes sp^3-C hybridization and no clustering effects.

2.4.2. Infrared Spectroscopy

Infrared spectroscopy has been widely used to characterize amorphous hydrogenated carbon films [28, 31–33]. The main feature observed in the infrared spectra of diamond-like a-C:H films is a broadband in the 2900 cm^{-1} wave number range, which is due to the C—H stretching vibration of CH_n groups containing carbon atoms in all the hybridization states. Although it has been sometimes used to determine the hybridization state of carbon atoms, by deconvoluting the CH broad band into its constituent bands [32, 33], this analysis is taught to give only qualitative information.

Nitrogen incorporation into a-C:H films was reported to give rise to dramatic changes in the infrared spectrum, as first reported by Kaufman, Metin, and Saperstein [10]. As shown in Figure 24 [10], as the nitrogen content increases, the spectra show the appearance, with increasing intensities, of absorption bands located at 2300 and 3500 cm^{-1}. These two bands have been respectively ascribed to the C≡N and N—H stretching bands. In addition, a continuous decrease of the CH stretching band is also reported. A broadband, located in the 1500 cm^{-1} wave number range also emerges with increasing intensity for increasing nitrogen content.

This broad band at 1500 cm^{-1} was ascribed by Kaufman, Metin, and Saperstein [10], to an IR observation of the amorphous carbon Raman D and G bands. This is forbidden by the selection rules, and has been attributed to the symmetry breaking introduced by the presence of CN bonds in the amorphous network. As carbon and nitrogen have different electronegativities, the formation of CN bonds gives the necessary charge polarity to allow the IR observation of the collective C=C vibrations in the IR spectrum. This conclusion was stated by the comparison of spectra taken from films deposited from $^{14}N_2$ and $^{15}N_2$. In the $^{15}N_2$-film spectrum, no shift was observed for the 1500-cm^{-1} band, whereas all other bands shifted as expected from the mass difference of the isotopes. Figure 25 compares

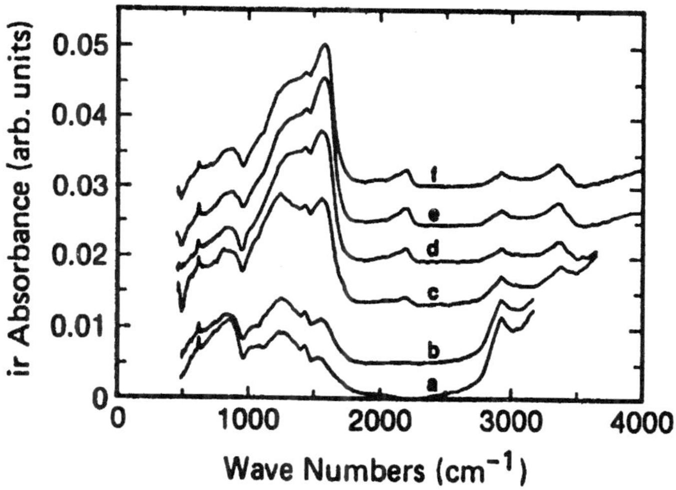

FIG. 24. Infrared absorption spectra taken from a-C(N):H films with several N contents. N increases from *a* to *f*. (Reproduced from [10].)

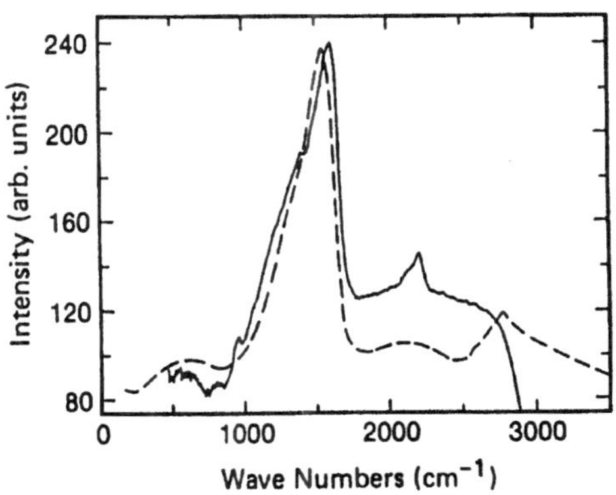

FIG. 25. Comparison of a Raman spectrum with an IR spectrum of an a-C(N):H film. (Reproduced from [10].)

Raman and IR spectra taken from the same a-C(N):H film [10], showing an almost perfect overlapping of the two bands.

The variation of the IR band intensities upon nitrogen incorporation for RF plasma-deposited a-C(N):H films is shown in Figure 26, as reported by Schwan et al. [53]. As mentioned before, the intensity of the C—H stretching band decreases upon nitrogen incorporation, at the same time that an increase in the N—H stretching band intensity is observed. This suggests that hydrogen preferentially

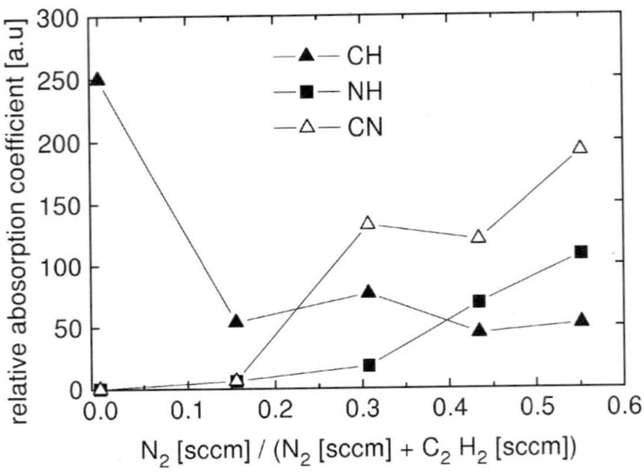

FIG. 26. Variation of the intensities of the IR CH, C≡N, and NH bands, for films obtained with several N_2 partial pressures in C_2H_2-N_2 plasmas. (Reproduced from [53].)

bonds to nitrogen in a-C(N):H films. Since the hydrogen content does not decrease at a rate higher than that of the nitrogen content (see Table I), it may be supposed that most of the nitrogen atoms would be hydrogenated, and thus at a terminal site of the amorphous network. The CN bonds shown in the IR spectra are also terminal bonds. As pointed out by Kaufman, Metin, and Saperstein [10], the IR spectra show that nitrogen atoms which are present in the IR-active groups may be located at the periphery of condensed aromatic rings.

Work on a-C(N):H films, deposited by Ar^+ ion-beam sputtering of graphite and ion-beam assistance by $N_2{}^+/H_2{}^+$ ions [80] lead basically to the same conclusions of Kaufman, Metin, and Saperstein [10]. In addition, these authors evidenced that nitrogen incorporation is not the only factor affecting the presence of the Raman band in the IR spectrum of a-C(N):H films. Hydrogen also plays a role, as evidenced by the theoretical simulation of the vibrational spectra of model organic molecules, and by the experimental comparison of Raman and infrared spectra of hydrogenated and nonhydrogenated carbon–nitrogen films. It was found that only for the hydrogenated films Raman and IR spectra were coincident, which was also corroborated by the model calculations.

2.4.3. Electron Energy Loss Spectroscopy

The hybridization of carbon atoms is the major structural parameter controlling DLC film properties. Electron energy loss spectroscopy (EELS) has been extensively used to probe this structural feature [5, 6]. In a transmission electron microscope, a monoenergetic electron beam is impinged in a very thin sample, being the transmitted electrons analyzed in energy. Figure 27 shows a typical

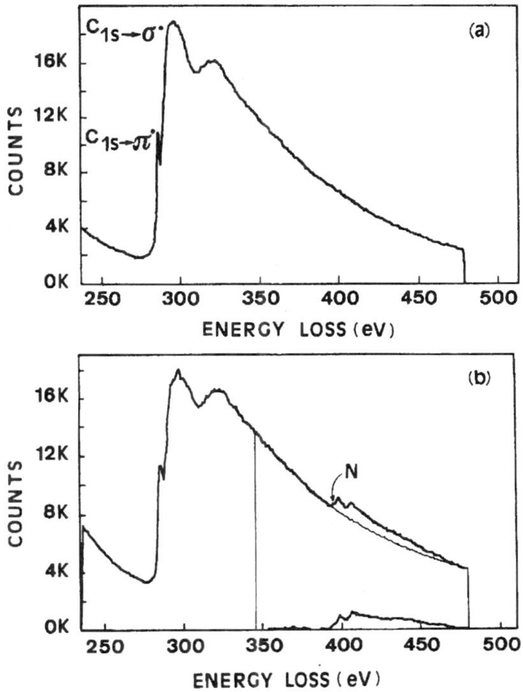

FIG. 27. Electron energy loss spectra taken from (a) an a-C:H film; and (b) an a-C(N):H film. (Reproduced from [56].)

EELS spectra from an a-C(N):H film [56]. The energy loss peaks shown in this figure correspond to the carbon and nitrogen K-edges, each one being composed of peaks corresponding to 1s-σ^* and 1s-π^* electronic transitions. The carbon and nitrogen sp^2-fraction estimates are obtained by comparing the π to σ peak area ratio to the one of a standard all-sp^2 sample, like highly oriented pyrolitic graphite HOPG [81].

The sp^2 C atom fraction variations against nitrogen content for a-C(N):H films deposited by RFPECVD from methane–nitrogen and methane ammonia atmospheres [11] are shown in Figure 28. In both cases, nitrogen incorporation results in a continuous increase of the sp^2-carbon atom fraction from about 50 to about 80%. Such a variation is in accordance with the previously discussed Raman spectroscopy results, which showed an increase in the size of graphitic domains present in the sample. The reported variations indicate a very strong effect, as compared to the relatively low nitrogen incorporation. Such kinds of effects are also reported for nitrogen incorporation in highly tetrahedral amorphous carbon [82], and seems to be inherent to the nitrogen incorporation process in DLC films.

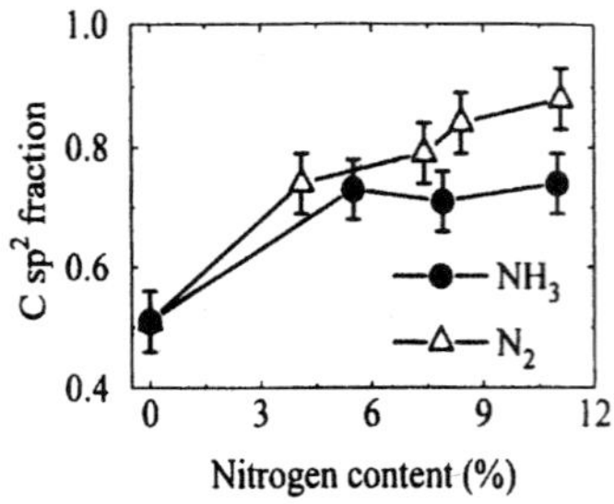

FIG. 28. Variation of the C sp^2 fraction as a function of N content in film, for films deposited from CH_4-N_2 and CH_4-NH_3. (Reproduced from [11].)

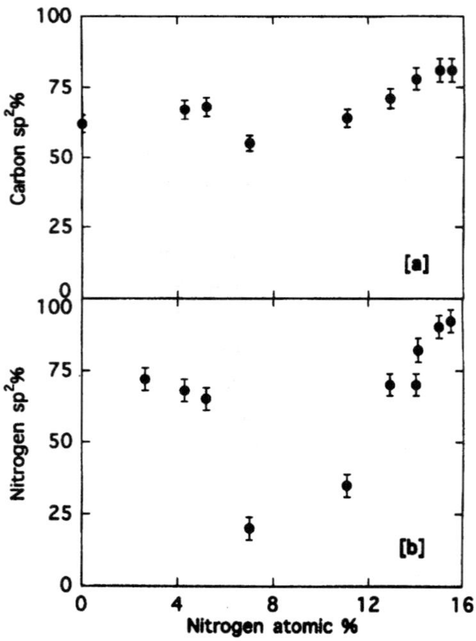

FIG. 29. C and N sp^2 fractions as functions of the nitrogen content in films deposited by magnetically enhanced RFPECVD in CH_4-He-N_2 atmospheres. (Reproduced from [14].)

However, different variations were also reported. A small decrease of the sp^2 fraction was observed for a 7-at.% nitrogen incorporation in films deposited by magnetically enhanced RFPECVD from CH_4-N_2 mixtures highly diluted in He [15], as shown in Figure 29. At the same time, an N atom sp^2 fraction decrease from 75 to about 20% in the same nitrogen content range was observed. For higher nitrogen content, the C sp^2 fraction increased again, resulting in a variation from 75%, for the nitrogen-free film, to about 90%, for the 16-at.% N containing film.

It is important to stress that nitrogen incorporation in a-C:H films always result, at least above a certain level, in a strong decrease in the tetrahedrally bonded carbon atom fraction. Raman spectroscopy also gives support to this observation, because the increase in the size of graphitic clusters only can proceed with also increasing sp^2 fraction.

2.4.4. X-RAY PHOTOELECTRON SPECTROSCOPY

X-ray photoelectron spectroscopy (XPS) is based on the measurement of the binding energy of core-level electrons ejected from the sample surface, as a consequence of monochromatic X-ray irradiation. The knowledge of the photoelectron production cross sections of the elements allows the estimation of sample chemical composition. This is done by comparing the corrected integral photoelectron intensities from each element present. In addition, it is expected that changes in the chemical environment of an atom may give rise to a shift in the photoelectron binding energy, due to charge transfer effects arising from chemical bonding [44]. However, the interpretation of the results is generally controversial [3], and shall be treated with care. The chemical shifts of XPS lines are small, strongly dependent on experimental factors such as surface charging effects or the presence of contaminants, and very often cannot be surely assigned to specific structural features, due to the lack of suitable model calculations.

In the following paragraphs, reports on XPS studies of a-C:N:H films are discussed. Most of the work that is discussed is not related to usual plasma deposition, but is related to ion-beam deposition, ion-beam assisted deposition, or other methods, because only a very few wide scope XPS studies on plasma-deposited a-C(N):H films were done up to this moment. Nevertheless, these results bring useful information on the role of hydrogen in the structure of a-C(N):H films.

Mansur and Ugolini [45] reported an XPS study on ion-beam deposited a-C(N):H films using a Kaufman-type ion source fed by CH_4-N_2 mixtures. They studied the changes in the C1s and N1s photoemission lines, for samples containing up to 10-at.% N (considering only C and N) (Fig. 30). Upon nitrogen incorporation, the C1s binding energy (BE) was found to shift from 284.8 to 285.2 eV. The corresponding linewidths (full width at half-maximum FWHM) were also found to increase, from 1.8 to 2.1 eV, being the broadening asymmetric toward higher BEs. At the same time, no changes were observed in the N1s line position or breadth, which remained fixed at 399 ± 0.1 and 2.4 eV, respectively. The shift of the C1s line to higher BEs was ascribed to the higher electronegativity of nitrogen atoms, as compared to carbon. Therefore, carbon–nitrogen bonding should result in charge transfer from carbon to nitrogen, and thus leading to the increase in the C1s binding energy.

Based on the IR spectroscopy, evidence that nitrogen incorporation into a-C:H films may occur in the periphery of condensed aromatic rings [10], Mansur and

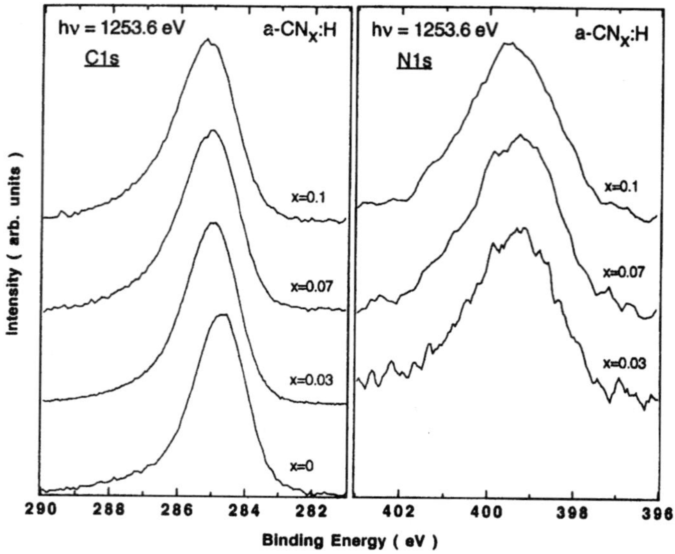

FIG. 30. C1s and N1s X-ray photoelectron spectra taken from a-C(N) films with several N contents, deposited by CH_4/N_2 ion beams. (Reproduced from [45].)

Ugolini compared the C1s spectra of the a-C(N):H films with that taken from benzene (C_6H_6) and pyridine (C_5H_5N). Data from benzene agreed well with that of pure a-C:H film (284.5-eV BE and 1.8-eV FWHM), but data from pyridine showed much higher BE (286.6 eV) than that shown by the 9-at.% N film. Since in pyridine, the nitrogen content (16.7%) is higher than that of the most nitrogenated film (9.0%), they made a linear extrapolation of the BEs to the actual film nitrogen content. The obtained shift, relative to the C1s BE of the a-C:H film, was still higher (2.5 times) than that observed for the 9-at.% N films, thus leading to discarding the nitrogen content as the only source of the observed difference. Taking into consideration the asymmetry of the line broadening, a two-phase model was proposed. Accordingly, the C1s line was decomposed by fitting two Gaussian lines. This procedure resulted for the carbon–nitrogen and pure carbon environments BEs of 286.2 and 285 eV, respectively, a 1.2-eV shift in a reasonably good agreement with the 1.4-eV shift observed for benzene and pyridine. The area ratio between the a-C:H and a-C(N):H features of the spectra was found to be 28%.

The bigger widths of the a-C(N):H N1s lines, as compared to that of pyridine, were also ascribed to a two-phase structure. As hydrogen is less electronegative than nitrogen, N—H bonding should result in a decreased BE, as compared to that of nitrogen bonded only to carbon. The authors were also able to fit the N1s lines with two Gaussian lines, the one at lower BE being ascribed to N atoms bonded to C and H, whereas that at higher BE was ascribed to N bonded only to C.

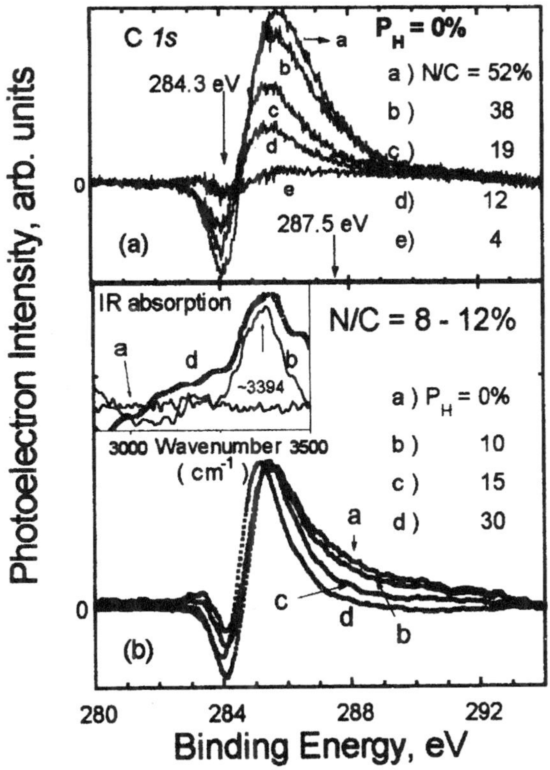

FIG. 31. XPS C1s difference spectra, as explained in the text. (Reproduced from [83].)

Souto and Alvarez [83] closely studied the role of hydrogen in the XPS spectra of a-C(N):H films. They studied a-C(N):H films produced by reactive RF sputtering of a graphite target in Ar-N_2-H_2 atmospheres, onto unbiased substrates, maintained at 285 °C during deposition. Their study was done by the analysis of the core-level C1s and N1s spectra obtained from hydrogenated and nonhydrogenated CN_x films. The C1s line analysis was performed by subtracting the C1s spectrum of a pure carbon film from (a) the spectra obtained from hydrogen-free CN_x films, and (b) the spectra obtained from hydrogenated CN_x films (respectively, Fig. 31a and b). Figure 31a shows that nitrogen incorporation into a-C films results in the growing of a contribution at higher BEs, which was ascribed to charge transfer from carbon to nitrogen, as discussed before. On the other hand, hydrogenated films showed decreasing intensities in the higher energy tail of the difference spectra, for increasing hydrogen incorporation (Fig. 31b). This effect was attributed to the preferential bonding of hydrogen to nitrogen atoms, which

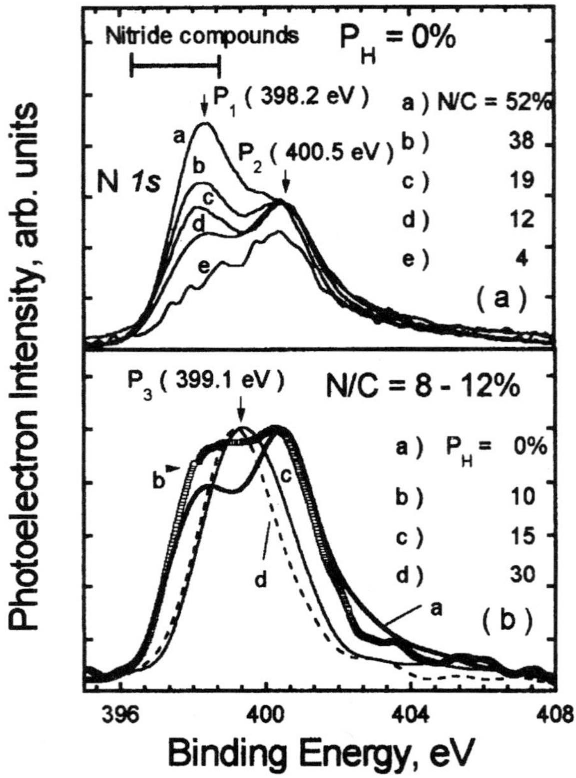

FIG. 32. XPS N1s difference spectra, as explained in the text. (Reproduced from [83].)

leads to the decrease of the number of CN bonds. Additional support for this hypothesis is given by the N1s photoelectron spectra, shown in Figure 32. Figure 32a shows an increased contribution of the peak located at 398.2 eV, named P1 in the figure, which was ascribed to nitrogen atoms sharing all of their valence electrons with carbon atoms. On the other hand, peak P2 at 400.5-eV binding energy was ascribed to N probably bonded to only one C atom. Figure 32b shows that hydrogen incorporation causes the collapse of P1 and P2 peaks, giving rise to the peak P3 at about 399 eV, which was identified to be probably due to a $C{-}NH_2$ bonding. This is additional evidence of preferential bonding of hydrogen to nitrogen atoms, even in a so different deposition method like reactive magnetron sputtering. The main effect of hydrogen incorporation is to avoid the formation of a C$-$N extended network.

A more detailed approach was used by Hammer, Victoria, and Alvarez [60], in the analysis of a-C(N):H films deposited by Ar^+ ion-beam sputtering of graphite

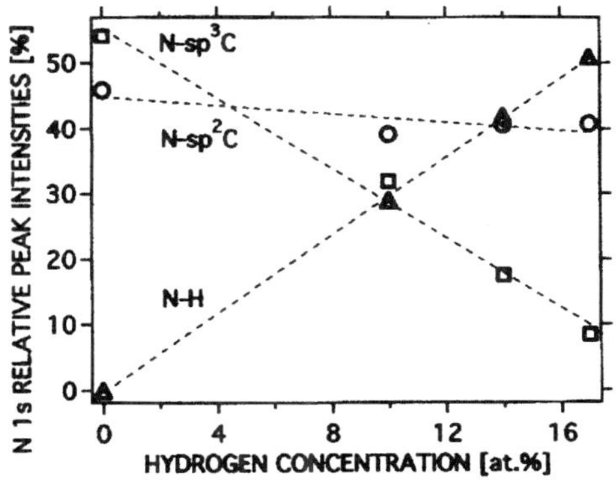

FIG. 33. Result of a-C(N):H films N1s peak decomposition. (Reproduced from [66].)

targets, and N_2^+ and H_2^+ ion-beam assistance. They used four Gaussian lines to fit the N1s peaks of the hydrogen free CN_x film (398, 400.6 eV, and two low-intensity components at 402.6 and 404.6 eV). The assignment of these lines was performed by numerical simulations applied to representative N-containing molecules with several sp-, sp^2-, and sp^3-carbon atom arrangements, in agreement with previous extensive XPS work on a-CN_x films. The 398-eV line was ascribed to N bonded to sp^3 C with isolated lone pair electrons. The peak at 400.6 eV was attributed to N atoms in graphite-like configurations, with the lone pair involved in p-bonding. The smaller peaks at higher BEs were not clearly identified, but were supposed to be due to N—N bonds. For the analysis of the hydrogenated films, the previously mentioned 399.1-eV peak due to N—H bonding also had to be included. The result of the fitting procedure applied to samples with different hydrogen contents is shown in Figure 33. This figure shows deep changes in film structure, with a strong decrease on the N-sp^3 C bonding, and the expected increase in the N—H bonding, all these occurring with an almost constant N-sp^2 fraction.

The previously discussed XPS results, despite being obtained from films deposited from deposition methods other than usual plasma deposition, give valuable information on a-C(N):H film structure. Results reported in the works of Mansur and Ugolini [45]; Souto and Alvarez [83]; and Hammer, Victoria, and Alvarez [60] are in total accordance with the picture that emerges from IR results: hydrogen bonds preferentially to nitrogen atoms, which are bonded mainly to sp^2-carbon atoms.

2.4.5. ^{13}C NUCLEAR MAGNETIC RESONANCE SPECTROSCOPY

Solid-state nuclear magnetic resonance (NMR) on ^{13}C nuclei has been used for a-C:H characterization [19, 85]. It was shown that ^{13}C NMR is well suited for hybridization studies on amorphous hydrogenated carbon films, since it shows well-defined peaks for sp^3- and sp^2-carbon atoms. Its use, in conjunction with the knowledge on the sample chemical composition lead to the determination of the detailed carbon atom coordination in plasma-deposited a-C:H films with varying self-bias [19]. The combination of NMR on ^{13}C and ^{1}H nuclei, allowed the detailed determination of all the protonation states of sp^2- and sp^3-carbon atoms, in addition to the determination of the bound-to-unbound hydrogen fraction in a-C:H films [84], in which concerns to hybridization determination, ^{13}C NMR spectroscopy is advantageous over other techniques, like electron energy loss spectroscopy and infrared spectroscopy. The high-energy electron beam used in EELS may damage the sample and may give rise to misleading results. Infrared spectroscopy can detect only IR-active bonds. Despite its power, NMR spectroscopy is scarcely used in a-C:H film characterization due to its low sensitivity. Samples as large as 70 mg in weight are needed [84], which may correspond to an about 350 cm^2 in area, 1-μm thick a-C:H film.

A ^{13}C-NMR study was on a-C(N):H RFPECVD deposited films in butadiene–nitrogen or butadiene–ammonia atmospheres was reported [85]. As the results, about a 65% sp^3 fraction was obtained for both the nitrogen-free film and the nitrogen-containing film deposited from nitrogen gas, while a 73% sp^3 fraction was obtained from the NH_3-derived film. The Auger electron spectroscopy estimates of the nitrogen content of the ammonia and nitrogen-derived films were about 5- and 2-at.% N, respectively. Such quite impressive sp^3-fraction enhancement was not confirmed by later studies on N_2/NH_3 comparison as nitrogen precursors [11]. The observed increase in the sp^3 fraction upon nitrogen incorporation with ammonia perhaps may be due to a higher hydrogen fraction, which may occur with the use of ammonia as the precursor gas [56].

Lamana et al. [86] also reported a ^{13}C-NMR study on a-C(N):H films. Despite being related to films deposited on the grounded electrode of an RFPECVD deposition reactor (no substrate bias at all), the obtained results are useful in the discussion of nitrogen modification of a-C:H film structure. For the nitrogen-free films, the authors observed the broad resonance bands related to the sp^2- and sp^3-hybridized carbon (Fig. 34). However, for the nitrogen-containing films sharp peaks were observed, superimposed on the broad ones, characterizing an ordering behavior. The authors fitted the spectrum to model structures composed of nitrogen-containing fused aromatic rings, terminated by NH_2 groups. No references were done by the authors on the evolution of the broad peaks in the spectra 3 and 4 (respectively, low nitrogen content and nitrogen-free film) shown in Figure 34. However, it must be noted that nitrogen incorporation leads to a decrease of the sp^3-carbon peak at 40 ppm, if we compare the spectrum 3 to that of the nitrogen-free sample, characterizing an increase in the sp^2-C atom fraction. For the higher nitrogen content samples, the model structures adopted for spectrum

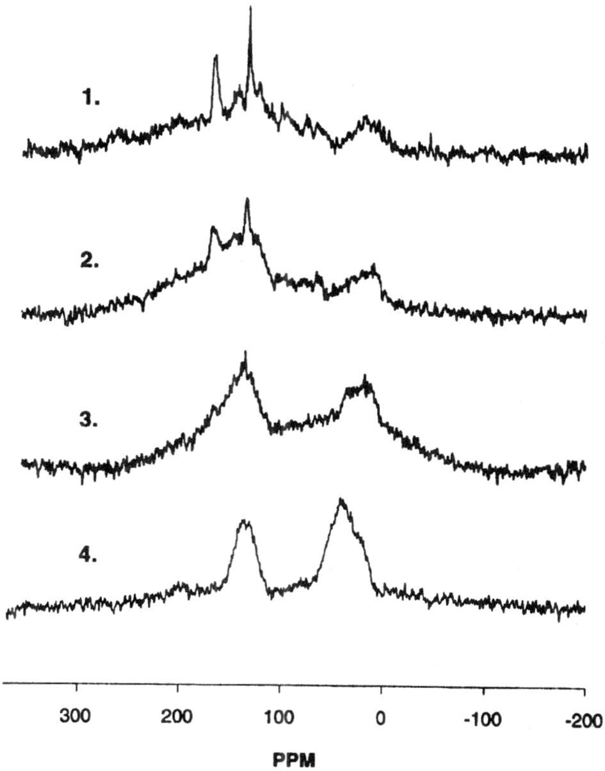

FIG. 34. NMR spectra taken from (1) a highly nitrogenated a-C(N):H film as deposited; (2) a highly nitrogenated a-C(N):H film annealed; (3) a low nitrogen content a-C(N):H film as deposited; (4) an a-C:H film as deposited. (Reproduced from [86].)

fitting are very similar to the previously discussed structure changes caused by nitrogen incorporation.

2.4.6. Overall Discussion

The results discussed earlier on the characterization of plasma deposited a-C(N):H films leads to a clear picture of the structure changes. Unless for a few exceptions, nitrogen incorporation leads to a strong increase of the C sp^2 fraction, as shown by electron energy loss spectroscopy. In parallel, Raman spectroscopy revealed that the graphitic clusters increase in size upon nitrogen incorporation. On the other hand, IR spectroscopy suggests, and XPS studies confirm, the preferential bonding of hydrogen to nitrogen, which bring two consequences. The first one is the location of N atoms in the border of the graphitic domains. The second one is to make N atoms as network terminators, impeding the formation of an extended C−N network. IR also shows the appearance of C≡N groups, which are network terminators as well.

At this point, we may suggest what would be the consequence of this microstrutural changes on the amorphous network as a whole. The foregoing modifications would lead to a lower cross-linking level, which is equivalent to a more open structure. This is in fact confirmed by other experimental techniques. Dopler-broadening observation of positron annihilation in a-C(N):H films [87] showed that nitrogen incorporation increases the open volume fraction in the film, what we would expect from the breaking of network interconnections. Hydrogen thermal effusion studies showed, for the same nitrogen-incorporated films, low temperature (about 200 °C) effusion peaks. As the compact a-C:H films usually show only high temperature (about 600 °C) hydrogen effusion peaks, the low temperature peaks present in a-C(N):H films were assigned to the presence of open voids [88]. This may lead, as discussed in the next section, to great changes in the film mechanical properties, as a consequence at the loss of connectivity.

It will be interesting to discern if the preceding transformations are a consequence of electronic effects due solely to nitrogen incorporation, or if there is some influence of the growth process itself. Different theoretical approaches were used in the study of hydrogen-free carbon–nitrogen solids.

Weich, Widany, and Frauenheim [89] reported a molecular dynamics simulation study on the formation of carbon–nitrogen solids. They studied the nitrogen incorporation on amorphous carbon within several density ranges, and they found that the presence of nitrogen induces the continuous increase of the sp^2-C atom fraction in all the cases. Hu, Yang, and Lieber [90] studied the semiempirical methods of the nitrogen incorporation effects in carbon clusters formed by diamond cells, and they found that for nitrogen contents greater than about 12 at.% a transition to the sp^2 state is likely to occur. Both findings at least support the possibility that the sp^2-C atom increase observed in a-C(N):H films may be due to a chemical bonding effect. Both visions match with examples coming from our previous discussion: the more or less continuous hybridization changes observed in films deposited from CH_4-N_2 and CH_4-NH_3 (see Fig. 28), and the apparently sudden structural evolution, as revealed by Raman spectroscopy, for C_2H_2-N_2 deposited films [58].

In plasma deposition of a-C(N):H films, as the nitrogen precursor partial pressure increases, the bombardment by N_2^+ ions is strongly enhanced. On the other hand, it is known that ion bombardment governs the establishment of a highly cross-linked network. However, in addition to deposit energy upon incidence on film surface, nitrogen ions promotes chemical sputtering of carbon atoms. As pointed out by Marton et al. [91], this kind of damage by ion impact may act as a source for trigonal bond formation in carbon–nitrogen films, and thus may be playing a role in the development of a-C(N):H film structure.

2.5. Mechanical Properties

2.5.1. HARDNESS AND STRESS

Some of the first works on the mechanical properties of plasma-deposited a-C(N):H reported strong stress decrease upon nitrogen incorporation, without appreciable changes in the mechanical hardness [12, 13]. Such observation attracted attention to a-C(N)H films because high internal stress is a stringent limitation to the application of thin films as mechanical protective coatings. Table II shows results on hardness and stress variation upon nitrogen incorporation, into a-C:H films together with deposition details and hardness measurement method.

Analysis of Table II shows discrepancies in the hardness and stress behavior of a-C(N) H films. Although all the works reported a clear stress reduction upon nitrogen incorporation, the hardness sometimes is quoted as almost constant, or on the other hand clearly decreasing. In addition to the possible effect of different deposition methods and conditions, it can be easily seen that the differences in hardness testing methods are the major source for discrepancies. Constant hardness behavior is only reported with the use microindentation methods, like Vickers and Knoop microhardness. On the other hand, the use of low-load nanoindentation methods always led to a nitrogen-induced decrease in hardness. This is basically the consequence of two factors. The first one is the higher penetration

TABLE I
MECHANICAL HARDNESS AND INTERNAL STRESS OF A-C(N):H FILMS

Reference	Deposition method gas mixture	Maximum N content (at.%)	Hardness method	Stress range (GPa)	Hardness range (Gpa)
Wood, Wyedeven, and Tsuji [57]	RFPECVD/ Hydrocarbon/N_2	16	Vickers microhardness	0.4–0.8	10–20
Jacobsohn et al. [54]	RFPECVD/C_2H_2-N_2	22	Nanoindentation	1.4–2.9	9.2–17.7
Schwan et al. [53]	RFPECVD/C_2H_2-N_2	20	Knoop microhardness	1.4–0.5	9–17
Franceschini, Achek, and Freire [12]	RFPECVD/CH_4-N_2	11	Vickers microhardness	1.5–2.4	19–21
Freire and Franceschini [56]	RFPECVD/CH_4-NH_3	11	Vickers microhardness	1.3–2.8	13–15
Houert et al. [92]	RFPECVD dc bias C_5H_{10}-N_2-Ar	8.2	Nanoindentation	—	13–23
Chan et al. [49]	ECRRF CH_4-N_2-Ar	12	Nanoindentation	—	9.5–4.5

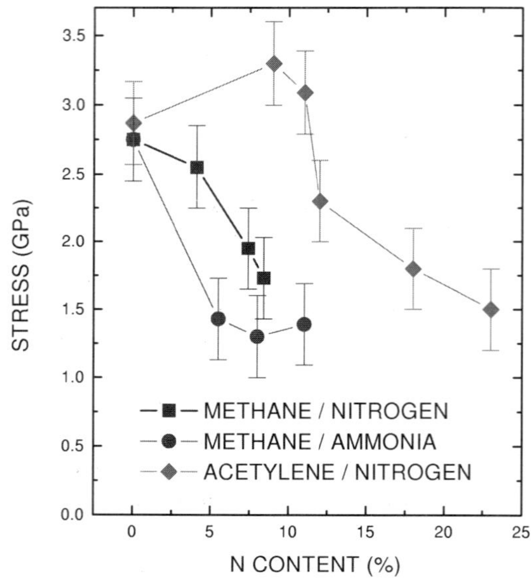

FIG. 35. Internal stress as a function of nitrogen content for several precursor mixtures. Data taken from [12, 54, 56].

depth owned by high-load microindentation, which allows the substrate mechanical properties to interfere in the measurement. The second one is the high elastic recovery of a-C(N):H films, which was found to be independent of nitrogen content [54]. In microindentation techniques, the penetration depth is determined by measuring the indentation dimensions after the load–unload process. On the other hand, nanoindentation methods measure the penetration depth during the load–unload process, allowing the suppression of elastic recovery effects from the hardness determination.

As mentioned earlier, nitrogen incorporation always leads to stress reduction. This is a straightforward consequence of the structural changes observed in a-C(N):H films. The combined increase in the sp^2-carbon atom fraction, the enlargement of the condensed aromatic ring clusters, the preferential bonding of H to N atoms, and the introduction of the terminal C≡N bonds may contribute to the decrease of network connectivity. This is equivalent to a decrease on the mean coordination number and hence to a reduction in the degree of overconstraining, thus leading to stress relief.

Figure 35 shows the internal stress variation for a-C(N):H films deposited by RFPECVD under similar conditions, from different precursor mixtures (CH_4-N_2 [12], CH_4-NH_3 [56], and C_2H_2-N_2 [54]). Despite all the nitrogen-free films showed about the same stress, the variations upon nitrogen incorporation are

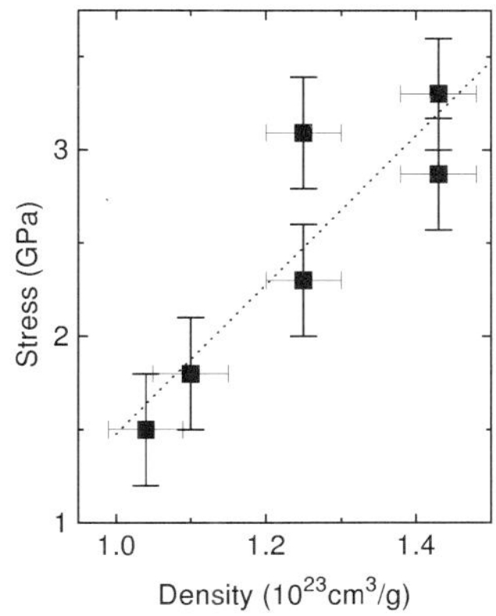

FIG. 36. Internal stress as a function of film atomic density for films deposited from a C_2H_2-N_2 mixture. (Reproduced from [56].)

different. Even for the CH_4 derived samples, which almost showed the same variation of carbon atom sp^2 fraction (see Fig. 28), which was attributed to the differences in the hydrogen content variation of the samples [11]. On the other hand, the C_2H_2 derived film shows a wide range of constant stress, followed by a strong decrease. This behavior closely follows the sp^2-carbon bonding variation upon nitrogen incorporations, as revealed by Raman spectroscopy on these films [54]. In this case, the Raman spectra parameters showed almost constant behavior for the region of constant stress.

This strong sensitivity to all of the parameters involved in the network connectivity, make internal stress measurement a useful diagnostic tool for the overall a-C(N):H film structure. This can be verified by the strong correlation showed between the internal stress and the atomic density, which is shown in Figure 36 [54]. The same structural evolution that led to the stress decrease, may also strongly affect the mechanical hardness. As pointed out by Robertson [23], structural features that lead to a decrease of network connectivity do not contribute to network rigidity. Thus, the hardness variation must be strongly correlated with internal stress, as shown in Figure 37 [54].

2.5.2. FRICTION AND WEAR

Relatively few works considered the tribological properties of plasma-deposited a-C(N):H films. First, considering the coefficient of friction, nitrogen incorporation was not found to strongly modify it. Prioli et al. [93] studied the friction coefficient of a-C(N):H films against a Si_3N_4 probe in an atomic force microscope, in ambient air. Nitrogen incorporation up to about 10 at.% into the

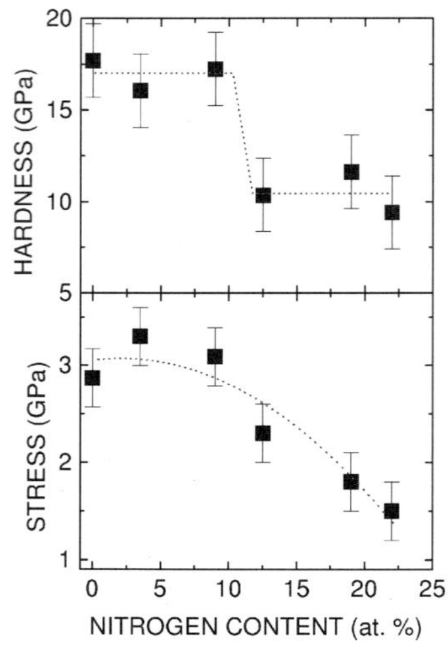

FIG. 37. Internal stress and mechanical hardness as functions of nitrogen content, and for films deposited from a C_2H_2-N_2 mixture. (Reproduced from [56].)

films (RFPECVD deposited in CH_4-NH_3 atmospheres) resulted in no appreciable changes of the friction coefficient, which remained in the 0.22–0.25 range. In addition, they also measured the film surface roughness, which was found to be very low, and to increase upon nitrogen incorporation, in the 0.11- to 0.24-nm range. Chan et al. [49] obtained similar behavior in a-C(N):H films deposited by ECRRF in CH_4-N_2-Ar atmospheres, with up to 12-at.% N incorporation. They essentially obtained constant friction coefficients in the 0.11 to 0.16 range, while film roughness increased from 0.2 to 0.7 nm. This roughness behavior was also reported by Silva, Amaratunga, and Barnes [94], and was ascribed to erosion effects arising from N_2^+ ion bombardment [74].

Considering the wear behavior of a-C:H films, it was reported that wear does not depend so strongly on film hardness as could be expected [95]. Wear behavior of a-C:H also depends on other factors, like the nature of the transfer layer—the layer modified by the wear process—and on the chemical reactions carried out by the ambient atmosphere. So, it could be expected that despite the strong drop in hardness, a-C(N):H films could have good wear properties. However, the few results reported on a-C(N):H films did not show an optimist wear performance for this material.

Dekempeneer et al. [96] studied the wear behavior of a-C(N):H films deposited by RFPECVD in CH_4-N_2 atmospheres, with up to 13-at.% N. The wear tests were done in a ball-on-disk tribometer, under air, at fixed 50% relative humidity. The initial friction coefficient was about 0.2 for all samples, and the wear life of the film was determined by the time necessary to achieve a friction coefficient equal to 0.5. Tests performed at 1- and 5-N normal loads showed wear lives greater than 24 h. On the other hand, tests performed with 10-N normal loads showed a strong decrease on the wear life, achieving a 12-h lifetime for the 13-at.% N sample. As should be expected, the studied samples also showed decreasing hardness and stress, and changes in the Raman spectra indicative of increased size or number of sp^2-C condensed ring domains. The same kind of behavior was reported by Wu et al. [97], in a study of friction and wear of ion-beam deposited a-C(N):H films. Despite the very low friction coefficient in ultrahigh vacuum conditions (about 0.02), the wear life was found to be very short, as compared with nitrogen-free a-C:H films.

2.6. Optical and Electrical Properties

As a semiconductor, the electrical and optical properties of a-C:H films are controlled by their electronic band structure. The electronic density of states of a-C:H films, as in a-Si:H and other amorphous semiconductors, does not show sharp edges and well-defined bandgaps as crystalline semiconductors. The disorder in the lattice potential gives rise to electronic localized states, which contribute to a tail in the density of states, leading it to expand into the gap [98]. The band is divided in two regions, one populated by the higher energy extended states like in crystalline semiconductors, and the tail localized states, which are separated from the extended states by a mobility edge. Since both conduction and valence bands have mobility edges, a mobility gap is thus defined. In addition to the tail states, disorder also gives rise to coordination defects, e.g., dangling bonds, thus introducing electronic defect states in the midgap.

In a-C:H, the tail states are dominated by π electrons, which results, as pointed out by Robertson [99, 100], in an enhanced localization as compared to a-Si:H, giving rise to higher band tail density of states and also to higher defect density in the midgap.

2.6.1. Optical Properties and Electron Spin Resonance

In the infrared (IR) wavelength range of the light spectrum, the a-C(N):H films are transparent, unless for the localized absorption bands due to CH, NH, C≡N and, if the films are nitrogenated enough, the IR Raman band. In the ultraviolet-visible (UV-vis) wavelength range, optical absorption is dominated by electronic transitions between valence band and conduction band states. So the parameter of most interest is the energy bandgap, or the optical bandgap. In amorphous semiconductors, because of the lack of sharp band edges, E_G is determined by the extrapolation of the bands. The usually adopted procedure is to assume parabolic

TABLE II
OPTICAL (TAUC) GAP, URBACH ENERGY, AND SPIN DENSITY OF A-C(N):H FILMS

Reference	Maximum nitrogen content (at.%)	E_g (eV)	Urbach tail width (meV)	Spin density (spins/cm^3)
Han and Feldman [8] (optical)	18	2.5–2.0	480–270	2×10^{18}–6.4×10^{16} (spins/g)
Lin et al. [101] (ESR)				
Amir and Kalish [9]	9	0.8–0.4		5×10^{20}–$<10^{18}$
Silva et al. [14]	16	1.8–2.15–1.7	260–170–200	4×10^{20}–$<10^{18}$
Schwan et al. [53]	20	1.5–0.8	100–250	—
(several deposition		1.3–0.7	100–250	
pressures)		1.0–0.4	200–250	

band edges, which leads to the so-called Tauc relation,

$$\left[\hbar\omega\alpha(\hbar\omega)\right]^{1/2} = A(\hbar\omega - E_{\mathrm{G}})$$

which allows the determination of the optical gap width from measurements of the absorption coefficient α, which is in turn derived from optical transmittance and reflectance spectra.

Several optical gap variations upon nitrogen content in a-C(N):H films are displayed in Table III, together with deposition details and other optical and electrical parameters. With only one exception, all results quoted in Table III showed a clear reduction of the gap width upon nitrogen incorporation. The nitrogen-free film gaps spanned a relatively wide range of values, indicating different initial structures, which might be expected by the different deposition techniques and deposition parameters. As discussed in Section 2.2, nitrogen incorporation was found to more frequently lead to an increase in the C sp^2 fraction and in the size of the π clusters. Both modifications in a-C:H films are expected, as stated by Robertson and O'Reilly [21] and Robertson [22], to result in the gap shrinking, as observed. Unfortunately, data on the hybridization state of C atoms of these films are not always available.

The exception to the regularity discussed before is the work of Silva et al. [14], which reported a gap width passing through a maximum at 7-at.% N, with a remarkable variation in the 1.7- to 2.8-eV range. In accordance, a slight decrease of the sp^2 fraction of carbon atoms was reported around 7-at.% N, and also a strong decrease, of the sp^2 nitrogen atom fraction in the same nitrogen content range. The N sp^2 fraction varied from 75 to about 20%, and after increased to about 90%, for the higher nitrogen content.

In amorphous semiconductors, information about the width of the band tail states (or disorder) may also be extracted from the optical absorption spectra. For photon energies near bandgap energy, the optical absorption coefficient of amorphous semiconductors exhibit an exponential dependence on the photon energy, following the so-called Urbach relationship:

$$\alpha(\hbar\omega) = \alpha_o \exp\left[(E - \hbar\omega)/E_0\right]$$

It was found that in this energy range, α follows the same behavior of the joint (valence + conduction) density of states. Thus, E_0, may be interpreted as a measure of the structural disorder [98], as it represents the inverse of band tail sharpness.

Only three results of Urbach width variation are shown in Table III. Schwan et al. [53] reported a clear increase of the Urbach energy for the whole nitrogen incorporation range, typically from 100 to 250 meV, for several deposition pressures. Han and Feldman [8] reported a decrease of the Urbach energy in the 480- to 270-meV range. Silva et al. [14] reported a decrease of E_0 from 270 to 160 meV up to 7-at.% N, which was followed by a small increase to about 200 meV for the higher nitrogen content, indicating, as the results of Han and Feldman, a clear decrease in disorder.

This disorder decrease on alloying is in principle unexpected, because alloying introduces chemical disorder in addition to bond disorder. However, Silva et al. [14] reported results on the joint density of states as determined by EELS which showed a clear decrease in the density of tail states at 7-at.% N and showed a further increase for higher nitrogen content.

The variation of the Tauc gap and the Urbach energy obtained by Silva et al. are shown in Figure 38, together with the variation of the refraction index.

Information on the evolution of the defect states in the midgap complement the information on the tail states. Electron spin resonance is a useful tool to probe spin-carrying defects (such as dangling bonds) in amorphous semiconductors. All the reports on ESR spin density quoted in Table I show a remarkable decrease upon nitrogen incorporation. Lin et al. [101] reported the spin density only for two samples, nitrogen-free and 18-at.% N. Amir and Kalish [9] reported an almost constant spin density for a wide range of nitrogen content, being the decrease observed only for the 9.0-at.% sample. In all the cases, the spin signal decreased by more than 2 orders of magnitude. So, in fact, in two of the cases shown in Table III, nitrogen incorporation has led to reduction in the disorder and in the defect density of states.

2.6.2. Electrical Properties

Nitrogen incorporation is also found to increase the electrical conductivity of a-C(N):H films. Wood, Wyedeven, and Tsuji [57] reported a 4 orders of magnitude

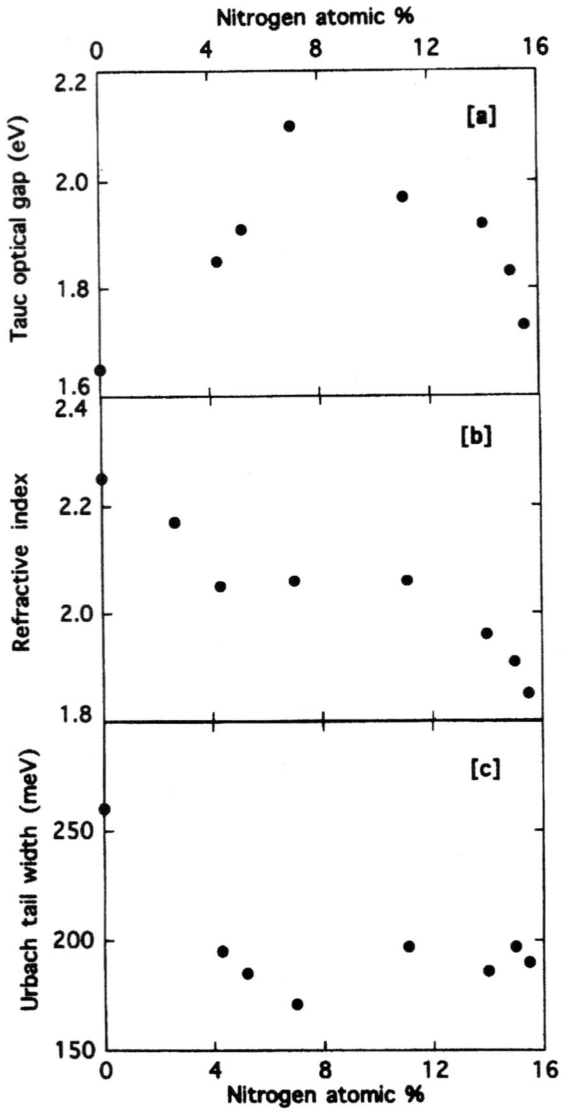

FIG. 38. Tauc gap, refractive index, and Urbach width of a-C(N):H films. (Reproduced from [14].)

decrease in the room temperature volume resistivity, for a 16-at.% N incorporation. Equivalent results were reported by Schwan et al. [53]. What is argued is that it is caused by electronic doping or by the increase in the sp^2 fraction, in parallel with the increase in the size of graphitic domains.

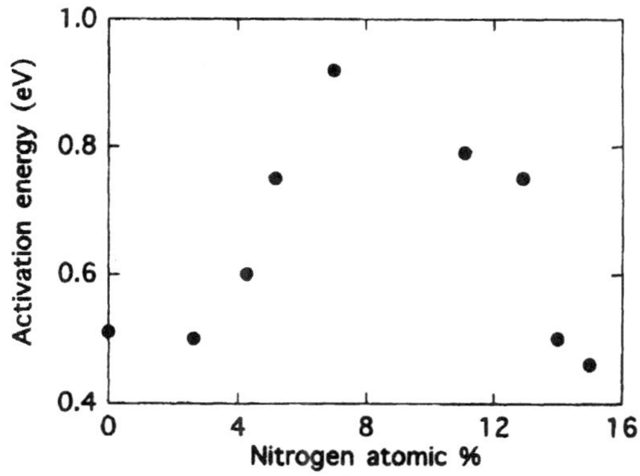

FIG. 39. Electrical conductivity activation energy vs nitrogen content. (Reproduced from [14].)

Evidence on this question may be taken by the behavior of the electrical conductivity σ as a function of temperature. A thermally activated process (T^{-1} dependence on $\log(\sigma)$, Arrhenius plot) is expected if doping takes place, whereas a $T^{-1/4}$ dependence, characteristic of a variable range hopping at the Fermi level is expected for a nondoping situation.

Schwan et al. [53], Amir and Kalish [9], Lin et al. [101], and Stenzel et al. [102] reported a $T^{-1/4}$ behavior for the nitrogen-free as well as the a-C(N):H films conductivities. The only report of a thermally activated behavior in RFPECVD-deposited a-C(N):H films was from the work of Silva et al. [14]. In addition, they reported an N-content dependence on the activation energy of the thermally activated conductivity. Upon increasing the nitrogen content, the activation energy was found to increase up to 7-at.% N from 0.5 to more than 0.9 eV, and again reducing to about 0.5 when the nitrogen content increases to 16-at.% N, as shown in Figure 39.

This behavior led Silva et al. to suggest that nitrogen acts as a weak electron donor in these films. The variation of the activation energy was ascribed to the Fermi level moving up in the bandgap [14]. It must be noted that if this process is really identified as a doping process, its efficiency is very low (7-at.% N for the compensation point). In addition, one must take into consideration that the effect occurs after relatively strong nitrogen-induced structure changes.

In fact, an apparent doping effect was also reported by Schwan et al. [39] in a-C(N):H films deposited by the highly ionized plasma beam deposition technique in C_2H_2-N_2 atmospheres. Schwan et al. also observed thermally activated behavior for the conductivity. As reported by Silva et al. [14], they also observed increasing optical gap, and decreasing ESR spin signal, but the Urbach energy was found to increase.

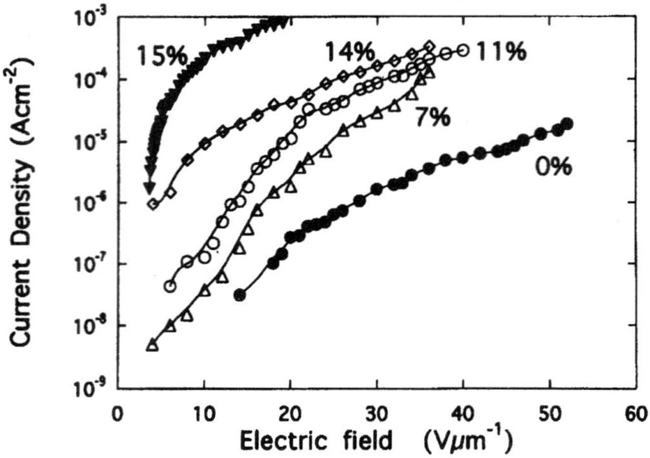

FIG. 40. Electron emission current density as a function of applied electric field, for a-C(N):H films with several N contents. (Reproduced from [15].

An interesting additional feature of the a-C(N):H films discussed in the previous paragraph [14] is their nitrogen-driven enhancement of the electric field-induced electron emission. Electron field emission is one of the alternatives considered for high-quality flat panel displays. Due to their presumed negative electron affinity, carbon-based materials are one of the possible choices for making emission cathodes. Nitrogen incorporation into a-C:H films was found to remarkably increase emitted electron current density, and to reduce the threshold field by electron emission [15], as shown in Figure 40. Such behavior was also reported by other workers [16], and was in fact used to improve the emission performance of Mo-tip emitters, by coating them with a-C(N):H films [16]. Further works emphasized the importance of surface plasma treatment [94], film thermal treatment after deposition [103], film thickness [104], and sp^2-C clusters concentration and size [105] on the improvement of emission properties.

References

1. A. Y. Liu and M. L. Cohen, *Science* 245, 841 (1989).
2. D. Marton, K. J. Boyd, and J. W. Rabalais, *Int. J. Mod. Phys. B* 9, 3527 (1995).
3. S. Muhl and J. M. Mendez, *Diamond Relat. Mater.* 8, 1809 (1999).
4. Patents: K. S. Goodson, S. Z. Gornicki, A. S. Lehil, C. K. Prabhakara, and W. T. Tang, Western Digital Corporation, US5855746-A; L. P. Franco, J. H. Kaufman, S. Metin, D. D. Palmer, D. D. Saperstein, and A. W. Wu, IBM Corporation, EP844603-A1.
5. J. Robertson, *Prog. Solid State. Chem.* 21, 199 (1991).

6. A. Grill, *Diamond Relat. Mater.* 8, 428 (1999).
7. D. I. Jones and A. D. Stewart, *Philos. Mag. B* 46, 423 (1982).
8. H. X. Han and B. J. Feldman, *Solid State Commun.* 65, 921 (1988).
9. O. Amir and R. Kalish, *J. Appl. Phys.* 70, 4958 (1991).
10. J. H. Kaufman, S. Metin, and D. D. Saperstein, *Phys. Rev. B* 39, 13053 (1989).
11. D. F. Franceschini, F. L. Freire Jr., and S. R. P Silva, *Appl. Phys. Lett.* 68, 2645 (1996).
12. D. F. Franceschini, C. A. Achete, and F. L. Freire Jr., *Appl. Phys. Lett.* 60, 3229 (1992).
13. S. Metin, J. H. Kaufman, D. D. Saperstein, J. C. Scott, J. Heyman, and E. Haller, *J. Mater. Res.* 9, 396 (1994).
14. S. R. P. Silva, J. Robertson, G. A. J. Amaratunga, B. Raferty, L. M. Brown, J. Schwan, D. F. Franceschini, and G. Mariotto, *J. Appl. Phys.* 81, 2026 (1997).
15. G. A. J. Amaratunga and S. R. P. Silva, *Appl. Phys. Lett.* 68, 2529 (1996).
16. E. J. Chi, J. Y. Shim, H. K. Baik, H. Y. Lee, S. M. Lee, and S. J. Lee, *J. Vac. Sci. Technol. B* 17, 731 (1999).
17. J. C. Angus and F. Jansen, *J. Vac. Sci. Technol. A* 6, 1778 (1988).
18. M. F. Thorpe, *J. Non-Cryst. Solids* 182, 135 (1995).
19. M. A. Tamor, W. C. Vassel, and K. R. Carduner, *Appl. Phys. Lett.* 58, 592 (1991).
20. A. Grill, *Thin Solid Films* 335–336, 189 (1999).
21. J. Robertson and E. P. O'Reilly, *Phys. Rev. B* 35, 2946 (1987).
22. J. Robertson, *Adv. Phys.* 35, 317 (1986).
23. J. Robertson, *Phys. Rev. Lett.* 68, 220 (1992).
24. S. Eisenberg and R. Chabot, *J. Appl. Phys.* 42, 2953 (1971).
25. Y. Lifishitz, S. R. Kasi, and J. W. Rabalais, *Phys. Rev. Lett.* 68, 620 (1989).
26. J. Robertson, *Diamond Relat. Mater.* 3, 361 (1994).
27. G. J. Vandentrop, M. Kawasaky, R. M. Nix, I. G. Brown, M. Salmeron, and G. A. Somorjay, *Phys. Rev. B* 41, 3200 (1990).
28. T. Schwarz-Selinger, A. Von Keudell, and W. Jacob, *J. Appl. Phys.* 86, 3988 (1999).
29. W. Jacob, *Thin Solid Films* 326, 1 (1998).
30. Y. Catheryne and C. Couderc, *Thin Solid Films* 144, 265 (1986).
31. A. Bubenzer, B. Dischler, G. Brandt, and P. Koidl, *J. Appl. Phys.* 54, 4590 (1983).
32. J. W. Zou, K. Reichelt, K. Schmidt, and B. Dischler, *J. Appl. Phys.* 65, 3914 (1989).
33. J. W. Zou, K. Schmidt, K. Reichelt, and B. Dischler, *J. Appl. Phys.* 67, 487 (1990).
34. A. Grill, "Cold Plasmas for Materials Fabrication," IEEE Press, New York, 1994.
35. A. Grill, *IBM J. Res. Dev.* 43, 147 (1999).
36. S. C. Kuo, E. E. Kunhardt, and A. R. Srivatsa, *Appl. Phys. Lett.* 59, 2532 (1991).
37. J. H. Kim, D. H. Ahn, Y. H. Kim, and H. K. Baik, *J. Appl. Phys.* 82, 658 (1997).
38. M. Weiler, S. Sattel, T. Giessin, K. Jung, V. Veerasamy, and J. Robertson, *Phys. Rev. B* 53, 1594 (1996).
39. J. Schwan, V. Batori, S. Ulrich, H. Ehrhardt, and S. R. P. Silva, *J. Appl. Phys.* 84, 2071 (1998).
40. H. W. Song, F. Z. Cui, X. M. He, and W. Z. Li, *J. Phys. Condens. Matter* 6, 6125 (1994).
41. M. A. Tamor and W. C. Vessel, *J. Appl. Phys.* 76, 3823 (1994).

42. M. A. Tamor, J. A. Haire, C. H. Wu, and K. C. Hass, *Appl. Phys. Lett.* 54, 123 (1989).
43. A. Grill, V. Patel, and S. Cohen, *Diamond Relat. Mater.* 3, 281 (1994).
44. L. C. Feldman and J. W. Mayer, "Fundamentals of Surface and Thin Film Analysis," North-Holland, Amsterdam, 1996.
45. A. Mansur and B. Ugolini, *Phys. Rev. B* 47, 10,201 (1993).
46. R. Kalish, O. Amir, R. Brener, A. Spits, and T. E. Derry, *Appl. Phys. A* 52, 48 (1991).
47. F. L. Freire Jr., D. F. Franceschini, and C. A. Achete, *Nucl. Instrum. Methods Phys. Res. B* 85, 268 (1994).
48. A. Grill and V. Patel, *Diamond Films Technol.* 2, 61 (1992).
49. W.-C. Chan, M.-K. Fung, K.-H. Lai, I. Bello, S.-T. Lee, and C.-S. Lee, *J. Non-Cryst. Solids* 124, 180 (1999).
50. S. Rodil, N. A. Morrinson, W. I. Milne, J. Robertson, V. Stolojan, and D. N. Jayavardane, *Diamond Relat. Mater.* 9, 524 (2000).
51. J. Seth, P. Padiyath, and S. V. Babu, *Diamond Relat. Mater.* 3, 221 (1994).
52. M. M. Lacerda, D. F. Franceschini, F. L. Freire Jr., and G. Mariotto, *Diamond Relat. Mater.* 6, 631 (1997).
53. J. Schwan, W. Dworschak, K. Jung, and H. Ehrardt, *Diamond Relat. Mater.* 3, 1034 (1994).
54. L. G. Jacobsohn, F. L. Freire Jr., M. M. Lacerda, and D. F. Franceschini, *J. Vac. Sci. Technol. A* 17, 545 (1999).
55. K.-R. Lee, K. Y. Eun, and J.-S. Rhee, *Mater. Res. Soc. Symp. Proc.* 356, 233 (1995).
56. F. L. Freire Jr. and D. F. Franceschini, *Thin Solid Films* 293, 236 (1997).
57. P. Wood, T. Wyedeven, and O. Tsuji, *Thin Solid Films* 258, 151 (1995).
58. J. Seth, P. Padiyath, and S. V. Babu, *Diamond Relat. Mater.* 3, 210 (1994).
59. D. F. Franceschini, F. L. Freire Jr., C. A. Achete, and G. Mariotto, *Diamond Relat. Mater.* 5, 471 (1996).
60. P. Hammer, N. M. Victoria, and F. Alvarez, *J. Vac. Sci. Technol. A* 16, 2941 (1998).
61. L. R. Shaginyan, F. Fendrych, L. Jastrabik, L. Soukup, V. Yu. Kulikovsky, and J. Musil, *Surf. Coat. Technol.* 116–119, 65 (1999).
62. T. Inoue, S. Ohshio, H Saitoh, and K. Kamata, *Appl. Phys. Lett.* 67, 353 (1995).
63. K. J. Clay, S. P. Speakman, G. A. J. Amaratunga, and S. R. P. Silva, *J. Appl. Phys.* 79, 7227 (1996).
64. S. Bhatasharyya, A. Granier, and G. Turban, *J. Appl. Phys.* 86, 4668 (1999).
65. M. Zhang, Y. Nakayama, T. Miyazaki, and M. Kume, *J. Appl. Phys.* 85, 2904 (1999).
66. C. Hopf, K. Letourneur, W. Jacob, T. Schwarz-Selinger, and A. von Keudell, *Appl. Phys. Lett.* 74, 3800 (1999).
67. H. Saitoh, H. Takamatsu, D. Tanaka, N. Ito, S. Ohshio, and H. Ito, *Jpn. J. Appl. Phys.* 39, 1258 (2000).
68. P. Hammer and W. Gissler, *Diamond Relat. Mater.* 5, 1152 (1996).
69. P. Hammer, M. A. Baker, C. Lenardi, and W. Gissler, *Thin Solid Films* 291, 107 (1996).
70. J. Hong and G. Turban, *Diamond Relat. Mater.* 8, 572 (1999).
71. J. Hong and G. Turban, *J. Vac. Sci. Technol. A* 17, 314 (1999).
72. F. D. A. Aarão Reis and D. F. Franceschini, *Appl. Phys. Lett.* 74, 209 (1999).
73. S. Todorov, D. Marton, K. J. Boyd, A. H. Al Bayati, and J. W. Rabalais, *J. Vac. Sci. Technol. A* 12, 3192 (1994).

74. F. D. A. Aarão Reis and D. F. Franceschini, *Phys. Rev. E* 61, 3417 (2000).
75. R. O. Dillon, J. A. Woolam, and V. Katkanant, *Phys. Rev. B* 29, 3482 (1984).
76. J. Schwan, S. Ulrich, V. Batori, H. Ehrhardt, and S. R. P Silva, *J. Appl. Phys.* 80, 440 (1996).
77. A. C. Ferrari and J. Robertson, *Phys. Rev. B* 61, 14,095 (2000).
78. G. Mariotto, F. L. Freire Jr., and C. A. Achete, *Thin Solid Films* 241, 255 (1994).
79. F. L. Freire Jr., *Jpn. J. Appl. Phys.* 136, 4886 (1997).
80. N. M. Victoria, P. Hammer, M. C. dos Santos, and F. Alvarez, *Phys. Rev. B* 61, 1083 (2000).
81. S. D. Berger, D. R. Mackenzie, and P. J. Martin, *Philos. Mag. Lett.* 57, 285 (1988).
82. V. S. Veerasamy, J. Yuan, G. A. J. Amaratunga, W. I. Milne, K. W. R. Gilkes, M. Weiller, and L. M. Brown, *Phys. Rev. B* 48, 17,954 (1993).
83. S. Souto and F. Alvarez, *Appl. Phys. Lett.* 70, 1539 (1997).
84. C. Donnet, J. Fontaine, F. Lefèbvre, A. Grill, V. Patel, and C. Jahnes, *J. Appl. Phys.* 85, 3264 (1999).
85. J. Seth, A. J. I. Ward, and V. Babu, *Appl. Phys. Lett.* 60, 1957 (1992).
86. J. LaManna, J. Bradok Wilking, S. H. Lin, and B. J. Feldman, *Solid State Commun.* 109, 573 (1999).
87. F. L. Freire Jr., D. F. Franceschini, and C. A. Achete, *Phys. Status Solidi B* 192, 493 (1995).
88. D. F. Franceschini, F. L. Freire Jr., W. Beyer, and G. Mariotto, *Diamond Relat. Mater.* 3 (1993).
89. F. Weich, J. Widany, and Th. Frauenheim, *Phys. Rev. Lett.* 78, 3226 (1997).
90. J. Hu, P. Yang, and C. M. Lieber, *Phys. Rev. B* 57, 3185 (1998).
91. D. Marton, K. J. Boyd, J. W. Rabalais, and Y. Lifishitz, *J. Vac. Sci. Technol. A* 16, 455 (1998).
92. R. Hauert, A. Gisenti, S. Metin, J. Goitia, J. H. Kaufman, P. H. M. Loosdrecht, A. J. Kellok, P. Hoffman, R. L. White, and B. D. Hermsmeier, *Thin Solid Films* 268, 22 (1995).
93. R. Pricli, S. I. Zanette, A. O. Caride, D. F. Franceschini, and F. L. Freire Jr., *J. Vac. Sci. Technol. A* 14, 2351 (1996).
94. S. R. P. Silva, G. A. J. Amaratunga, and J. R. Barnes, *Appl. Phys. Lett.* 71, 1477 (1997).
95. A. Grill, *Surf. Coat. Technol.* 94–95, 507 (1997).
96. E. H. A. Dekempeneer, J. Maneve, J. Smeets, S. Kuypers, L. Eersels, and R. Jacobs, *Surf. Coat. Technol.* 68–69, 621 (1994).
97. R. L. C. Wu, K. Myioshi, W. C. Lanter, J. D. Wrbanek, and C. A. DeJoseph, *Surf. Coat. Technol.* 120–121, 573 (1999).
98. R. A, Street, "Hydrogenated Amorphous Silicon," Cambridge Univ. Press, Cmabridge, U.K., 1991.
99. J. Robertson, *J. Non-Cryst. Solids* 198–200, 615 (1996).
100. J. Robertson, *Diamond Relat. Mater.* 6, 212 (1997).
101. S. Lin, K. Noonan, B. J. Feldman, D. Min, and M. T. Jones, *Solid State Commun.* 80, 101 (1991).
102. O. Sterzel, M. Vogel, S. Pönitz, R. Petrich, T. Wallendorf, C. V. Borczyskowski, F. Rozploch, Z. Krasolnik, and N. Kalugin, *Phys. Status Solidi A* 140, 179 (1993).

103. A. P. Burden, R. D. Forrest, and S. R. P. Silva, *Thin Solid Films* 337, 257 (1999).
104. R. D. Forrest, A. P. Burden, S. R. P. Silva, L. K. Chea, and X. Shi, *Appl. Phys. Lett.* 73, 3784 (1998).
105. J. D. Carey, R. D. Forrest, R. U. A. Khan, and S. R. P. Silva, *Appl. Phys. Lett.* 77, 2006 (2000).

Subject Index

Recent Volumes In This Series

Maurice H. Francombe and John L. Vossen, *Physics of Thin Films*, Volume 16, 1992.

Maurice H. Francombe and John L. Vossen, *Physics of Thin Films*, Volume 17, 1993.

Maurice H. Francombe and John L. Vossen, *Physics of Thin Films, Advances in Research and Development, Plasma Sources for Thin Film Deposition and Etching*, Volume 18, 1994.

K. Vedam (guest editor), *Physics of Thin Films, Advances in Research and Development, Optical Characterization of Real Surfaces and Films*, Volume 19, 1994.

Abraham Ulman, *Thin Films, Organic Thin Films and Surfaces: Directions for the Nineties*, Volume 20, 1995.

Maurice H. Francombe and John L. Vossen, *Homojunction and Quantum-Well Infrared Detectors*, Volume 21, 1995.

Stephen Rossnagel and Abraham Ulman, *Modeling of Film Deposition for Microelectronic Applications*, Volume 22, 1996.

Maurice H. Francombe and John L. Vossen, *Advances in Research and Development*, Volume 23, 1998.

Abraham Ulman, *Self-Assembled Monolayers of Thiols*, Volume 24, 1998.

Subject and Author Cumulative Index, Volumes 1–24, 1998.

Ronald A. Powell and Stephen Rossnagel, *PVD for Microelectronics: Sputter Deposition Applied to Semiconductor Manufacturing*, Volume 26, 1998.

Jeffrey A. Hopwood, *Ionized Physical Vapor Deposition*, Volume 27, 2000.

Maurice H. Francombe, *Frontiers of Thin Film Technology*, Volume 28, 2001.

Maurice H. Francombe, *Non-Crystalline Films for Device Structures*, Volume 29, 2002.